A NEW DEAL
FOR AN EFFECTIVE EUROPEAN RESEARCH POLICY

A NEW DEAL FOR AN EFFECTIVE EUROPEAN RESEARCH POLICY

The Design and Impacts of the 7th Framework Programme

by

Ugur Muldur, Fabienne Corvers, Henri Delanghe,
Jim Dratwa, Daniela Heimberger, Brian Sloan, and Sandrijn Vanslembrouck

With a Preface by Commissioner for Science and Research Janez Potočnik

Springer

A C.I.P. Catalogue record for this book is available from the Library of Congress.

ISBN-13 978-90-481-7390-7
ISBN-10 1-4020-5551-x (e-book)
ISBN-13 978-1-4020-5551-5 (e-book)

Published by Springer,
P.O. Box 17, 3300 AA Dordrecht, The Netherlands.

www.springer.com

Printed on acid-free paper

CONTENTS

LIST OF FIGURES AND TABLES

PREFACE

More than a century ago Louis Pasteur observed that "Science knows no country, because knowledge belongs to humanity, and is the torch which illuminates the world". His remark has all the more resonance today with the increasing globalisation of science, technology, and commerce. But it also underlines an important truth: that science has always advanced most rapidly when it is a collective endeavour, with a strong circulation of knowledge.

When it comes to forging cooperation in research, Europe has an impressive track record. This year, the European Union's Framework Programme for Research and Technological Development celebrates its 20th birthday. It is the world's largest programme in support of transnational research, promoting cooperation, coordination, and mobility across borders. The year 2007 will see the launch of the 7th Framework Programme, the biggest and most ambitious so far. This book will show how the new Framework Programme was put together and explain why it took the shape it did. It will also set out its potential impacts and the conditions necessary for it to be a success.

In the challenging new global economy, knowledge is Europe's most valuable resource, and I am convinced that, with its unprecedented scale and scope, the 7th Framework Programme is the right strategy for a Europe that wants to progress towards a knowledge-based economy.

However, the Framework Programme is only a very modest part of total public R&D spending in Europe, the bulk of which is carried out by national governments. They must continue their efforts to raise levels of research investment, but this is not an easy task. National budgets are tight, and there are many competing priorities. We must face facts. Public resources for research remain scarce across the EU, and it will take time for this to change. It is therefore essential to ensure that we make the most of those funds we have. Inefficiencies in the European research system mean that we are not yet investing this money to greatest effect: efforts are poorly coordinated across national boundaries, there is much duplication, and resources are thinly spread.

The European Research Area initiative was introduced to tackle these problems. It has become a cornerstone of European research policy and its further realisation remains vital. However, in the last few years, in spite of some important advances, it has become clear that we are moving too slowly towards the ERA, and indeed too slowly towards the Lisbon goal of the knowledge-based economy. Time is pressing.

I believe that we need a new bolder approach for increasing the efficiency and impact of the EU research system. We need to go further than just coordinating and linking up the existing research structures: we need a more dynamic method which would re-design the very architecture of the public research system in Europe. I am talking about re-examining public research activities and making sure that they are managed and implemented at the level where they can be most effective – be it regional, national, EU, or intergovernmental. We need to be more courageous in looking at how research can be most efficiently distributed across these levels. We need to base these decisions less on tradition and history, and more on hard evidence of added value and potential impact. And when it comes to impact, research policies need to be more geared towards our major strategic objectives such as growth, employment, and a sustainable and healthy society.

What this implies, as the title of this book suggests, is a "New Deal" for research in Europe, involving decisive action by two committed partners – the Member States and the Union – in consultation with universities and industry. We must share more information about research schemes across the Union and about their relative efficiency and effectiveness. Only in this way can we move towards a clear and efficient division of labour between different policy levels and ensure that our R&D investment achieves the greatest impact.

Janez Potočnik
Commissioner for Research and Science

ACKNOWLEDGEMENTS

This book draws upon a wide variety of inputs and would not have been possible without the contribution of many.

Among those to be mentioned are, first of all, the European Commission colleagues from the Directorates General which form the so-called "Framework Programme family". We would like to thank in the first place our in-house colleagues from the Directorate General for Research, but also those from DG Enterprise, DG Transport and Energy, DG Information Society, DG Fisheries, and the Joint Research Centres, who all provided useful input and feedback in the course of writing the Impact Assessment Report accompanying the 7th Framework Programme proposal.

Many other colleagues have generously commented on earlier drafts of the manuscript for this book. For their time, expertise, and constructive criticism, we are grateful to (in alphabetical order): Michel André, Roman Arjona, Jean-Michel Baer, Paraskevas Caracostas, Bénédicte Caremier, Francesca Cervelli, Vincent Duchêne, Jocelyne Gaudin, Peter Haertwich, Dermot Lally, Célina Ramjoué, Neville Reeve, Keith Sequeira, Clara de la Torre, and Kurt Vandenberghe.

We also wish to thank Pablo Amor and Michel Claessens, as well as Valentina Fratto from OPOCE, for their efficient collaboration. They provided valuable advice in all aspects regarding the publication of this book. Valérie Dierckx deserves a special mention for her secretarial support.

Finally, without the constant support of our hierarchy it would not have been possible to write a study on the Framework Programme as we have done here. Our sincere thanks go out to Richard Escritt, Director of Directorate A (Coordination of Community Activities) in DG Research, to José Manuel Silva Rodríguez, the Director General, and to the Commissioner for Research and Science, Janez Potočnik.

Nevertheless, any errors are to be attributed to the authors only. This book does not represent an official position from the European Commission.

LIST OF ABBREVIATIONS

AIDS	Acquired Immuno-Deficiency Syndrome
ATP	Advanced Technology Programme
BERD	Business Expenditure on R&D
BSE	Bovine Spongiform Encephalopathy
CIS	Community Innovation Survey
CMD	Congenital Muscular Dystrophy
CoR	Committee of Regions
CORDIS	Community Research and Development Information Service
COREPER	Comité des Représentatives Permanents
COST	European Co-operation in the Field of Scientific and Technical Research
CRAFT	Cooperative Research Action for Technology
CREST	Scientific and Technical Research Committee
DEST	Department of Education, Science and Training
ECCP	European Climate Change Programme
EDCTP	European and Developing Countries Clinical Trials Partnership
EESC	European Economic and Social Committee
EFTA	European Free Trade Association
EPO	European Patent Office
ERA	European Research Area
ERA-NET	European Research Area Network
ERC	European Research Council
ESF	European Science Foundation
EURAB	European Research Advisory Board
FET	Future and Emerging Technologies
FP	Framework Programme
GDP	Gross Domestic Product
HIV	Human Immunodeficiency Virus
ICT	Information and Communication Technology
INCO	International Cooperation
IP	Integrated Projects
IPM	Interactive Policy Making
ITRE	Committee on Industry Research and Energy (European Parliament)
JPO	Japanese Patent Office
JRC	Joint Research Center
JTI	Joint Technology Initiative

NBER	National Bureau of Economic Research
NEST	New and Emerging Science and Technology
NoE	Networks of Excellence
NUTS	Nomenclature of Territorial Units for Statistics (EU territorial units)
OECD	Organization for Economic Co-operation and Development
PhD	Doctor of Philosophy
PPS	PPS
R&D	Research and Development
RTD	Research and Technological Development
S&T	Science and Technology
S&E	Science & Engineering
SARS	Severe Acute Respiratory Syndrome
SEPSDA	Sino-European Project on SARS Diagnostic and Antivirals
SME	Small and Medium Sized Enterprises
STREP	Specific Targeted Research Project
TP	Technology Platform
TEC	Treaty of the European Communities
TSER	Targeted Socio-economic Research
UEAPME	European Association of Craft, Small and Medium Sized Enterprises
USPTO	US Patent Office

COUNTRY CODES ABBREVIATIONS

AT	Austria
BE	Belgium
CY	Cyprus
CZ	Czech Republic
DE	Germany
DK	Denmark
EE	Estonia
EL	Greece
ES	Spain
EU	European Union
HU	Hungary
FR	France
FI	Finland
IE	Ireland
IT	Italy
LT	Lithuania
LU	Luxembourg
LV	Latvia
NL	The Netherlands

MT	Malta
PL	Poland
PT	Portugal
SE	Sweden
SK	Slovakia
Sl	Slovenia
UK	United Kingdom
US	United States

INTRODUCTION

This is a book about the design of European Union (EU) research policy. It focusses on the preparation of the 7th Framework Programme, which will be the EU's main instrument for funding research and technological development for the period 2007–2013. It describes some of the key factors that helped to shape the new Framework Programme, and looks at its expected impacts. In the process, it also draws upon past experience of previous Framework Programmes and evidence from evaluations to pose some broader questions in relation to EU research policy and to provide some longer-term perspectives.

Making research policy at European level is not easy. There are many actors to be consulted with different needs and different visions of what is required. The EU institutional machinery is impressive and necessary, but can sometimes be unavoidably complicated and time-consuming. And there is an unresolved debate about how to achieve the best articulation between EU intervention in research and regional and national support schemes. In short, designing EU research policy is a very complex task. This book tries to guide the reader through this complex process, describing how the 7th Framework Programme was put together, and explaining why it took the shape it did.

The 7th Framework Programme has been drawn up at a time when society and governments expect science and technology to deliver solutions to an increasingly long list of problems. It will be the most ambitious Framework Programme to date, with a much larger budget and new support mechanisms, and signals a stronger role for EU research. But the 7th Framework Programme represents just a small part of total government spending on R&D in the EU, the bulk of which is carried out by the national and regional authorities. Clearly, the 7th Framework Programme on its own cannot hope to solve all Europe's problems. What is needed, this book argues, is a "New Deal" for research in Europe[1] involving the mobilisation of policy actors across all levels – regional, national and European – and their commitment to developing a more effective research system in Europe based on taking actions at the level where they can be shown to have the greatest impact.

The originality of the book lies in its perspective, its transparency and its objectivity. It is the first to present, from a viewpoint inside the European

[1] The term "New Deal" historically stands for an economic reform programme in the US during the Great Depression (1933–37) under President Franklin D. Roosevelt. It has been subsequently used as a political metaphor, epitomising a new, large-scale approach to a political project, based on the partnership of all actors involved.

Commission, the nuts and bolts of how EU research policy is actually constructed. It also provides a comprehensive analysis, on the basis of factual evidence, not only of the positive impacts of EU research, but also of the various criticisms that have been made of the Framework Programme.

This degree of openness and objectivity is the direct result of another innovative aspect of the book: It represents the first fruits of a new approach towards policy-making, based on ex-ante impact assessment – a structured, evidence-based method of evaluating policy options and their expected impacts. The 7th Framework Programme was the first to be accompanied by an impact assessment report, and the material presented here draws extensively on this work.[2]

Impact assessment is a new approach for improving the transparency and quality of policy design.[3] It informs decision-makers of the likely consequences of policy choices by answering a common set of questions. It assesses the issues at stake and the objectives to be pursued by the policy proposal. It examines the views of the main stakeholders that will be affected by the policy. It identifies the main policy options for achieving the objectives, and analyses their likely economic, environmental and social impacts. And it outlines the advantages and disadvantages of each option, as well as synergies, trade-offs and risks. Fig. 0.1 shows how these elements – most of which are closely interlinked – feed into the policy-making process, be it directly or channelled through the impact assessment.

As can be seen from the figure, the book itself follows the steps of the impact assessment for the 7th Framework Programme.

The first phase involved identifying what the problems are that needed to be tackled by the policy, and setting the main objectives. This is the subject of the first three chapters. In Chapter 1, an analysis is made of the main economic, social and environmental challenges facing Europe and of the potential role that science and technology can play in addressing these challenges. However, Chapter 2 goes on to show that Europe will not be able to meet these challenges unless it addresses a number of structural weaknesses which currently prevent it from achieving its full Science and Technology (S&T) potential. It is against this backdrop of high expectations for research in tackling society's problems, offset by important weaknesses in Europe's S&T system, that the 7th Framework Programme was prepared. But there was another key factor influencing its design: the political context. Chapter 3 argues that a new policy context has emerged based on the

---

2 *http://ec.europa.eu/research/future/pdf/comm_sec_2005_0430_1_en.pdf*

3 Since 2003, every legislative proposal by the European Commission must be accompanied by an in-depth impact assessment report. The introduction of impact assessment was one of the concrete actions taken by the Commission within the framework of its 2001 *European Strategy for Sustainable Development* and its 2002 *Better Regulation Action Plan*, in order to increase the transparency and quality of its policy design. It replaces existing requirements for business impact assessment, gender impact assessment, environmental assessment, SME assessment, trade impact assessment, and so on. Moving away from this partial approach towards one integrated impact assessment helps policy-makers to assess trade-offs and compare different scenarios when shaping a particular proposal.

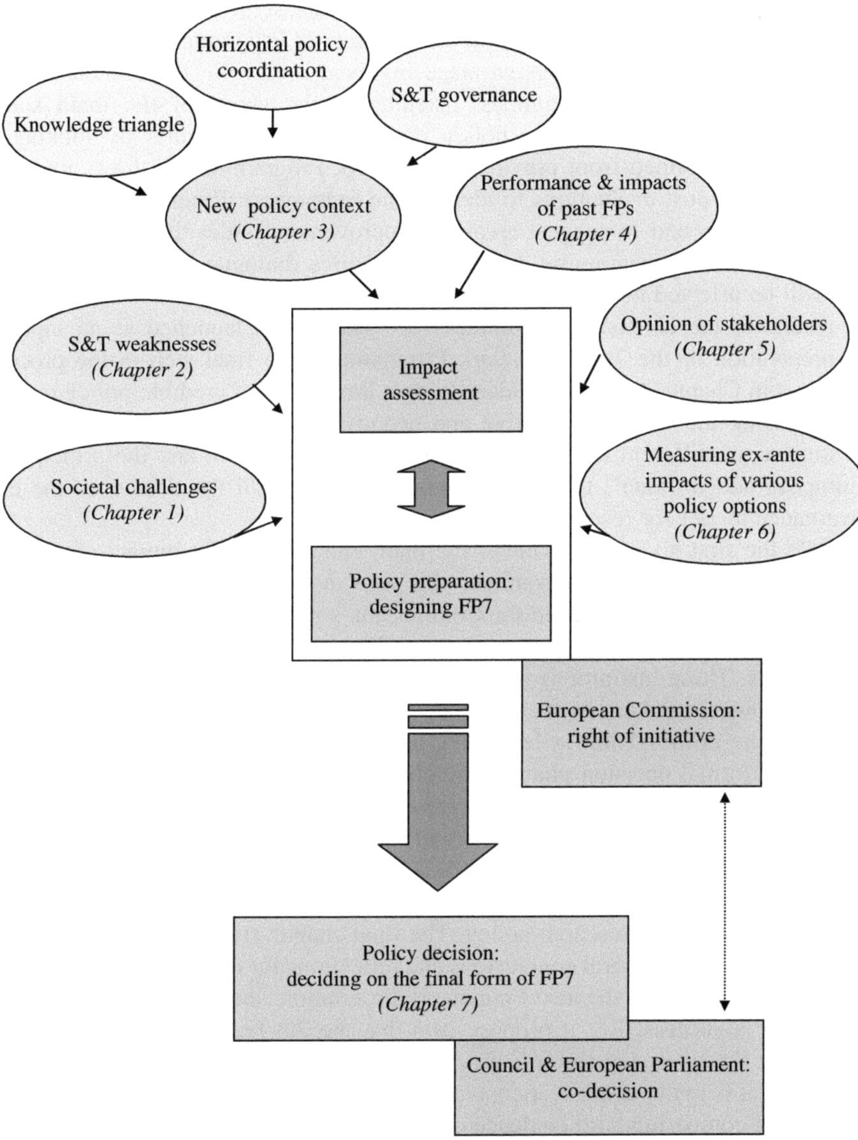

Fig. 0.1. View of the policy-making process of FP7
Source: DG Research

knowledge economy with research high on the agenda, but founded too on the realisation that the Lisbon method of soft inter-governmentalism will not resolve all the EU's problems. This has in turn made it possible to move towards a new Framework Programme, much more ambitious in scale and scope than its predecessors.

Having taken stock of these factors, the second phase of the impact assessment process focussed on how best to build the Framework Programme so as to respond to these issues. This policy design stage involved reviewing the lessons learned from past Framework Programmes, listening to the views of the main stakeholders, and assessing concrete policy options. Chapter 4 starts by looking at the experiences gained from previous Framework Programmes. It draws upon the evidence of ex-post evaluations to identify the Framework Programmes' strengths and weaknesses, and to suggest areas for improvement under the 7th Framework Programme. Good programme design also requires dialogue with the key actors who will be affected or involved. This is the subject of Chapter 5, which analyses the results of the extensive consultation of stakeholders launched as an input to the preparation of the 7th Framework Programme. The final step in the process, described in Chapter 6, involved identifying a limited set of credible policy options and assessing the expected positive and negative impacts of each. By analysing the impacts of different degrees of EU research intervention, and the principle of "European added value", this chapter goes to the heart of the debate on the best governance model for research in Europe.

While the first six chapters cover the main elements of the impact assessment for the 7th Framework Programme, this was not the end of the story. The impact assessment accompanied the Commission's proposal for the 7th Framework Programme, which was then presented to the European Parliament and the Council of Ministers. These institutions must examine the content of the proposal, propose their amendments and arrive at an agreement on the final text – in other words, they transform the proposal into a legal decision. Chapter 7 describes what happened during this formal decision phase, spotlighting the various actors, explaining how decisions are prepared and taken, and presenting the main themes debated.

By the end of the first seven chapters the reader should understand more clearly how and why the 7th Framework Programme evolved into its final shape. But what also emerges from these analyses is a set of broader questions concerning the future direction of EU research policy. The final chapter tries to draw this material together, with some general conclusions and reflections for the future. It summarises the expected impacts of the next Framework Programme, and sets out the conditions for it to be a success. But it proposes too that the 7th Framework Programme is only a first tentative step towards a new governance model for European research policy. What is really needed, it concludes, is a "New Deal" for research, based on sharing and comparing hard evidence of the impact of research policies at regional, national and EU levels. Only in this way will Europe be able to identify what is done best at each level, and how to get the most out of the public resources it invests.

CHAPTER 1

CHALLENGES TO THE EUROPEAN EXCEPTION: WHAT CAN S&T DO?

The 7th Framework Programme (FP) was designed against the background of Europeans feeling anxious that Europe is turning from a positive into a negative "exception" at global level. The continent is confronted with a number of important economic, social and environmental challenges. These undermine the European exception in the positive sense, namely what Europeans are most proud of: the progress made so far towards the achievement of valuable societal objectives (e.g. raising standards of living and quality of life). While some (e.g. environmental) difficulties are common to all major world economies, other troubles (e.g. low economic growth and high unemployment) appear to be unique to Europe, turning it into a global exception in the negative sense.

The purpose of this chapter is twofold. It is first of all, and without pretending completeness, to explore in greater detail some of the societal issues affecting Europe. A second objective is to investigate the role that S&T can play in addressing these challenges. This potential role is sizeable, as this chapter will demonstrate. In order for S&T to realise its potential, however, a number of substantial S&T weaknesses will have to be remedied, as will be discussed in the next chapter (Chapter 2).

An analysis of the difficulties society has to deal with, and of the potential role of S&T in addressing them, is key to the development of any S&T policy and programme. The magnitude and nature of the challenges can, for instance, affect decisions on the size and thematic coverage of the programme, and its associated work programmes.

1. EUROPE: TURNING FROM A POSITIVE INTO A NEGATIVE EXCEPTION

Europeans have set themselves ambitious societal objectives:

> The Community shall have as its task . . . to promote throughout the Community a harmonious, balanced and sustainable development of economic activities, a high level of employment and of social protection, equality between men and women, sustainable and non-inflationary growth, a high degree of competitiveness and convergence of economic performance, a high level of protection and improvement of the quality of the environment, the raising of the standard of living and quality of life, and economic and social cohesion and solidarity among Member States.[1]

---

[1] Treaty Establishing the European Community (Consolidated Text), *Official Journal C 325 of 24 December 2002.*

A quick review of the available evidence shows, however, that, while great strides have been made over the past few decades towards the achievement of these goals, Europe is facing significant challenges in most if not all of these areas. Economic growth is slow. Europe's competitive position is feeble. There are not enough jobs, and not enough of them are high-level. Europe is still characterised by significant poverty and regional inequality. An important demographic challenge is emerging. Europeans' health is affected by serious lifestyle and contagious diseases. And the environment is being degraded.

This is undermining what Europeans are most proud of and turning Europe into a negative exception at global level. The term "European exception" is most often used to refer to a European country not acting in accordance with what most other European countries are doing, whatever the field. Sometimes, however, the vocabulary is also used to refer to how Europe behaves differently from other advanced world economies. Usually, reference to the European exception has a positive tone to it. Europeans are proud of their commonly held values, their social model based on egalitarianism and solidarity, their high level of environmental awareness and protection and so on.

However, Europe appears to be the only advanced economy suffering from chronic low growth and high unemployment, and an unceasing lack of dynamism. Its levels of poverty and of individual and regional income inequality are not that far removed from US levels. And this makes Europeans feel anxious, and unsure of themselves, their future and further European integration.

1.1. Slow economic growth

Significant change has characterised the world economy over the past few decades. World trade has been liberalised as both formal and informal trade barriers have been reduced significantly, or disappeared altogether. Capital roams the planet freely in search of the best investment opportunities as barriers to capital mobility have been eliminated. Global communication and transportation networks have become denser and better integrated through a combination of technological and organisational innovation. The speed of technological change has accelerated while technologies are standardised more rapidly and use is made of modular production systems. As the combination of these factors has made it possible to locate the production of goods and services anywhere on the planet and still serve global markets, the global production system is in the process of being reconfigured.

The new international division of labour not only provides both developing and developed countries with ample opportunities, it also has shady sides. On the one hand, low-, medium- and to an increasing extent high-technology manufacturing and services industries are under threat from delocalisation or so-called off-shoring and outsourcing, resulting in at least short-term disruption and unemployment. Employment is also under threat from rapid process innovation leading to productivity increases.[2] On the other hand, rapid product innovation provides developed

[2] European Commission, *Employment in Europe 2004*, Luxembourg, 2004, p. 77.

countries with opportunities to improve competitiveness and serve global markets by fleeing forward as it were. The race to upgrade the economy is never-ending, however, and innovation-based advantages are fleeting and unsustainable as rapid standardisation and modular production techniques quickly allow the production process to move partially or completely to developing countries.

As reflected in its lacklustre economic growth performance, Europe has not yet adapted to the rules of this new game. In the first half of the post-war period, the European economy grew as fast as the world economy (Fig. 1.1).[3]

In the second half of the post-war period, however, the decline in economic growth was more pronounced in Europe than in the United States, Japan and other OECD economies (Figs 1.1 and 1.2). In the last 15 years or so, Europe has done worse than the United States, while Japan has once again started to outperform Europe, and the large BRIC (Brazil, Russia, India, China) economies and smaller East Asian economies continue to grow rapidly.[4] The growth of output amounted to 1.3 per cent in the Euro area in 2005, substantially lower than the 3.5 per cent in the United States and the 2.7 per cent in Japan, and the 4.8 per cent at world level. Output is projected to grow by a higher 2.0 per cent in the Euro area in 2006, still

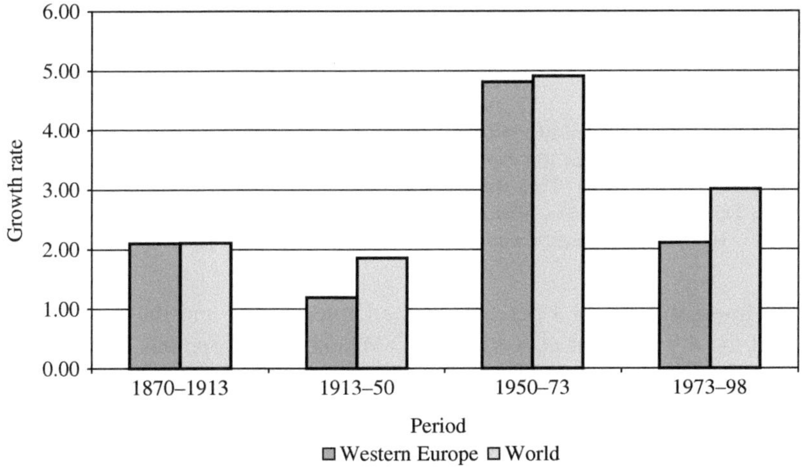

Fig. 1.1. Slow European economic growth in the second half of the post-war period compared to the rest of the world (annual average compound growth rate of GDP, 1870–1998)
Source: DG Research
Data: Maddison, 2001

[3] Angus Maddison, *The World Economy: A Millennial Perspective*, Paris, 2001.

[4] OECD, *Economic Survey of the Euro Area 2005: Outlook and Challenges*, Paris, 12 July 2005: Economic growth in the euro area has been lagging that of the best performing OECD countries since the mid-1990s. It should be acknowledged, however, that some EU countries have performed rather well economically in the past decade. This group includes the Member States formerly classified as cohesion countries (especially Ireland), as well as Finland, the Netherlands and the UK.

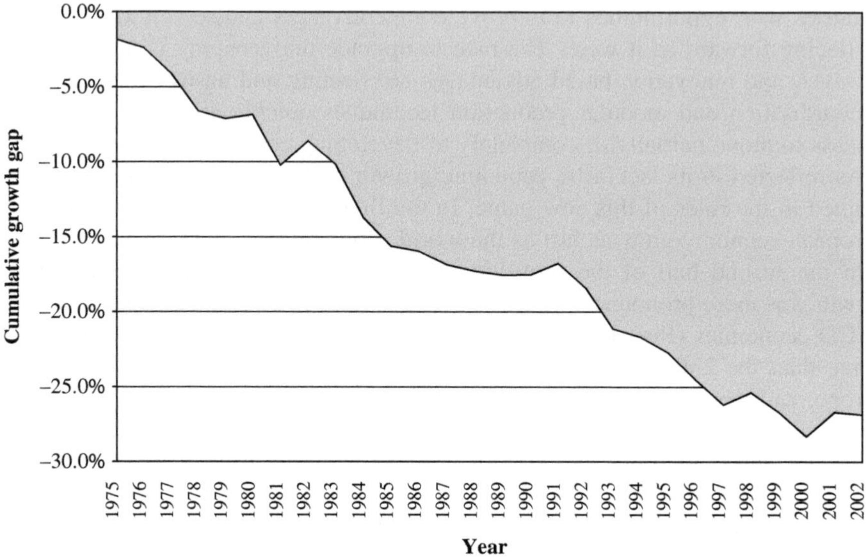

Fig. 1.2. Slow European economic growth in the second half of the post-war period compared to other industrialised countries (cumulative economic growth gap between the EU and the other industrialised countries (current prices and current PPPs))

Source: DG Research

Data: OECD

Note: For both the EU-15 and the non-EU-15 OECD countries, 1974 GDP at current prices and current PPPs (billions of dollars) was taken as 100. For all following years, GDP growth in percentages relative to the 1974 amount was calculated. Then the series for the non-EU15 OECD countries (Australia, Canada, Iceland, Japan, Korea, Mexico, New Zealand, Norway, Switzerland, Turkey, US) was set to 100 and the difference with the series for the EU-15 calculated.

significantly lower than the 3.4 per cent in the United States and the 2.8 per cent in Japan, and the 4.9 per cent at world level.[5] Whenever Europe has been able to increase productivity in the past it has suffered in the field of employment, and vice versa, pointing to the existence of structural barriers to growth.[6]

1.2. A feeble competitive position

Underlying Europe's lacklustre economic growth performance is its weak competitive position. The most common definition of competitiveness refers to the overall capacity to improve standards of living in a sustainable way.[7] By this standard, European competitiveness is not improving. Europe caught up with the United

[5] International Monetary Fund, *World Economic Outlook April 2006*, Washington, DC, 2006.

[6] *An Agenda for a Growing Europe. Making the EU Economic System Deliver*, Report of an Independent High-Level Study Group Established on the Initiative of the President of the European Commission, Brussels, 2003, pp. 27–28.

[7] European Commission, *European Competitiveness Report 2004*, Luxembourg, 2004, p. 7.

States during the 1950s and 1960s. But since the 1970s, European standards of living have not increased relative to the United States (Fig. 1.3).[8]

Labour productivity is another common measure of competitiveness. Though, except for a few countries, the productivity gap was never closed in the end, for most of the post-war period the EU somehow caught up on average with the United States.[9] This catch-up has now stopped and is even being reversed. Since 1995, for the first time in three decades, growth in US labour productivity has outstripped that of the Union (Fig. 1.4).[10] This EU productivity downturn is of a structural nature and mainly due to an outdated and inflexible industrial structure slow to adapt to the intensifying pressures of globalisation and rapid technological change.[11]

Deindustrialisation is often taken as a further sign of Europe's deteriorating competitiveness. The fear is that slow labour productivity growth, high labour costs,

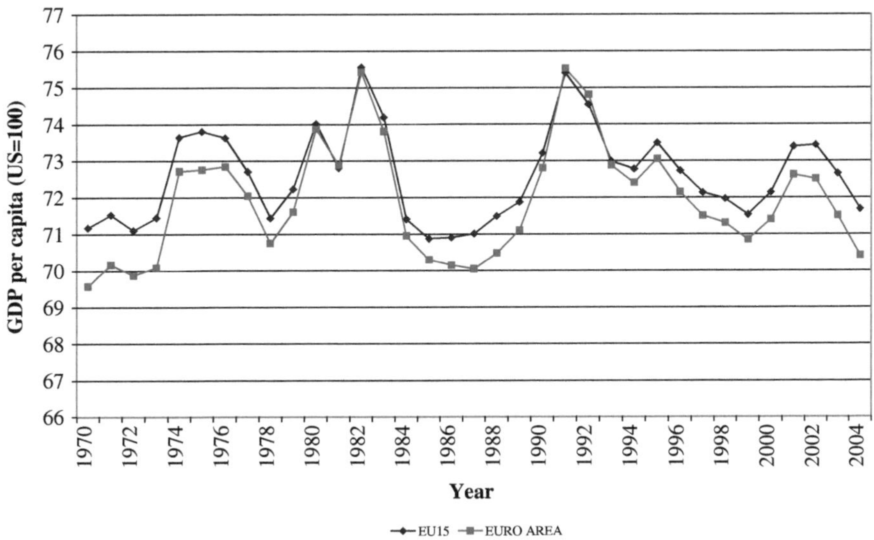

Fig. 1.3. European standards of living are much lower than US ones, and not catching up (GDP per Capita (US=100), EU-15 and Euro Area, 1970–2004 (per head at the price levels and PPPs of 2000 (US Dollars)))
Source: DG Research
Data: OECD

[8] International Monetary Fund, *Euro Area Policies*, 2 July 2004, Washington, DC, p. 14.

[9] Europe has reached past productivity increases to a large extent in a forced manner, by pushing low-skilled labour out of the labour market: *An Agenda for a Growing Europe*, p. 28.

[10] Mary O'Mahony and Bart van Ark (Eds.), *EU Productivity and Competitiveness: An Industry Perspective – Can Europe Resume the Catching-up Process*?, Luxembourg, 2003.

[11] Cécile Denis *et al., The Lisbon Strategy and the EU Structural Productivity Problem*, European Economy, European Commission, Directorate-General for Economic and Financial Affairs, Economic Papers, N° 221, February 2005, p. 4.

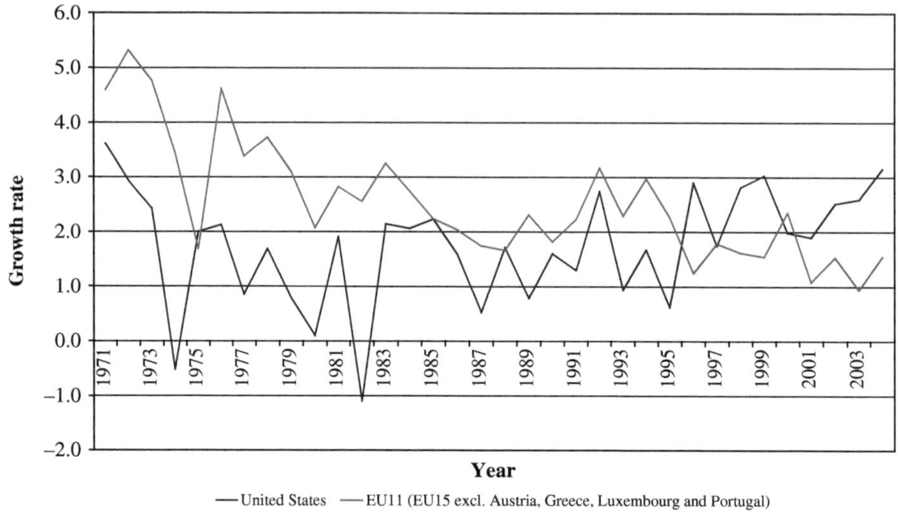

Fig. 1.4. European productivity is no longer catching up with the US one (GDP per hour worked, annual growth rates, EU-11 and US, 1970–2004)
Source: DG Research
Data: OECD

and short and inflexible working hours drive entire industries to low-cost, high-tech countries in Eastern Europe and Asia. The evidence for deindustrialisation is not clear-cut. Some analyses point out that industry still accounts for the same important share of Gross Domestic Product in terms of volume as in the past, while the declining share in terms of value added and employment is due simply to decreasing prices because of productivity gains and exposure to competition higher than that for services. Should it occur, the impact of deindustrialisation would indeed be worrying: the existence of many services depends on the presence of industry; industry pays better wages than services, even for low-skilled jobs; industry accounts for most innovations and technological revolutions; and industry has an important strategic role.[12]

Europe's feeble competitive position is also clear from its weak trade performance, especially that at the high-tech end. Europe's most dynamic export products are generally not those one would closely associate with the knowledge-based economy. The top three products with the fastest growing market share are floor coverings, pork and poultry fat, and hemp. On the other hand, if one looks at products for which market share is in major decline (> 10 per cent loss in market share), the EU has many more (345 product groups) than the United States (65) or Japan (90). What is more, in Europe many technological products are among

[12] CEPII-CIREM, *European Industry's Place in the International Division of Labour*, Report Prepared for DG Trade of the European Commission, July 2004.

them (e.g. air launchers, turbines, insulating glazing, drugs containing alkaloids or hormones, telephones, photographic film).[13] High-tech manufacturing exports represent a much smaller proportion of total manufacturing exports in Europe than in the United States or Japan (in 2002, 19.7 per cent vs. 28.5 per cent and 26.5 per cent respectively).[14] Europe's share of global high-tech manufacturing exports, though increasing, is lower than that of the United States (in 2002, 16.7 per cent vs. 19.5 per cent respectively).[15] And Europe runs a structural deficit in high-tech manufacturing trade, whereas the United States and Japan run surpluses.[16]

1.3. Not enough jobs, and not enough of them high-level

The European employment input is significantly lower than that in the United States. First, though apparently catching-up, the European employment rate is still substantially lower than that of the United States (Fig. 1.5). In 2004, the EU-25 employment rate was 63.3 per cent and the EU-15 one 64.7 per cent, so 6 to 7 percentage points below the target under the Lisbon agenda, compared to 71.2 per cent in the United States.[17] This is mainly due to the limited participation of women, the young, and the elderly in the labour force. At 55.7 per cent and

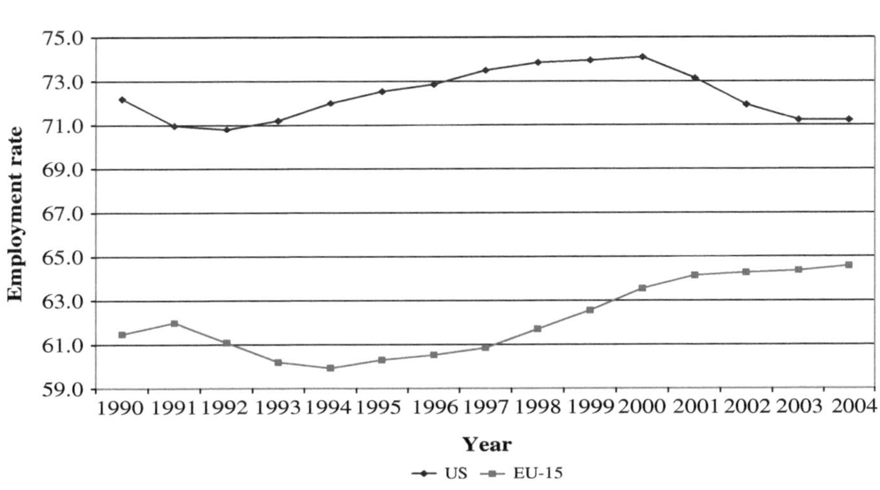

Fig. 1.5. Relatively fewer people are employed in Europe than in the US (employment rate, EU-15 and US, 1990–2004)
Source: DG Research
Data: OECD
Note: Share of persons of working age (15–64 years) in employment

---

[13] CEPII-CIREM, *European Industry's Place In the International Division Of Labour.*

[14] European Commission, *Key Figures 2005: Towards a European Research Area: Science, Technology and Innovation*, Luxembourg, 2005, p. 66.

[15] European Commission, *Key Figures 2005*, p. 67.

[16] European Commission, *Key Figures 2005*, p. 67.

[17] European Commission, *Employment in Europe 2005*, Luxembourg, 2005, p. 17.

41.0 per cent, the female and older people's employment rates were about 4 and 9 percentage points below the Lisbon targets for 2010.[18]

Second, Europe also scores lower than the United States in terms of the number of hours worked annually per employee (Fig. 1.6).[19]

For a long time, the low employment rate and number of hours worked annually per employee were explained with reference to the European emphasis on work-life balance. A growing number of authors draw attention to the existence of disincentives to work, however, the main one being the lack of employment opportunities.[20] This lack of employment opportunities is clear from the high unemployment rates. In 2004, about 19.4 million Europeans were out of work. This equalled 9.0 per cent of the labour force, some 4 percentage points higher than the rates in the United States and Japan (Fig. 1.7).[21] The proportion of high-level jobs is also considerably lower in Europe than in the United States.[22]

1.4. The cohesion and enlargement challenges: Substantial poverty and regional inequality

Though Europe likes to pride itself on its superior social model, poverty rates are rather high, and regional inequality is substantial. In 2004, the at-risk-of-poverty

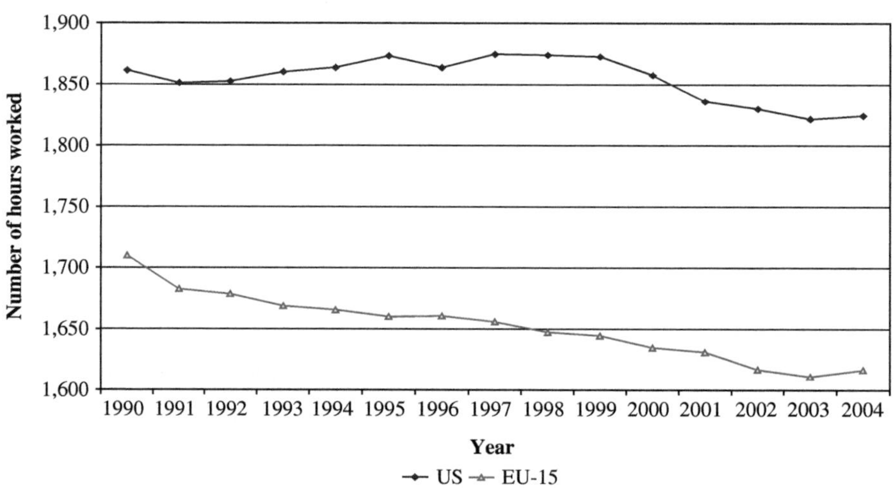

Fig. 1.6. An average European works far fewer hours than an average American (actual hours worked per year per person in employment, EU-15 and US, 1990–2004)
Source: DG Research
Data: OECD

[18] European Commission, *Employment in Europe 2005*, p. 9.
[19] International Monetary Fund, *Euro Area Policies*, p. 14.
[20] International Monetary Fund, *Euro Area Policies*, pp. 14 and 17.
[21] European Commission, *Employment in Europe 2005*, p. 24 and Statistical Annex.
[22] European Commission, *Employment in Europe 2004*, p. 128.

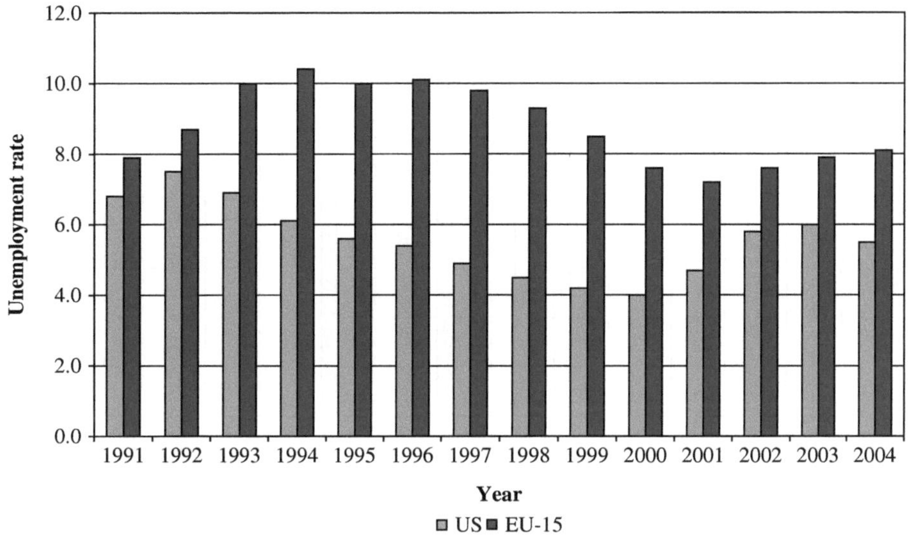

Fig. 1.7. Relatively more people are unemployed in Europe than in the United States (unemployment rate, EU-15 and US, 1991–2004)
Source: DG Research
Data: OECD

rate was 17 per cent in the EU-15, and 16 per cent in the EU-25.[23] In the same year, the 20 per cent of the population with the highest income (top quintile) received almost 5 times as much income as the 20 per cent of the population with the lowest income (lowest quintile) in both the EU-15 and the EU-25.[24] The Gini coefficient – a number between 0 and 1 used to express the degree of income inequality, where 0 corresponds to perfect income equality and 1 corresponds to perfect income inequality – was 0.30 in both the EU-15 and the EU-25 (Fig. 1.8).[25] The share of children living in households with income below the poverty line ranges from 7 per cent in Slovenia and 9 per cent in Denmark to 30 per cent in Slovakia.[26] The proportion of people aged 65 and over with income below the poverty line varies across the EU from 4 per cent in the Czech Republic and under 10 per cent in France, Hungary and Poland to 30 per cent in Spain, over 40 per cent in Ireland, and over 50 per cent in Cyprus.[27]

---

[23] Eurostat [The at-risk-of-poverty rate must be understood as the share of persons with an equivalised disposable income below the risk-of-poverty threshold, which is set at 60 per cent of the national median equivalised disposable income (after social transfers)].

[24] Eurostat (Income must be understood as equivalised disposable income).

[25] The Gini coefficient is a measure of inequality of a distribution.

[26] Applica *et al.*, *Network on Social Inclusion and Income distribution – Final Report,* European Observatory on the Social Situation (SSO) – Contract No VC/2004/0462, December 2005, pp. v and vi.

[27] Applica *et al.*, *Network on Social Inclusion and Income Distribution*, pp. v and vi.

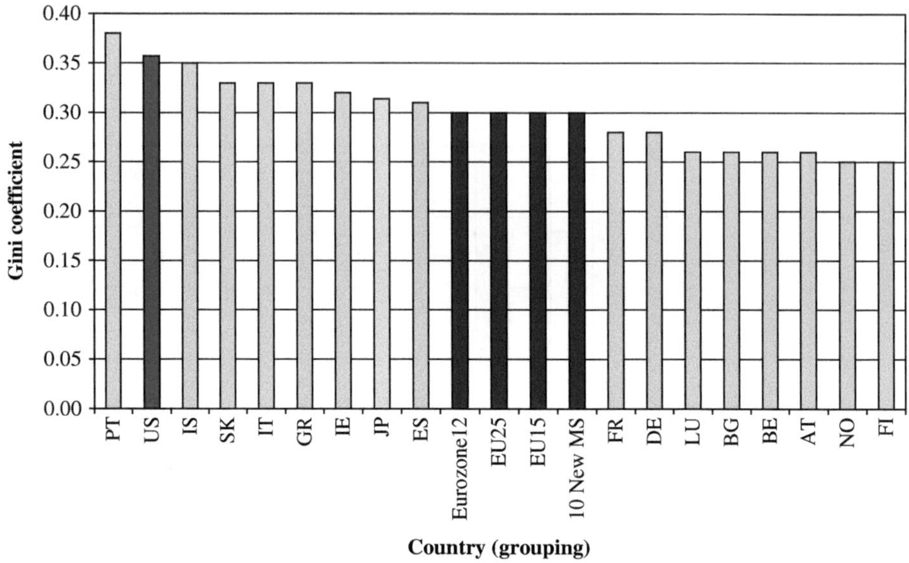

Fig. 1.8. The degree of income inequality is relatively high in Europe (GINI-Coefficient, 2004)
Source: DG Research
Data: Eurostat
Note: The Gini coefficient is a number between 0 and 1 used to express the degree of income inequality, where 0 corresponds to perfect income equality and 1 corresponds to perfect income inequality. Data for US and JP are for 2000 and taken from: Förster, M. and M. Mira d'Ercole (2005), "Income Distribution and Poverty in OECD Countries in the Second Half of the 1990s", OECD Social Employment and Migration Working Papers, No. 22, OECD Publishing. doi:10.1787/882106484586.

Substantial regional diversity and inequality characterises the European Union. In 2004, employment rates ranged from 39.5 per cent in Réunion (France) to 78.2 per cent, or almost twice as high, in Bedfordshire, Hertfordshire (United Kingdom). Unemployment rates, on the other hand, ranged in 2003 from 20.1 per cent in Sicily (Italy), 20.4 per cent in Mecklenburg-Vorpommern (Germany), and 25.5 per cent in the Polish region of Zachodniopomorskie to a mere 3.4 per cent in the Dutch region of Noord-Brabant, 3.7 per cent in Luxembourg and 4.0 per cent in the Budapest area.[28] The EU is also marked by substantial inequality in income levels. In 2002, Gross Domestic Product (GDP) per capita was below 75 per cent of the EU-25 average in 63 out of 254 NUTS 2 regions examined in the EU-25.[29] The highest

---

[28] Eurostat.

[29] At the beginning of the 1970s, Eurostat set up the 'Nomenclature of Statistical Territorial Units' (NUTS) as a single, coherent system for dividing up the European Union's territory in order to produce regional statistics for the Community. NUTS subdivides each Member State into a whole number of regions at NUTS 1 level. Each of these is then subdivided into regions at NUTS level 2, and these in turn into regions at NUTS level 3. Leaving aside

regional Gross Domestic Product per capita (Inner London – United Kingdom) was about 10 times the lowest one (Lubelskie – Poland).

Enlargement, for the European Union, is at one and the same time a challenge and an achievement, a "raison d'être" and a "façon d'être". It is a continuation of the historical process that started over 50 years ago with the Communities' inception, developed through several steps (in 1973, 1981, 1986, 1991, 1995), and reached a high point – albeit not an end-point – with the enlargement of the European Union to 10 countries of Eastern and Southern Europe on 1 May 2004. Preparation for that enlargement took several years and by the time they joined, the EU-10 had successfully transformed their economies from centrally planned to functioning free market ones. Compliance with the Copenhagen criteria for accession served as a powerful catalyst for change. This assessment is detailed in a recent stock-taking exercise in which the Commission services have provided strong evidence and analyses indicating that the 2004 enlargement constitutes an economic success for the "old" and the "new" Member States alike.[30] It has to be noted that enlargement has been a dynamic process rather than a discrete event and that its effects will become visible over time.

Figure 1.9 shows that convergence and catching up in real income have been at work throughout the period since the late 1990s. Per-capita incomes are now much closer to EU-15 levels than they were in 1997, the year in which enlargement prospects crystallized in the Commission's Agenda 2000. After the output collapse in the early years of transition, growth rates in the EU-10 have been higher than in the EU-15, but also more volatile. The key contributors to actual and potential economic growth in the EU-10 have been capital accumulation and technical progress (the so-called Total Factor Productivity, TFP), while the contribution of labour has been mostly negative (that being a reflection of weak employment growth and, to a lesser extent, of an ongoing decline in hours worked per employee). In general, and consistent with the convergence hypothesis, Member States with lower initial (1997) per capita income tended to grow faster in the intervening years.

1.5. The demographic challenge: Fertility decline and ageing

Birth rates continue to be low in Europe.[31] Everywhere, the fertility rate is below the threshold needed to renew the population (around 2.1 children per woman),

---

the local level (municipalities), the internal administrative structure of the Member States is generally based on two of these three main regional levels. This existing national administrative structure may be, for example, at NUTS 1 and NUTS 3 levels (respectively the Länder and Kreise in Germany, or at NUTS 2 and NUTS 3 (régions and départements in France, Comunidades autónomas and provincias in Spain).

[30] European Commission, *Enlargement, Two Years after – An Economic Success*, Communication from the Commission to the Council and the European Parliament, COM(2006) yyy final, 3 May 2006; European Commission, *Enlargement, Two Years after: An Economic Evaluation*, European Economy, Occasional Papers, No. 24, May 2006.

[31] Ageing also threatens a developing country like China, see Howard W. French, As China Ages, a Shortage of Cheap Labor Looms, In: *The New York Times*, 30 June 2006.

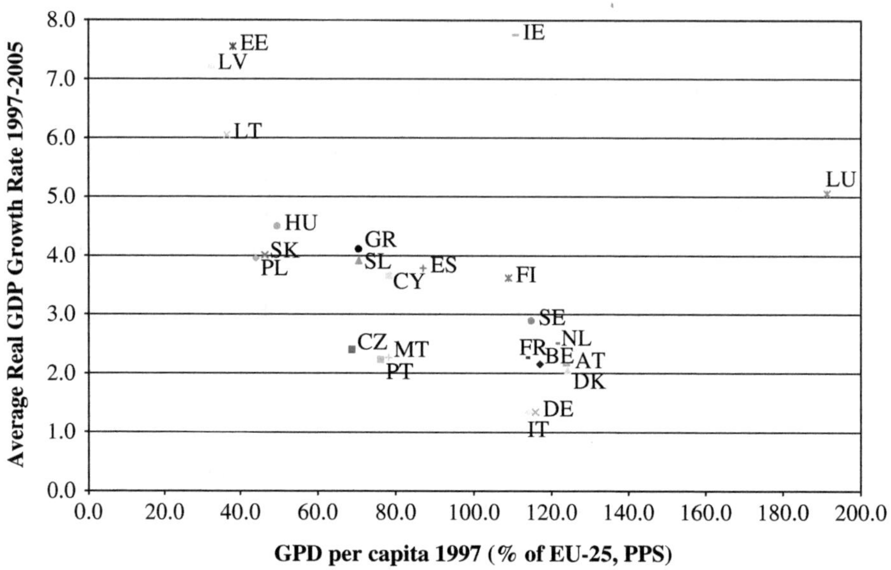

Fig. 1.9. The enlarged Europe is converging in terms of real income
Source: DG Research

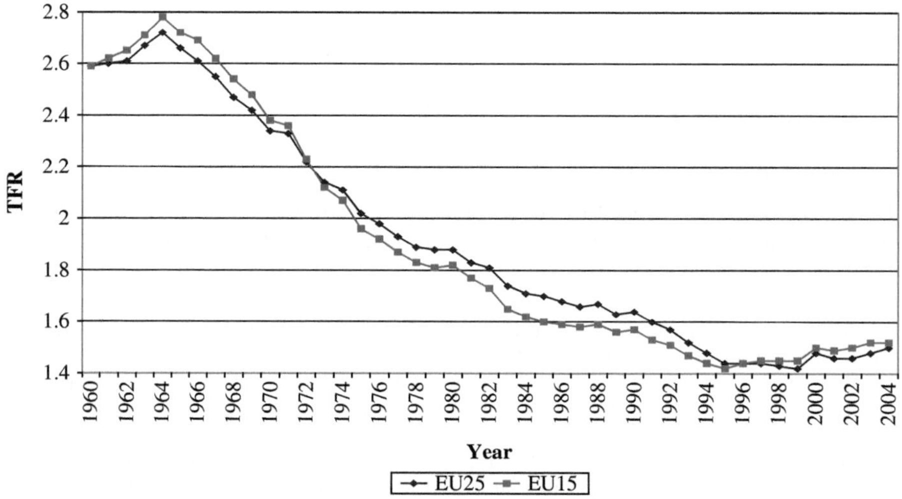

Fig. 1.10. European fertility is declining (total fertility rate, EU-15 and EU-25, 1960–2004)
Source: DG Research
Data: Eurostat

and in many Member States it has even fallen below 1.5 children per woman (Fig. 1.10). In 2003, the natural population increase in Europe was just 0.04 per cent per annum, and in many countries the population would shrink if it were not for immigration. Except for Cyprus and Malta, all new Member States already saw falling populations. The Union's population is set to grow just slightly up until 2025, thanks to immigration, before starting to drop: 458 million in 2005, 469.5 million in 2025, and 468.7 million in 2030. Between 2005 and 2030, the total working age population (15–64 years) is set to fall by 20.8 million. At the same time, as average life expectancies are increasing, the demographic dependency rate will rise from 49 per cent in 2005 to 66 per cent in 2030, putting pressure on pension and health care systems. Ageing could cause potential annual growth in Gross National Product in Europe to fall from about 2 per cent now to 1.25 per cent in 2040.[32]

1.6. The public health challenge: Lifestyle and communicable diseases

The health of European citizens is negatively affected by lifestyle as well as communicable diseases. For both men and women in the EU-25, circulatory diseases are the major cause of death. They accounted for 1.8 million or 42 per cent of all deaths in the EU-25 in 2002 (38 per cent of male deaths and 46 per cent of female deaths).[33] Cancer is the second most frequent cause of death in Europe. In 2002, it was responsible for a quarter of all deaths (29 per cent of deaths for men and 22 per cent of women).[34] In the same year, for every 100,000 men in the EU-25 311 new cases of cancer were reported, and for every 100,000 women 232 new cases.[35] The chronic disease diabetes (mellitus) and its complications have become a major public health problem in all countries. It causes significant physical and psychological morbidity, disability and premature mortality among those affected and imposes a heavy financial burden on health services. The prevalence of diabetes is rising. In the EU, there were an estimated 12.7 million diabetics in 1995 projected to rise to 19.6 million by 2010. In 2000, an estimated 4.1 per cent of the EU population were diabetics.[36]

Communicable diseases killed some 15 million people worldwide in 2002, of which some 225 thousand in Europe.[37] HIV/AIDS is the most important communicable disease of our times, followed by tuberculosis and malaria. In 2005, the number of people living with HIV, for instance, was estimated at 40.3 million worldwide, and that of people newly infected with HIV at 4.9 million. The number

---

[32] European Commission, *Green Paper "Confronting Demographic Change: A New Solidarity between the Generations"*, Communication from the Commission, COM(2005) 94 final, 16 March 2005.

[33] European Commission, *Health in Europe – Data 1998–2003*, Luxembourg, 2005, p. 63. European Commission, DG Health and Consumer Protection, *Public Health, Disease and Conditions Information Sheets: Cardiovascular Diseases*.

[34] European Commission, *Health in Europe*, p. 63.

[35] European Commission, *Health in Europe*, p. 33.

[36] European Commission, DG Health and Consumer Protection, *Public Health, Disease and Conditions Information Sheets: Diabetes*.

[37] WHO, *The WorldHealth Report 2004*, Geneva, 2004, Annex Table 1.2.

of AIDS deaths was estimated at 3.1 million.[38] In Europe, the number of newly reported HIV infections is increasing, while that of newly diagnosed AIDS cases is decreasing. In the 17 EU countries with data available for 1996 and 2003 for both HIV infections and AIDS cases, the number of newly reported HIV infections increased by almost 75 per cent (from 7641 to 13,257) while the number of newly diagnosed AIDS cases fell by over 55 per cent (from 4 085 to 1 772).[39] Europe is also affected by other communicable diseases including SARS and avian influenza.

1.7. The environmental challenge: Water, climate change, and biodiversity

One of the most worrying challenges for Europe, and indeed for the whole world, concerns the deterioration of the environment. European citizens overwhelmingly agree that the state of the environment influences their quality of life (72 per cent), that policy-makers should consider the environment to be just as important as economic and social policies (85 per cent), and that policy-makers should take into account environmental concerns when deciding policy in other areas such as the economy and employment.[40] "A high level of protection and improvement of the quality of the environment" is a European Community objective (see above). Europe has been implementing environmental action plans and pursuing sustainable development strategies at both national and European level for quite some time now. It plays a leading role in the fight against global warming.[41] And it occupies a strong position in the field of environmental technologies.

Yet, because of population growth; consumption patterns; market, policy and political failures; features of existing technologies; and world views and values, Europe and the world at large are still far removed from a development trajectory that is truly sustainable, that is, which satisfies the current needs of society (growth, competitiveness, employment, etc.) without compromising the needs of future generations.[42] European citizens worry most about water pollution (of seas, rivers, lakes, underground sources, etc.) (47 per cent); man-made disasters (major oil spills, industrial accidents, etc.) (46 per cent); climate change (45 per cent); and air pollution (45 per cent).[43] The Sixth Environment Action Programme of the European Community 2002–2012 (6th EAP) identifies four priority areas for urgent action: (1) climate change; (2) nature and biodiversity; (3) environment and health and quality of life; and (4) natural

---

[38] UNAIDS, *AIDS Epidemic Update: December 2005 – Global Summary of the AIDS Epidemic*, Geneva, December 2005.

[39] European Commission, *Health in Europe*, p. 35.

[40] European Commission, *Special Eurobarometer – The Attitudes Of European Citizens Towards Environment*, Luxembourg, April 2005, Annex Tables.

[41] Though, in contrast to its image in the outside world, Europe still relies to a large extent on coal. See for instance: Mark Landler, The Energy Challenge – Europe's Green Image Clashes with Reliance on Coal, In: *The New York Times*, 20 June 2006.

[42] Robert L. Olson and Jessica Biamonte, *New Global Agenda: An Analysis of Major Overview Studies of the Global Environment*, A White Paper Prepared for the Foresight and Governance Project, Woodrow Wilson International Center for Scholars, Washington, D.C., August 2004.

[43] European Commission, *Special Eurobarometer*, pp. 5–6.

resources and waste. The environmental objectives of the EU Sustainable Development Strategy include: (1) addressing climate change; (2) better management of natural resources; and (3) making transport more sustainable. A 2004 review of nine recent comprehensive analyses of global environmental problems (Table 1.1) showed near-unanimous agreement that the three problems posing the greatest threats to the global environment and continuing economic development include: (1) water quality and access; (2) climate change; and (3) loss of biodiversity.[44] Climate change forecasts indicate that, if the level of emissions is not curbed, the temperature level will rise and risks such as water shortage, malaria and hunger will increase and affect millions of people by 2080 (Fig. 1.11).

Addressing such environmental problems is highly complex. One of the premises of sustainable development is that environmental problems interact with each other, as well as with economic and social issues. Climate change affects agriculture, forestry, water availability, marine systems, terrestrial ecosystems, health and, last but not least, the economy. Forests and oceans act as climate regulators but also harbour a wide diversity of species. Decisively tackling the issue of biodiversity will require i.a. making forestry sustainable, addressing pollution, and dealing with climate change. Pollution negatively affects health, from allergies and infertility to cancer and premature death. In the mid-1990s damage costs to the EU caused by air pollution originated in the EU (see Table 1.2) were calculated to be around 2 per cent of EU GDP (ranging from 0.3 to 3.2 per cent) and damages to EU and non EU countries caused by air pollution originated within the EU were estimated to be 2.6 per cent of EU GDP (with ranges between 0.4 and 6.9 per cent), with health damages accounting for the largest share.[45] An animal and human health problem like aviary flu also constitutes a threat to biodiversity.

Environmental degradation contributes to the increase recorded in the number of disasters and, in relation to this, to a heightened sense of vulnerability (see Fig. 1.13 in the last section of this chapter). Disasters can be man-made or natural and include wildland fires, earthquakes, volcanic eruptions, landslides/debris flows, floods, extreme weather, tropical cyclones, sea and lake ice, coastal hazards including tsunamis, pollution events, and so on. During the period 1990–1999, disasters killed 500,000 people and caused 750 billion dollars of damage.

2. GREAT EXPECTATIONS OF S&T AS REGARDS TACKLING THE MULTITUDE OF CHALLENGES

Throughout history, the relation between science and society has been marked by both continuity and change.[46] The continuity is situated in the tension between the

[44] Robert L. Olson and Jessica Biamonte, *New Global Agenda*.

[45] This was demonstrated by a number of FP-funded projects In particular, GARP II funded under FP4 and GREENSENSE funded under FP5 applied the ExternE methodology to the calculation of national damage accounts.

[46] This and following paragraph based on Andrew Ede and Lesley B. Cormack, *A History of Science in Society – From Philosophy to Utility*, Toronto, 2004.

Table 1.1. What are the main environmental challenges?

Source	Major problems							
	Water quality and access	Climate change	Biodiversity	Indoor air quality	Air pollution	Food production/soil degradation	Over-fishing	Deforestation
World Resources Institute *et al.*, World Resources 2000–2001: People and Ecosystems. Washington, D.C.:WRI 2000.	■	■	■		♦	♦	♦	♦
Organisation for Economic Co-operation and Development, Environmental Outlook, Paris: OECD, 2001.	■	■	■		♦	♦	■	■
The World Bank. World Development Report 2003: Sustainable Development in a Dynamic World. New York: World Bank, 2003.	■	■	■		■	■	■	■
Ehrlich, Paul and Anne Ehrlich. One With Niveneh: Politics, Consumption, and the Human Future. Washington, DC: Island Press, 2004.	■	■	■		♦	♦	■	♦
Speth, James Gustave, Red Sky at Morning: America and the Crisis of the Global Environment. New Haven: Yale Univ. Press, 2004.	■	■	■		♦	♦	♦	

United Nation Environment Program. Global Environment Oultook: Past, Present, and Future Perspectives. London: Earthscan, 2002.	■	■	■			♦	♦	♦
Central Intelligence Agency, Global Trends 2015: A Dialogues About the Future with Nongovernmental Experts. Washington, D.C.: CIA, 2000.	■	■	■		♦	■		♦
National Advisory Council for Environmental Policy and Technology (NACEPT). The environmental Futures: Emerging Challenges and opportunities for EPA. Washington, D.C.: NACEPT, 2002.	■	■	■	■	♦			
Brown, Lester R. Plan B: Rescuing a Planet Under Stress and a Civilization in Trouble. New York: W.W. Norton, 2003.	■	■			♦	■	♦	♦

■ Problem is a serious threat to the global environment and continued economic development.
♦ Problem is a threat to the global environment.
Source: Robert L. Olson and Jessica Biamonte, New Global Agenda, 2004.

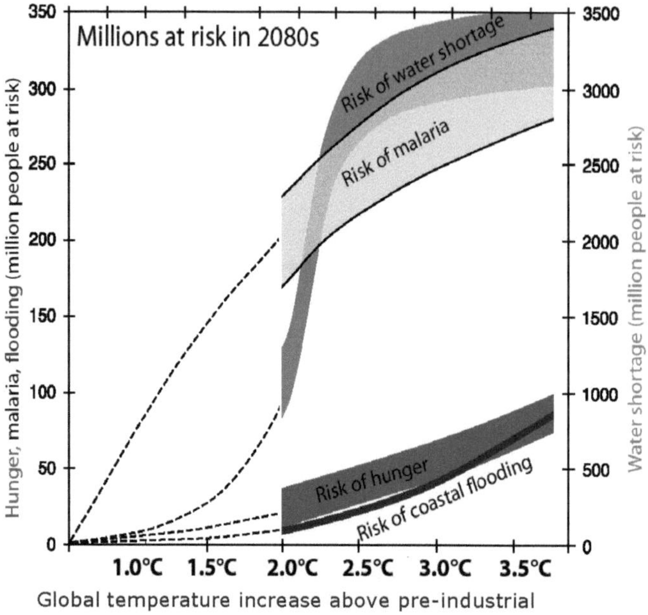

Fig. 1.11. The number of people at risk from global warming in 2080
Source: Parry, Martin *et al.*, Millions at Risk: Defining Critical Climate Change Threats and Targets, In: *Global Environmental Change*, 11, 3, 2001, pp. 181–183.

philosophical and intellectual pursuit of and search for knowledge on the one hand, and the desire of researchers and their supporters to make scientific knowledge useful and apply it on the other hand. This tension was first recognised by the ancient Greek philosophers, and has been reflected in recurring calls from philosophers and scientists throughout history, including today, for more "research for its own sake". Within the context of this tension, the change has been located in what has constituted or better what has been considered useful knowledge in each age, in other words in "the changing social expectation of science": "What counts as useful knowledge differed from patron to patron and society to society, so that Cosimo de Medici and the United States Department of Energy looked for quite different 'products' to be created by their clients, but both traded support for the potential of utility".

From century to century, societal expectations of S&T have not just changed. They have also increased. In the era of the ancient Greek philosophers, societal expectations of S&T were rather low. S&T was a highly controlled activity carried out by a small elite group of people for philosophical or religious objectives. At present, however, it is considered a powerful tool for political, economic, and social change. In between, S&T helped exploit worldwide resources as trade empires and colonies expanded (18th century); helped expanding and consolidating trade empires and colonies, and turn their natural resources into wealth, or make up for the lack of trade empires and colonies (19th century); helped fight wars (First World

Table 1.2. What air pollution is costing us

	Damage costs (Million euro)		Damage costs per GDP (%)	
Member States	To EU-15	To non EU-15	To Eu-15 + non-EU-15	TO EU-15
Austria	1200	1800	1.8	0.7
Belgium	4400	400	2.4	2.2
Denmark	1200	400	1.2	0.9
Finland	300	100	0.5	0.4
France	23200	2000	2.2	2.0
Germany	34400	17000	2.9	2.0
Greece	2100	3700	6.9	2.5
Ireland	700	0	1.5	1.5
Italy	15800	6800	2.6	1.8
Luxembourg	300	0	2.3	2.3
Netherlands	4900	500	1.8	1.7
Portugal	1600	0	2.1	2.1
Spain	13500	400	3.3	3.2
Sweden	500	300	0.4	0.3
UK	24700	1200	3.0	2.8
Tot. EU	128800	34600		
Average EU	16100	4325	2.6	2.0

Source: DG Research, GARP II, and GREENSENSE.

War and Second World War); and helped producing consumer goods, consumer medicines, exploring space, addressing environmental challenges, exploring the human genome, and so on (post-war period).

It is no exaggeration to say that as a result today societal expectations of S&T have never been higher in industrial countries. In the United States, the *Carnegie Commission on Science, Technology, and Government* listed in 1992 no less than 25 major societal goals to which S&T can contribute (Table 1.3). And a *National Academies* report noted in 2005 that "the nation increasingly looks to the scientific and engineering communities for solutions to some of its most intractable problems, from chronic disease to missile defence, to transportation woes, to energy security, to ensuring clean air and clean water. Expectations for S&T are perhaps higher than at any other time in our history and are placing unprecedented demands on leadership".[47] In Japan, the 2004 White Paper on Science and Technology noted that "what is needed... is for science and technology to respond flexibly to the

[47] Committee on Ensuring the Best Presidential and Federal Advisory Committee Science and Technology Appointments – Committee on Science, Engineering, and Public Policy – National Academy of Sciences, National Academy of Engineering and Institute of Medicine of the National Academies, *Science and Technology in the National Interest, Ensuring the Best Presidential and Federal Advisory Committee Science and Technology Appointments*, Washington, DC, 2005.

Table 1.3. To which societal goals can S&T contribute?

Quality of life, health, human development, and knowledge
Education and diffusion of knowledge
Personal and public health and safety
Personal development and self-realisation
Exploration and expansion of knowledge
High standard of living
Creation and maintenance of civic culture
Cultural pluralism and community harmony
Population stabilisation
A resilient, sustainable, and competitive economy
Economic growth
Full employment and workforce training
International competitiveness
Modernised communications and transportation
International cooperation and action
Environmental quality and sustainable use of natural resources
Worldwide sustainable development
Resource exploration, extraction, conservation, and recycling
Energy production and efficiency in use
Environmental quality and protection
Provisions for public recreation
Maintenance and enhancement of productivity of the biosphere
Maintenance of urban infrastructure
Energy security and strategic materials
Personal, national, and international security
Personal security and social justice
National and international security
Individual freedom
Worldwide human rights

Source: Carnegie Commission on Science, Technology and Government, *Enabling the Future: Linking Science and Technology to Social Goals*, 1992.

needs of society as they change over time, or in other words, to become a 'science and technology for society' ".[48]

Things are no different in Europe. In 2000, the European Commission remarked that "expectations of science and technology are getting higher and higher, and there are few problems facing European society where science and technology are not called upon, one way or another, to provide solutions".[49] Starting with that year's Lisbon European Council, this trend has if anything only strengthened. In the past few years, a great number of high-profile analyses have been carried

[48] Ministry of Education, Culture, Sports, Science and Technology, *Science and Technology and Society in the Future*, White Paper on Science and Technology 2004, Tokyo, 2004.

[49] European Commission, *Science, Society and the Citizen in Europe*, Commission Working Document, SEC(2000) 1973, 14 November 2000, p. 5; see also Biotechnology: Europeans Start to Believe the Health Benefits, In: *Europe Information Service*, European Report No. 3108, 20 June 2006.

out of the challenges Europe is facing, and recommendations have been made on how to address them. Time and again the same wide range of urgently to be addressed challenges is identified. The reports are also near-unanimous in the key role assigned to S&T in this respect, as will be seen in Chapter 3. In other words, great expectations are held of S&T as regards the tackling of the multitude of challenges Europe is facing. This will be developed in Chapter 3 as part of the new policy context that enabled the genesis of the Lisbon Strategy as well as of the 7th Framework Programme.

3. THE ROLE THAT S&T CAN PLAY

The role that S&T can play in addressing all these challenges is expected to be substantial. This section will show that S&T indeed has the potential to contribute to a range of economic, social and environmental challenges: it can improve economic performance, promote employment, improve public health, tackle demographic, cohesion and environmental challenges, and so on.

3.1. S&T, economic growth and competitiveness

Modern mainstream economic theory – whether neoclassical, endogenous or evolutionary – has recognised for quite some time now that technological progress and innovation are the main engines of economic growth. According to Baumol, innovation explains much of the extraordinary economic growth record under capitalism. The reason is that in important parts of the economy, competition is based on innovation rather than price. Firms are therefore forced by market pressure to support innovative activity systematically and substantially.[50] According to Romer, productivity growth is driven by innovation resulting in the creation of new though not necessarily improved product varieties.[51] And under the Schumpeterian paradigm, growth results from "quality improving innovations that render old products obsolete, and hence involves the force that Schumpeter called 'creative destruction' ".[52] Even basic research generates several direct economic benefits. It is a source of useful new information; it creates new instrumentation and methodologies. Those engaged in basic research develop skills which yield economic benefits when individuals move from basic research carrying codified and tacit knowledge. Through participation in basic research, access is granted to networks of experts and information. Those trained in basic research may be good at solving complex technological problems. And, finally, on the basis of basic research, spin-off companies are created.[53]

---

[50] William J. Baumol, *The Free-Market Innovation Machine. Analyzing the Growth Miracle of Capitalism*, Princeton, NJ and Oxford, 2002, viii–ix.
[51] Philippe Aghion and Peter Howitt, *Appropriate Growth Policy: A Unifying Framework*, 9 August 2005, pp. 2–3.
[52] Aghion and Howitt, *Appropriate Growth Policy: A Unifying Framework*, p. 3.
[53] Ben Martin *et al.*, *The Relationship between Publicly Funded Basic Research and Economic Performance,* A SPRU Review, Report Prepared for HM Treasury, July 1996, p. vii.

There is also empirical support for the contribution of S&T to economic performance (see tables and sources in annex). Estimates of private returns to firms' own investment in R&D still produce varying figures, but there is an emerging consensus that gross returns between 20 and 30 per cent are common and plausible (Table 1.4). Microeconomic studies confirm the existence of significant spillovers of knowledge from the firms that perform the R&D to other firms and industries. Taking account of measured spillovers typically raises the estimated gross rate of return on business investment into the range of 30 to 40 per cent (Tables 1.5–1.7). Macroeconomic studies, which by definition cover all sectors of the economy, also find significantly higher returns to R&D in OECD countries, with estimates ranging from 50 per cent to over 100 per cent. A recent Austrian report found that the rise of corporate spending on R&D from 0.8 per cent to 1.1 per cent of Gross Domestic Product in the second half of the 1990s produced a boost of three tenths of a per cent in growth.[54] Both microeconomic and macroeconomic studies find that an important source of productivity growth in all OECD countries comes from the international diffusion of technology. A country's ability to absorb those foreign technologies is enhanced by investment in education and by investment in own R&D.

3.2. S&T and employment

The economic literature is not conclusive on the employment effects of innovation, since process innovation (the introduction of labour-saving technologies) is likely to have a negative effect on employment, assuming all other factors remain constant, while product innovation creates new markets and employment opportunities.[55] But empirical evidence suggests that technological change promotes employment. Such evidence includes a recent study of the Directorate-General Employment which found that the rate of growth of total factor productivity (due to improvements

Table 1.4. Private rate of return to R&D

Author	Year	Rate of return (%)
Bernstein and Nadiri	1991	14–28
Bernstein and Nadiri	1988, 1989	9–27
Mansfield	1977	25
Nadiri	1993	20–30
Sherer	1982, 1984	29–43
Sveikauskas	1981	10–23
Terleckyj	1974	0–29

Source: Robert M. Margolis and Daniel M. Kammen, Evidence of under-investment in energy R&D in the United States and the impact of Federal policy, In: *Energy Policy* 27, 1999, 575–584.

[54] Federal Ministry of Education, Science and Culture, together with Federal Ministry of Transport, Innovation and Technology, and Federal Ministry of Economics and Labour, *Austrian Research and Technology Report 2004*, Vienna, 2004, p. 2.

[55] European Commission, *Employment in Europe 2004*, p. 77.

Table 1.5. Social rate of return to R&D

Author	Year	Rate of return (%)
Bernstein and Nadiri	1988, 1989	10–160
Bernstein and Nadiri	1991	56
Griffith, Redding and Van Reenen	2000	Canada: 57.2 Denmark: 67.9 Finland: 95.2 France: 54.9 Germany: 49.9 Italy: 71.6 Japan: 70.8 Netherlands: 49.6 Norway: 75.6 Sweden: 68.0 UK: 80.5 US: 41.7
Griliches	1964	35–40
Griliches	1994	30 (on own R&D)
Griliches and Lichtenberg	1984 (Interindustry…)	71 (30 on own R&D plus 41 on used R&D)
Griliches and Lichtenberg	1984 (R&D and…)	34 (on own R&D)
Hall	1995	33 (on own R&D)
Mansfield	1977	56
Nadiri	1993	50
Scherer	1982	103 (29 on own R&D plus 74 on used R&D)
Sveikaukas	1981	17 (on own R&D)
Terleckyj	1974	48–78
Terleckyj	1980	107 (25 on own R&D and 82 on used R&D)

Source: Rachel Griffith, *How Important Is Business R&D for Economic Growth and Should the Government Subsidise it?*; Charles I. Jones and John C. Williams, *Measuring the Social Return to R&D*, February 1997; Charles I. Jones and John C. Williams, Measuring the Social Return to R&D, In: *The Quarterly Journal of Economics*, November 1998; Robert M. Margolis and Daniel M. Kammen, Evidence of Under-Investment in Energy R&D in the United States and the Impact of Federal Policy, In: *Energy Policy* 27, 1999, pp. 575–584.

in the efficiency of production or to pure technological progress) has a positive impact on the employment rate, with a one-year lag, and that both in the short- and long-term, countries with higher than average total factor productivity growth tend also to have higher than average growth in employment.[56] Clear evidence exists that more computerised or R&D-intensive industries increased their demand for college-educated workers at a faster rate in the 1980s. Such high-skilled workers also command higher wages, as the consensus is that the increase in the schooling wage premium and the rise in wage inequality are driven by technological change.[57]

56 European Commission, *Employment in Europe 2004*, p. 80.

57 European Commission, *Human Capital in a Global and Knowledge-Based Economy – Final Report*, Luxembourg, 2003, pp. 14–15.

Table 1.6. Rate of return to public R&D

Author	Year	Rate of return (%)
Bredahl and Peterson	1976	36–47
Evenson	1979	45–130
Evenson and Welch	1964	55–60
Griliches	1958	20–40
Norton	1981	27–132
Peterson	1967	21–25
Schmitz and Seckler	1970	16–46
Smith, Norton, and Havlicek	1983	22–61
Sundquist, Cheng, and Norton	1981	97–118

Source: Keith Fuglie *et al.*, Economic Returns to Public Agricultural Research, In: *Agricultural Research and Development: Public and Private Investments under Alternative Markets and Institutions*, Agricultural Economics Report AER735, May 1996.

Support also comes from the observation that all Member States saw employment levels in the high technology sector rise between 1997 and 2002, leading to an increase of almost 2 million for the Union as a whole, with employment in high-tech services accounting for 1.4 million of this total (Fig. 1.12).[58]

Table 1.7. R&D and innovation

Author	Year	R&D and innovation
Mansfield	1991	11% of new products and 9 % of new processes could not have been developed in the absence of academic research without substantial delay
Mansfield	1998	15 % of new products and 11 % of new processes could not have been developed without a substantial delay in the absence of academic research
Beise and Stahl	1999	One-tenth of the firms that produced product or process innovations between 1993 and 1995 would not have done so without public research
Autant-Bernard	2001	Public research increases private innovation directly, and indirectly by increasing private research. These effects are geographically localised
Tijssen	2001	Approximately 20% of private sector innovations are partially based on public sector research

Source: Alister Scott *et al.*, *The Economic Returns to Basic Research and the Benefits of University-Industry Relationships – A Literature Review and Update of Findings*, Report for the OST by SPRU.

[58] European Commission, *Employment in Europe 2003*, Luxembourg, 2003, p. 42.

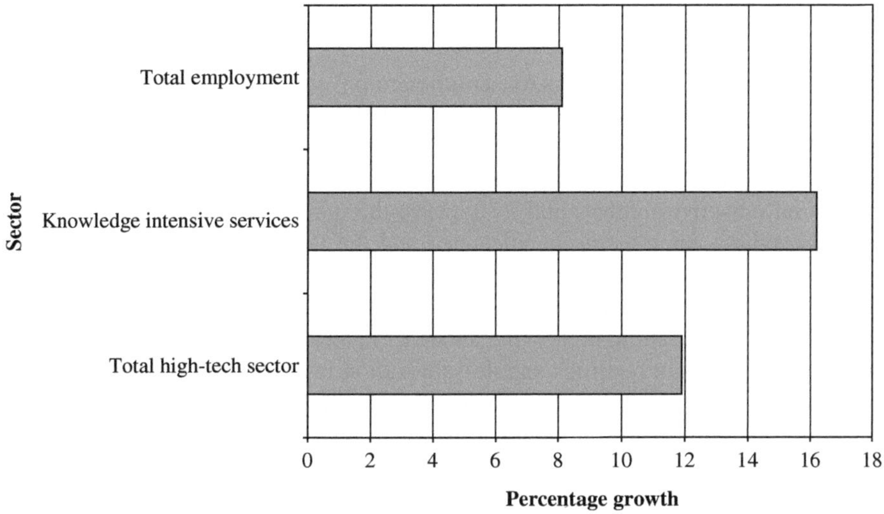

Fig. 1.12. Knowledge-intensive industries and services create more jobs (changes in employment in % (1997–2002, EU-15))
Source: DG Research
Data: European Commission, *Employment in Europe 2003*

3.3. S&T and the ageing and poverty challenges

Through its contribution to product, process and service innovation, productivity growth, and the creation of more and higher paid jobs, research and innovation can also help meet the challenges of ageing and cohesion. Higher employment rates and levels of productivity – to which S&T can contribute – would allow for maintaining or increasing living standards, and for the absorption of increasing medical and pension costs. Doubling the growth of productivity over the next few decades would allow for maintaining current levels of industrial production and average per capita income with some 40 million elderly in the EU.[59]

The best solution to poverty is investing in education.[60] For instance, in general the lower the illiteracy rate, the higher per capita income.[61] Higher levels of educational attainment enhance the chance of finding work and enjoying a decent standard of living. However, education is not yet accessible for everyone and often only to those who can afford it. Improving access to educations takes time and

[59] Paraskevas Caracostas and Ugur Muldur, *Society – The Endless Frontier*, Luxembourg, 1998, pp. 94–95.
[60] UNICEF, *The State of the World's Children 2005*, New York, 2005; World Bank, *World Development Report 2000/2001*, Washington, D.C., 2000.
[61] Per capita income in countries with a literacy rate less than 55 per cent averages about $600 whereas per capita income in countries with a literacy rate between 55 and 84 per cent is $2400. Source: World Bank, *World Development Report 2000/2001*.

effort. Education is, therefore, in its own right not powerful enough to solve the poverty problem.

In the meantime, contributions to a solution to poverty can also be expected from Science and Technology. Besides investing in education and developing skills, this means dedicating research programmes to find ways to fight inner-city poverty, to relieve the effects of urbanisation, to diminish the impacts of ever increasing mobility on our environment, and to improve the quality of life of the vulnerable groups in society, such as the handicapped and the ill, the elderly and the young. In developing countries this can take the form of helping to improve the productivity of natural and physical assets, for example, by protecting farmland against erosion and desertification, preserving an area's natural resources, building easy-to-maintain water storage facilities and de-salinisation installations, and strengthening farmers' diagnostic capabilities in relation to livestock diseases, to name a few.[62] That these advances have important impacts on farmers' income levels has been repeatedly demonstrated by the different targeted activities across the Framework Programmes.[63]

3.4. S&T and the public health challenge

Science and Technology can also make a large contribution to the improvement of public health. It can assist in prevention (e.g. through the development of vaccinations), it can play an important role in the quicker and more reliable diagnosis of diseases (e.g. through the further development of medical imaging), and it can find treatments for diseases or, in the absence of treatments, it can help finding ways to control them (e.g. HIV/AIDS retroviral drugs). S&T can also help to lessen the impact of disease. Furthermore, S&T can help to find new ways to deliver treatment (e.g. ambulant rather than hospital treatment) and can provide better tools for health care system management. A good illustration of the way in which Science and Technology can make a positive contribution to public health is the Article 169 EDCTP[64] initiative referred to in Chapter 4.

It is also useful to take a step back here. Globalisation in this regards also means the globalisation of infection transmission. As travel of people (and goods) intensifies, communicable diseases constitute challenges which it is increasingly difficult to confine. Interconnectedness is a defining feature of our modernity. As a case in point, healthcare systems are indeed organised as systems – which can lead to catastrophic failures such as the consequences of HIV-infected blood supplies that took a particular prominence in France but did in fact strike many countries. Ours is a vulnerable society. While that vulnerability is most strikingly epitomized by Ebola-type viruses, with diverse profiles of outbreaks, it is also revealed through

[62] World Bank, *World Development Report 2000/2001*.

[63] The International S&T cooperation with third countries (INCO) is one of those programmes which have been developed around the idea that poverty can be overcome by successfully developing human and institutional resources.

[64] European and Developing Countries Clinical Trials Partnership.

the rise of nosocomial infections (i.e. ills originating in the very places which are devised to heal). These further illustrate the flipside – or paradoxical unanticipated consequences – of healthcare as interconnected systems. Yet, while avian flu and SARS together with the above examples represent the globalisation of infection transmission, they also point to the globalisation of the means to tackle public health challenges. The relative containment of avian flu and SARS, and even more so the eradication of smallpox (the variola virus), constitute inspirational successes in that regard.

3.5. S&T and the environmental challenge

There is no doubt that the solution to the environmental challenge has to come first and foremost from elsewhere than from new technological development. Available technological best practices should first of all be disseminated as widely as possible. A change of mentality is also required leading to less consumption of more carefully selected resources and increased reuse and recycling within the limits of the current technological frontier.

Yet it does not seem unjustified to expect a contribution from new technological development. Technology is already used in a variety of ways when it comes to the environment, and everywhere there is great scope for improvement. Technology in the form of satellites is used to monitor the global environmental situation and change therein. Technology in the form of super computers is used to develop climate models and make predictions. Technological development has made industrial production less resource intensive. It has also reduced the energy consumption of machinery (e.g. cars). S&T has been successful at developing alternatives for harmful substances (e.g. within the context of fighting ozone depletion). Technological development has increased the extent to which a larger variety of goods can be recycled. The production of green energy is wholly dependent on technological development. And S&T is needed to mitigate the impacts of environmental degradation. This need for a joint undertaking – combining existing technologies, technological innovations, as well as political innovations – is illustrated in Fig. 1.13 in the case of climate change (the fight to curb greenhouse gas emissions, that is).

As the next chapter will further examine, S&T is not only an indispensable source for the evidence base on challenges such as environmental degradation, they are also taken to be one of the causes of such predicaments. One can undoubtedly point to the lack of societal controls on the use of S&T, to environmentally harmful production and consumption patterns, and to other types of failures in this regard. Nonetheless, the outlook can change fundamentally if one can conceive of S&T as part of the solution rather than the problem.

The "precautionary principle" is a useful notion to mark that double perspective. It can first be taken as stifling innovation in the name of environmental protection; but more interestingly, it can be understood as promoting innovations that take account of social and environmental difficulties, taking account of risks as well as benefits, taking account of less tractable, longer-term consequences. Its emphasis – even with its origin in German environmental legislation in the 1970s – was as

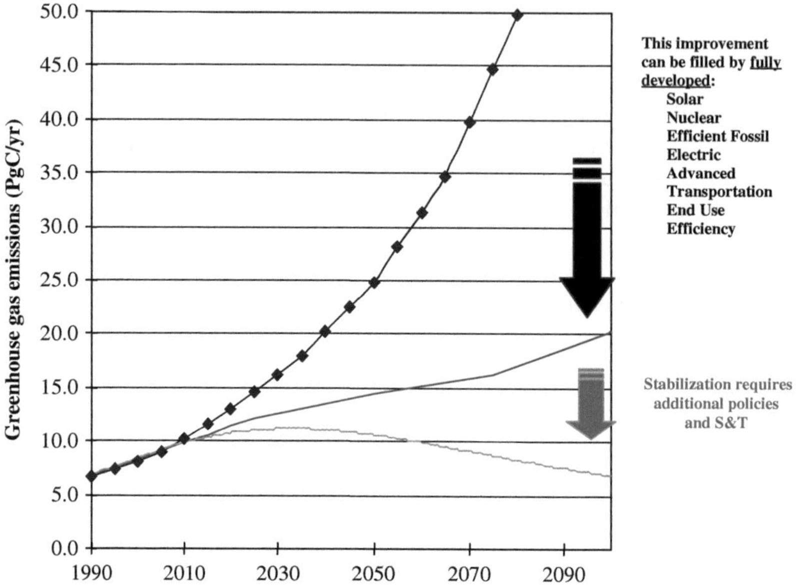

Fig. 1.13. Cutting back on greenhouse gas emissions – new technologies needed
Source: Joint Global Change Research Institute, *The Technological Challenge of Climate Change*, 2003

much on environmental protection as on gaining a competitive advantage through innovations on the backdrop of environmental regulation. Indeed, although this remains a fiercely debated question, a recent survey of the literature[65] indicates that a transparent and non-discriminatory regulatory framework, coupled with high environmental standards, is an engine for innovation and business opportunities. This engine functions notably through the creation of lead markets.[66] The story of the catalytic converters provides a compelling example of such R&D-based win-win.

A first step in that perspective consists in acknowledging the need to sever the link between economic growth and environmental degradation. The endeavour of a duly responsible polity – with a concern for the quality of life of present and future generations – is then to optimise the effects of its economic activity, that is to minimise adverse externalities without sacrificing part of its material well-being or endangering economic growth.

---

[65] Conseil d'Analyse Économique (D. Bureau *et al.*), *Politiques environnementales et compétitivité*, Paris, 2004. Knut Blind *et al.*, *New Products and Services: Analysis of Regulations Shaping New Markets*, Luxembourg, 2004.

[66] As discussed in European Commission, *Innovation Policy: Updating the Union's Approach in the Context of the Lisbon Strategy*, Communication from the Commission to the Council, the European Parliament, the European Economic and Social Committee and the Committee of the Regions, COM(2003) 112 final, 11 March 2003.

A second step consists not in ignoring the above "limits to growth" understanding, but in researching other links between development and sustainability. This move is at the heart of the role of S&T in relation to the environment – and is indeed at the heart of the Lisbon Strategy as underscored in the Conclusions of the 2001 Göteborg Summit. The potential of technology to create synergies between environmental protection and economic growth was emphasised by the October 2003 European Council. That well-established premise is taken to its most fruitful operational conclusions in the Environmental Technologies Action Plan.[67] More recently, the benefits of S&T for the economy and environment alike were further examined in the "Towards a more sustainable EU" report for the Dutch Presidency and indeed in the Kok report of November 2004.[68] In fact Europe occupies a strong position in the field of environmental technologies. Of course this also relates to the fragile but powerful synergies, introduced above, between environmental promotion/protection, S&T, and growth and competitiveness.

These potential benefits can also be of great importance for developing countries. With appropriate technology transfer they can provide these countries with affordable solutions for reconciling their desire for strong economic growth with the need to do so without increasing the pressure on the local – or the global – environment. This North–South dimension highlights the sustainable development predicament as differentiated yet common. The question of sustainable development can be posed along two main lines: a question of adapting – or otherwise innovating – appropriate "clean" technologies, and a question of redefining needs and lifestyles.

Now it is interesting to re-consider the climate change issue in the light of the above remarks. The European Union has taken a leading role in the international process to tackle global warming so as to promote environmentally responsible choices by all actors. The EU has ratified the Kyoto Protocol early on, joined by almost all of its international counterparts on this course – most recently Russia. Its successes are also the planet's successes. The EU is committed to meet its Kyoto emissions reduction targets[69] and continues to show leadership on this issue. The role of S&T is set to become even more central in the post-Kyoto (post-2012) regime, for which negotiations are starting now. The need for new and cleaner technologies as an indispensable means to tackle energy demands and CO_2 emissions was the main message of the latest yearly report of the International Energy Agency.[70]

---

[67] European Commission, *Stimulating Technologies for Sustainable Development: An Environmental Technologies Action Plan for the European Union*, Communication from the Commission to the Council and the European Parliament, COM(2004) 38 final, 28 January 2004.

[68] The September 2004 report for the Dutch Presidency and Environment Ministry *Towards a More Sustainable EU: The Need for Investments that Benefit Economy and Environment Alike* is available at: http://www.rivm.nl/en/milieu/internationaal/Towards_a_more_sustainable_EU.jsp.

[69] For a prospective state of play, see European Environment Agency, *Greenhouse Gas Emission Trends and Projections in Europe 2004 – Progress by the EU and Its Member States Towards Achieving Their Kyoto Protocol Targets*, Luxembourg, 2004.

[70] OECD and IEA, *World Energy Outlook 2004*, Paris, 2004.

More widely, S&T plays an important part in the EU's capacity to shape – and implement – international agreements.

By way of conclusion, it is worthy of note that the answers which science and technology can bring to environmental problems are increasingly judged with reference to the changes they bring in society. They demand choices of policies and governance, the impact of which on economic and social groups must be measured in terms of effectiveness and efficiency, the spread of costs and benefits, and social or regional equity. This is only possible if research also seeks to develop the knowledge-base and methodologies needed by such analyses.

3.6. S&T and the Knowledge Society: The ultimate answer? The ultimate challenge?

As the previous discussion of the contribution of S&T to employment or environmental challenges has shown, it is not always clear-cut where problems start and where solutions end. Or to put these tangled matters even more simply in this case: the role that S&T can play is manifold. And nowhere is this manifoldness better encapsulated than in the predicament of the "Knowledge Society".[71] Here the challenges, the expectations, and indeed the role of S&T in eliciting and addressing them, are brought together in ways that it is most illuminating to examine.

First, this section probes the mutual shaping of science and culture. Second, it foregrounds some collateral features of the knowledge society, and in particular the vulnerability that accompanies its emergence. This will lead up, in Chapter 2, to a discussion of our modernity – or modernities –as characterised by a distribution of goods but also of ills or risks, and of knowledge or claims thereon. Indeed, in this subsequent chapter, the problematic and ambivalent relations between S&T and the public at large will be considered in the perspective of the weaknesses of European S&T. But firstly we must examine the crucial place of S&T within our Knowledge Society in the making.

The mutual shaping of culture and S&T The examples in this chapter have already shown how profoundly our culture is marked by S&T developments. At the same time as S&T shapes our society, they are themselves produced, taken up, reconfigured, shaped by society. That is one (double) way in which culture is decidedly scientific culture, and thus in which S&T is at the heart of this nearly eponymic "Knowledge Society". But to allow all sections of society to benefit from those advances – as well as to take part in that shaping process – individuals need to be provided with the appropriate equipment, in terms of education, skills, awareness, and appreciation for the stakes in S&T endeavours. Vital for a democratic society

[71] On the genesis of this labile notion, besides the seminal work of Peter Drucker (Peter F. Drucker, *The Age of Discontinuity. Guidelines for a Changing Society*, London, 1969), see Michael Gibbons *et al.*, *The New Production of Knowledge: The Dynamics of Science and Research in Contemporary Societies*, London, 1994, and European Commission, *Building the Knowledge Society: Social and Human Capital Interactions*, Commission Staff Working Paper, SEC(2003) 652, 28 May 2003.

in this day and age, such demands point towards another crucial sense for scientific culture, also exposing the acute need for it to be developed. Actions to foster a thorough public grasp of what is science and how it contributes to society are thus *sine qua non* to a full-fledged democratic society.

Importantly, S&T developments accompany and affect lifestyle changes in societies. In this respect the taking up of mobile phones or GSM provides interesting illustrations.[72] The GSM has strikingly changed the way people communicate with their loved ones, organize their work and outings, and live everyday. As regards research, innovation, and competitiveness, the rise of the GSM standard provides an inspiring example of European leadership.[73] In effect, new information and communication technologies open up opportunities for new lifestyles and new ways of working.[74] Remote working or online trading decouples economic activity from a particular geographic location (be it the office, capital cities or structurally favoured regions). Moreover, such technologies can facilitate access to employment – and other forms of social inclusion/participation[75] – among sections of society (people with physical disabilities, the elderly) who may otherwise be excluded. Key to achieving those benefits is ensuring that people are equipped with the necessary skills to get involved. Much information society literature[76] also hypothesises that "eWork" (remote working) may contribute to environmental sustainability as, in addition to other dematerialisations, travelling to work is reduced. On the other hand, transport technologies themselves – from the wheel through to the airplane – continue to have a central role in society, for example in enabling communication.

The quality of human life is made up of many more components than the ones already mentioned: greater access to knowledge, better nutrition and health services, more secure livelihoods, clean air to breathe, security against crime and physical violence, satisfying leisure hours, political and cultural freedoms and sense

---

[72] International Development Research Centre, *Information and Communication Technologies for Development in Africa: Trends and Overview*, 2001. United Nations Research Institute for Social Development, *Information and Communication Technologies and Social Development in Senegal: an Overview/Les technologies de l'information et de la communication et le développement social au Sénégal: un état des lieux*, 2000.

[73] It is thanks to the political determination of the EU that this unassuming technical standard – in fact this far-reaching technical and commercial and political endeavour – was brought to fruition. GSM now stands for "Global System for Mobile Communications", it originates in the acronym for the "*Groupe Spécial Mobile*" hosted by the European Conference of Postal and Telecommunications Administrations, and its specifications where defined by the European Telecommunications Standards Institute in the late 1980s. Commercial operation began – and the world's first GSM phone call was made – in 1991 with Radiolinja in Finland.

[74] European Foundation for the Improvement of Living and Working Conditions & PREST (Manchester Business School), *The Knowledge Society Delphi: EUFORIA Project Report*, 2004.

[75] e-Government Strategic Support Unit (Office of the Deputy Prime Minister, UK), *Engaging the Community in e-Government*, 2004. European Commission, DG JRC – Institute for Prospective Technological Studies (C. Centeno, R. van Bavel and J.C. Burgelman), *eGovernment in the EU in the Next Decade: Vision and Key Challenges*, 2004.

[76] European Commission, DG INFSO, *Impact of ICT on Sustainable Development*, 2004.

of participation in community activities. S&T can contribute to improvements and bring lasting solutions in each of these areas. For example, investment in research and new technologies to achieve sustainable transport solutions generates desirable impacts on the quality of life worldwide: less energy consumption; fewer air pollution; less respiratory diseases; lower noise levels; increased space and security for pedestrians and cyclists resulting in more friendly cities for children and older people; less congestion; fewer road accidents; and so on. Besides, it is S&T which makes possible the novel lifestyles – and indeed the novel society – discussed above.

It may be that, in solving some age-old problems, S&T has created the possibility for new problems to emerge. Yet even to address these new problems we can hardly do without S&T. But we can – and rightfully do – concern ourselves with the consequences of the solutions we devise.

The vulnerable society and the knowledge society S&T has brought a mix of benefits and risks. In the modern world heightened wellbeing and security are accompanied by increased vulnerability and insecurity. This vulnerability can take many forms, from loneliness or travelling accidents to industrial disasters or the twisting of human rights in a totalitarian state. Fig. 1.14 provides an illustration of the rising challenge represented by disasters. Here "disasters" include both technological and natural events.[77]

The dramatic increase shown on the graph may be due not only to the consequences of concentrated urbanisation, climate change, and so on, but also

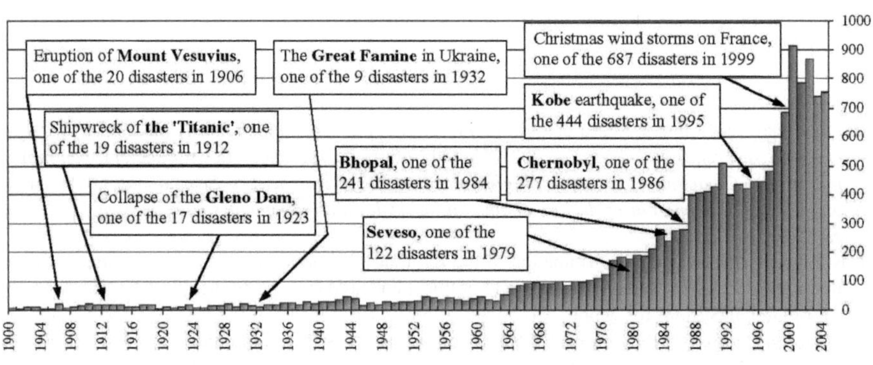

Fig. 1.14. The number of disasters reported is increasing (total number of disasters reported (1900–2004))
Source: DG Research
Data: EM-DAT – The OFDA/CRED International Disaster Database

[77] For a disaster to be entered into the database and thus shown on the figure, at least one of the following criteria must be fulfilled: 10 or more people reported killed; 100 people reported affected; declaration of a state of emergency; call for international assistance.

to a heightened sense of vulnerability and risk, together with a better ability to measure disasters.

Hence the emerging knowledge society will have its problems too. Besides, it will not depend solely on S&T but also on governance and on the citizens who will make up our society – and shape it. Yet it is characterized by an increasingly pivotal role for S&T. The knowledge society requires a revolution in our understanding of knowledge: not only with regard to S&T researchers, but also concerning a democratisation or broadening of knowledge production.[78] This has profound implications for decision-making, for the lay-expert divide, for the handling of risks and uncertainties, and indeed for the relations between citizens and institutions of governance, as every individual should be recognized as – and given the means to be – a *person of knowledge*. Europe finds itself in a peculiar situation in this regard, and the following chapter will unpack the paradoxical relations between S&T and its citizen.

CONCLUSION

This chapter has explored in greater detail some important economic, social and environmental challenges Europe is facing, the expectations held of S&T in addressing these challenges, and the role that S&T could potentially play. The 7th Framework Programme was designed against the background of Europeans feeling anxious because the continent is experiencing a number of important economic, social and environmental challenges – or indeed against the background of a Europe turning from a positive into a negative "exception" at global level. Economic growth is slow. Europe's competitive position is feeble. There are not enough jobs, and not enough of them are high-level. Europe is still characterised by significant poverty and regional inequality. An important demographic challenge is emerging. Europeans' health is affected by serious lifestyle and contagious diseases. And the environment is being degraded.

As will be further examined at the end of Chapter 2 and in Chapter 3, expectations of S&T have never been higher than they are now. Such expectations held of S&T are partially justified. S&T can indeed play an important role in addressing societal economic, social and environmental challenges. S&T is the engine of economic growth and competitiveness. The employment effects of S&T are positive. S&T can play a major role in addressing the consequences of ageing, and the cohesion and public health challenges. S&T can play a key role in addressing environmental challenges. S&T is part and parcel of our lives, be they framed in a Knowledge Society or otherwise, and they are the linchpin of the latter's emergence.

However, as will be seen in the next chapter, for S&T to be able to realise its potential, some serious S&T weaknesses will have to be addressed.

---

[78] Jim Dratwa, Taking Risks with the Precautionary Principle: Food (and the Environment) for Thought at the European Commission, In: *Journal of Environmental Policy and Planning* (Special Issue on Risk and Governance), Vol. 4 No. 3, 2002, pp. 97–213.

CHAPTER 2

ARE EXPECTATIONS TOO HIGH? WEAKNESSES OF EUROPE'S S&T SYSTEM

Across the globe, modern society expects R&D to deliver solutions to an increasingly long list of problems (see Chapter 1). Arguably, however, these expectations have become somewhat inflated and out of line with reality. There is a gulf between what citizens (and sometimes governments) think R&D can achieve and what it is actually capable of delivering today in the real world, with limited resources and complex systems of governance.

In Europe, the gulf between hopes and reality is particularly pronounced. EU citizens recognise that they live in one of the most advanced economies of the world, and expect R&D to be able to tackle the challenges they face. EU governments meanwhile, reflecting the concerns of citizens, have set an increasingly ambitious and necessary agenda for change centred on the creation and exploitation of knowledge. This is typified by the comprehensive Lisbon goals, which emphasise the key role of the transition to the knowledge-based economy in securing sustainable growth, more and better jobs and greater social cohesion.

Such ambitions and such a vision of the future are vital if policy in Europe is to remain on track and reflect major societal concerns. However, the part that R&D can play in this process will be constrained if a number of key factors, which currently prevent Europe from achieving its full S&T potential, are not adequately addressed. This chapter identifies and analyses these weaknesses of the European S&T system. It is organised as follows. Section 1 will demonstrate that Europe is not investing sufficiently in R&D compared with its international partners. At the same time, the system of governance for research in the EU needs to be adapted to address problems of fragmentation and lack of coordination, and this will be the topic of Section 2. It also needs to do more to link research policy with other complementary policies which are necessary to achieve the goals Europe has set itself (Section 3). And finally, Section 4 will deal with the problem of public distrust of science which has focussed in recent years around certain issues and which Europe must tackle if advances are to be made.

1. INSUFFICIENT INVESTMENT IN R&D

A factor contributing to Europe's S&T weakness in recent years is the lack of sufficient investment in R&D. If Europe is to tackle the tough economic, social and environmental challenges it faces at the beginning of the twenty-first century, it will need to spend more on research.[1]

1.1. Europe's low R&D intensity

The United States and Japan not only invest more of their GDP in R&D than the EU (2.67 and 3.20 per cent respectively in 2003 compared with 1.90 per cent for the EU), but have also increased their R&D intensity since the mid-1990s, leaving Europe seriously lagging behind (Fig. 2.1).[2] Meanwhile, a number of Asian countries either are ahead of the EU (e.g. Korea at 2.63 per cent in 2003) or have recorded dramatic increases in their R&D intensity since the mid-1990s (e.g. China and Singapore) (Fig. 2.2). China is now the third largest research spender, with its R&D growing at just under 20 per cent per year from 2000 to 2003, and it aims to reach an intensity of 1.5 per cent by 2005.

What is all the more worrying for Europe is that, while its partners have been boosting their funds for R&D, its own R&D intensity has remained stable for more than a decade, in spite of the increasing pace of technological innovation and the rising costs of R&D. In 2004, the R&D intensity of EU-25 was 1.86 per cent, only marginally above its level of 1.81 per cent in 1995, while the level of the EU-15 countries (1.92 per cent) was almost exactly the same as it was in 1991.

The steps taken at the Lisbon European Council in 2000 were an explicit reaction to these growing concerns about Europe's under-investment in the knowledge economy, and this was further reinforced at the Barcelona summit in 2002, where the EU set itself the objective of reaching an R&D intensity of 3 per cent by 2010. However, the growth in its R&D spending since 2000 has been insufficient to attain this target. Between 2000 and 2003 the average annual growth of EU-25 R&D intensity was just 0.7 per cent, a trend which, if continued, would lead to an intensity of only about 2.2 per cent in 2010.[3]

---

[1] Europe is underinvesting not only in R&D but in knowledge more generally too. According to the recent EC study, the United States is investing roughly \$200 billion more annually on its knowledge economy than the EU. This important knowledge investment gap is due to European underinvestment in R&D, education and innovation. See European Commission, *The Economic Costs of Non-Lisbon – A Survey of the Literature on the Economic Impact of Lisbon-type Reforms*, Commission Staff Working Document, SEC(2005) 385, 15 March 2005.

[2] Many of these issues have already been explored in various studies at national and international level. See for example: European Commission, *Third European Report on Science and Technology Indicators*, Luxembourg, 2003; OECD, *Science, Technology and Industry Outlook*, Paris, 2005 and *Science, Technology and Industry Scoreboard 2005 – Towards a knowledge-based economy*, Paris 2005; National Science Board, National Science Foundation, Division of Science Resources Statistics, *Science and Engineering Indicators 2006*, Arlington, VA, 2006; Observatoire des Sciences et des Techniques, *Indicateurs de Sciences et de Technologies, Edition 2004 – Rapport de l'OST*, Paris, 2004; UK Treasury, Dept. of Trade and Industry and Dept. of Education's, *Science & Innovation Investment Framework 2004–2014*, London, 2004.

[3] European Commission, *Key Figures 2005: Towards a European Research Area: Science Technology and Innovation*, Luxembourg, 2005.

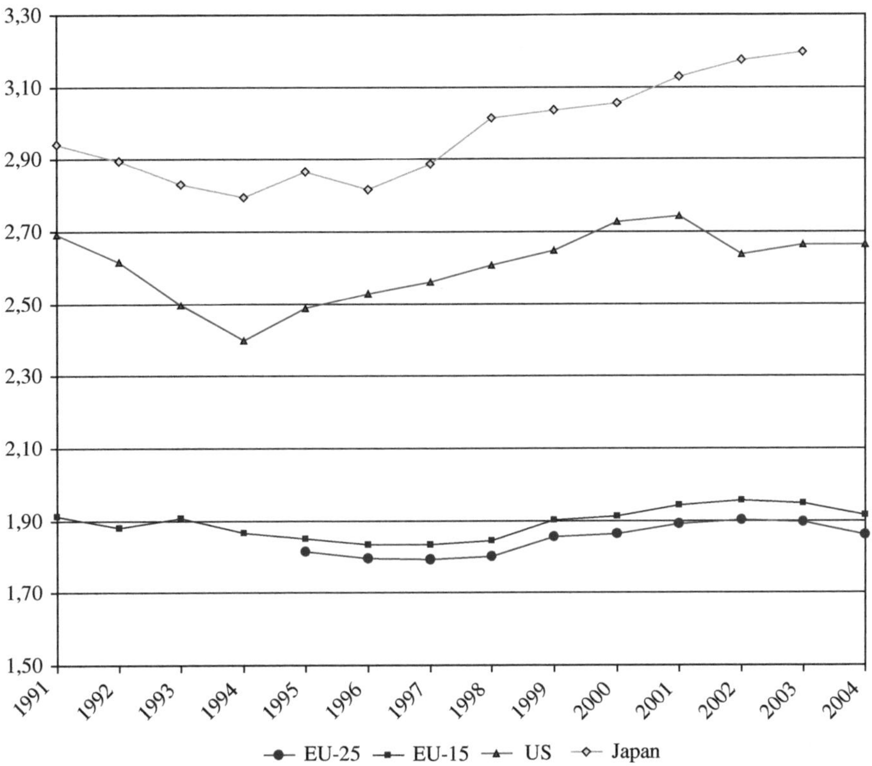

Fig. 2.1. Lagging behind in R&D intensity (Total R&D spending as a percentage of Gross Domestic Product)
Source: DG Research
Data: Eurostat, OECD

This situation led a recent expert group chaired by Mr. Esko Aho (Former Prime Minister of Finland and President of the Finnish national fund for research and development) to remark upon the large gap between "the rhetoric of a political system that preaches the knowledge society and the reality of budgetary and other priorities that have shown little shift in preparing to engage with it".[4]

The extent to which individual Member States have been able to raise their R&D spending has varied considerably (Fig. 2.3). Although, the four largest Member States (Germany, the United Kingdom, France and Italy), which account for over 70 per cent of EU research expenditure, saw their R&D expenditure rise by a yearly average of 2.2 per cent between 2000 and 2003, the same as the rate of increase for the EU-25 as a whole, a number of new Member States appear to be

[4] *Creating an Innovative Europe*, Report of the Independent Expert Group on R&D and Innovation Appointed following the Hampton Court Summit, Brussels, 2006.

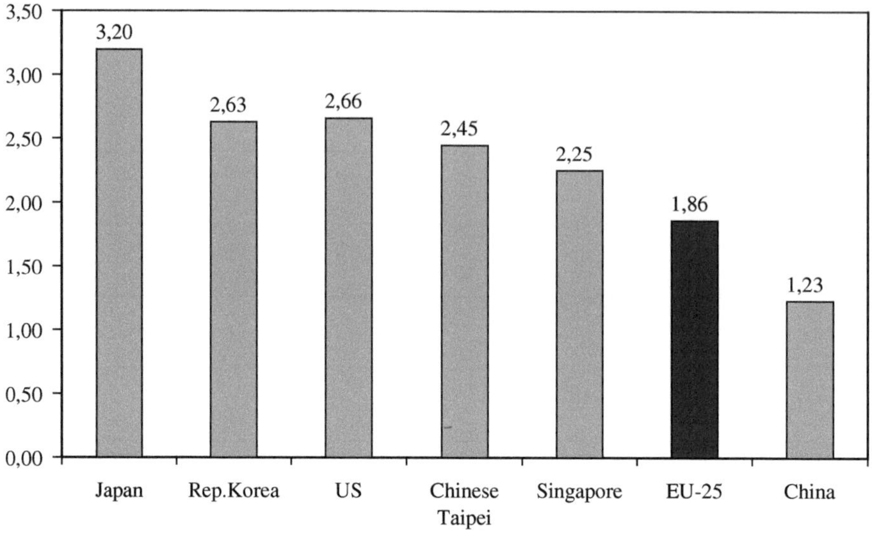

Fig. 2.2. The United States and Asia have become more R&D Intensive (Total R&D spending as a percentage of gross domestic product, 2004)
Source: DG Research
Data: Eurostat, OECD
Note: Japan, Rep. Korea and Chinese Taipei=2003

on a rapid track of "catching-up" (Estonia, Cyprus, Lithuania, Hungary, Slovenia and Czech Republic) and have recorded a significant rise in their R&D effort (6.2 per cent per year on average). Research spending for the four other new Member States (Poland, Malta, Slovakia and Latvia) was markedly less buoyant, notably for Poland and Malta, which even registered a decline in real terms. As to the countries which are planning to join the Union in 2007, both Bulgaria and Romania remain far behind the EU's average R&D spending with 0.5 per cent respectively 0.4 per cent of Gross Domestic Product in 2003. However, Bulgaria's R&D has risen significantly, growing by an annual average of 8.2 per cent.[5] A group of what one might term 'cohesion' countries – comprising Spain, Ireland, Greece and Portugal – have shown a mixed performance over the same period. While R&D expenditure grew strongly in real terms in Spain and Ireland, it increased more modestly in Greece (just below the EU average) and levelled off in Portugal.

Finally, among the Scandinavian countries, Finland and notably Denmark continue to expand their R&D spending faster than the average Member State, but Sweden's expenditure appears to be levelling off recently. However, Finland and Sweden are already well above the 3 per cent in terms of R&D intensity.

[5] In real terms, 2001–2004. No figures on the R&D growth rate available for Romania. Data from EUROSTAT.

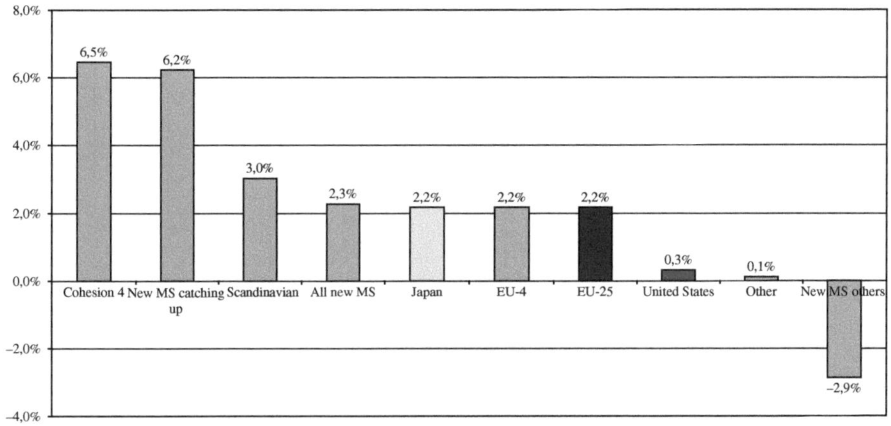

Fig. 2.3. Trends in R&D spending differ between Member States (Total R&D expenditure: average annual real growth 2000–2003)

Source: DG Research

Data: Eurostat, OECD

Notes: (1) "EU-4" = D, F, I, UK; "Cohesion 4" = Spain, Ireland, Greece, Portugal; "New MS catching up" = New EU Member States with growth above EU average (Estonia, Cyprus, Lithuania, Hungary, Slovenia, Czech Republic); "New MS others" = New EU Member States with growth below EU average (Latvia, Slovakia, Poland); "Scandinavian" = Denmark, Finland, Sweden; "Other" = Austria, Luxembourg, Belgium, Netherlands (2) SE, GR estimated for 2000.

1.2. The R&D investment gap between Europe and its competitors

The gap between the EU and the United States in terms of investment in R&D has been the subject of much attention.[6] Between 1991 and 2001, the United States invested nearly € 600 billion more (in real terms)[7] than the EU. By 2001, the annual gap had reached € 131 billion, more than 60 per cent higher in real terms than its level in 1995 (Fig. 2.4). Since then the gap has shown little sign of decreasing in real terms.

Investigating the EU–US gap in more detail, it becomes apparent that about 80 per cent of it is due to the difference in industry expenditure on R&D between the Union and the United States (Fig. 2.5). This is the reason for the target set by the European Council at the Barcelona summit for two-thirds of European R&D to be financed by industry by the year 2010.[8] Military R&D is a second important factor explaining the R&D gap with the United States. The gap between the EU-25 and the United States in government spending on defence-related R&D was € 45

[6] See for example the conclusions of the European Councils at Lisbon (March 2000) and Barcelona (March 2002) (http://www.europarl.europa.eu/summits/).

[7] The data cited in this section come from latest Eurostat and OECD official statistics.

[8] Compared to only 55.6 per cent in EU-25 in 2002 (Key Figures 2005).

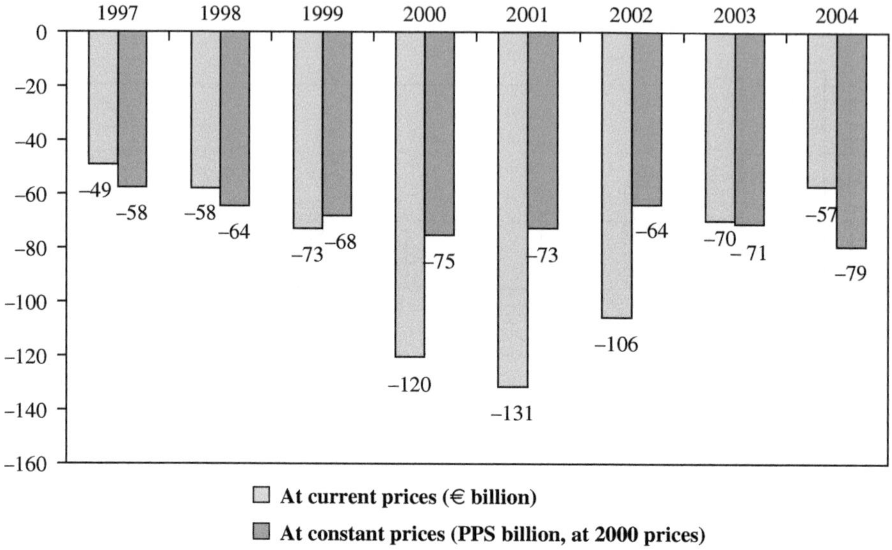

Fig. 2.4. The R&D investment gap between Europe and the United States (total R&D spending in € Billion and PPS at 2000 prices)
Source: DG Research
Data: Eurostat, OECD
Notes: (1) Data are for EU-25: Luxembourg and Malta are not included in the EU-25 total; (2) US: Most or all capital expenditure is excluded.

billion in 2003 (the United States invests € 56 billion in defence, versus € 11 billion by the EU).[9]

In terms of sectors and S&T fields, Europe's R&D spending in the more "traditional" areas (e.g. chemicals, automobiles) compares favourably with that of its partners. However, it lags behind the United States when it comes to investment in new technologies (Fig. 2.6). In the field of Information and Communication Technologies in particular, Europe spends far less on R&D than the United States and Japan.[10] In Space, the United States invests five times more than the EU. EU-15 has a public investment of about 0.06 per cent of its GDP in Space activities against an investment of 0.30 per cent of GDP in the United States. The investment gap is considerable in other promising sectors as well. In Nanotechnology the average level of public investment in 2003 for the EU-25 was € 2.4 per citizen (€ 2.9 for the EU-15), compared to € 3.7 for the United States and € 6.2 for Japan.[11]

---

[9] Source: Eurostat Newcronos database – GBAORD data.

[10] Didier Pouillot and Alain Puissochet, R&D Spending on ICT, In: *Communications & Strategies*, No. 48, 4th quarter 2002. Quoted from the Interim Report of the 5-Year Assessment on RTD in IST 1999–2003. June 2004.

[11] European Commission, *Towards a European Strategy for Nanotechnology*, Communication from the Commission, COM(2004) 338 final, 12 May 2004.

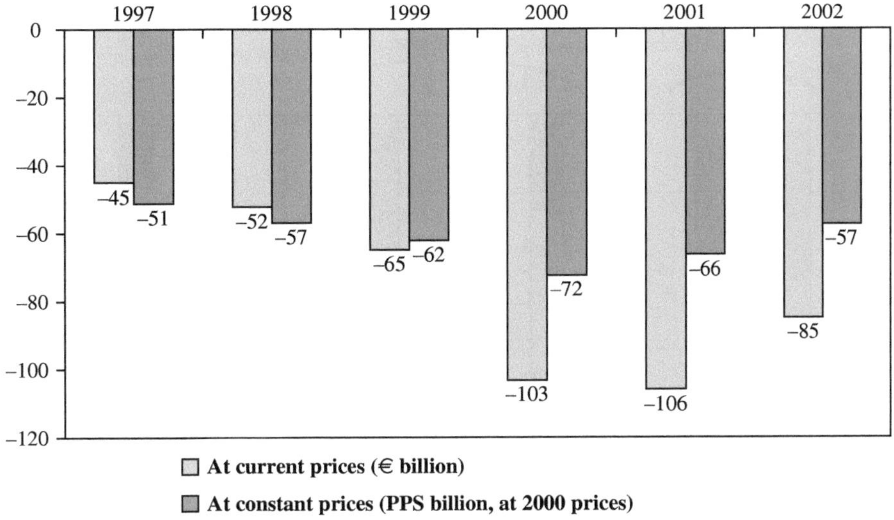

Fig. 2.5. The gap in industry spending on R&D between Europe and the United States (in € billion and billion purchasing power standards at 2000 prices)
Source: DG Research
Data: Eurostat, OECD
Notes: (1) EU-25: Luxembourg and Malta are not included in the EU-25 total; (2) US : Most or all capital expenditure is excluded; (3) EU-25: The values for Italy were estimated for the years 1997–2002 (1996 is the latest available year for Italy). The spending gap between Europe and US can be attributed in large part to differences in R&D spending by SMEs in the two regions. Higher R&D spending by European SMEs could therefore play a key role in helping to achieve the 3 per cent Barcelona target.

Looking ahead to the next ten years, America may find it difficult to reproduce its investment levels of the 1990s. The sizeable US budget deficit ($319 billion[12] – an estimated 2.6 per cent of GDP for fiscal year 2005) may negatively affect the level of public R&D financing and also give rise to higher interest rates, which would result in higher R&D financing costs for the private sector. In Europe, following the Barcelona summit, there is evidence of an enhanced resolve in most Member States to stop the decline in R&D spending (see Fig. 2.3), with several now setting national targets. One would therefore anticipate a slowing down or even reduction of the EU–US gap.

However, a new gap has now developed in relation to the dynamic and emerging economies of Asia. A number of countries in the Far East, including China and a recovering Japan, have been gradually intensifying their efforts to invest in the knowledge-based economy. During the 1990s, they were investing less than the EU

[12] Joint Statement of John W. Snow, Secretary of the Treasury, and Joshua B. Bolten, Director of the Office of Management and Budget, on Budget Results for Fiscal Year 2005. http://www.treasury.gov/press/releases/js2973.htm

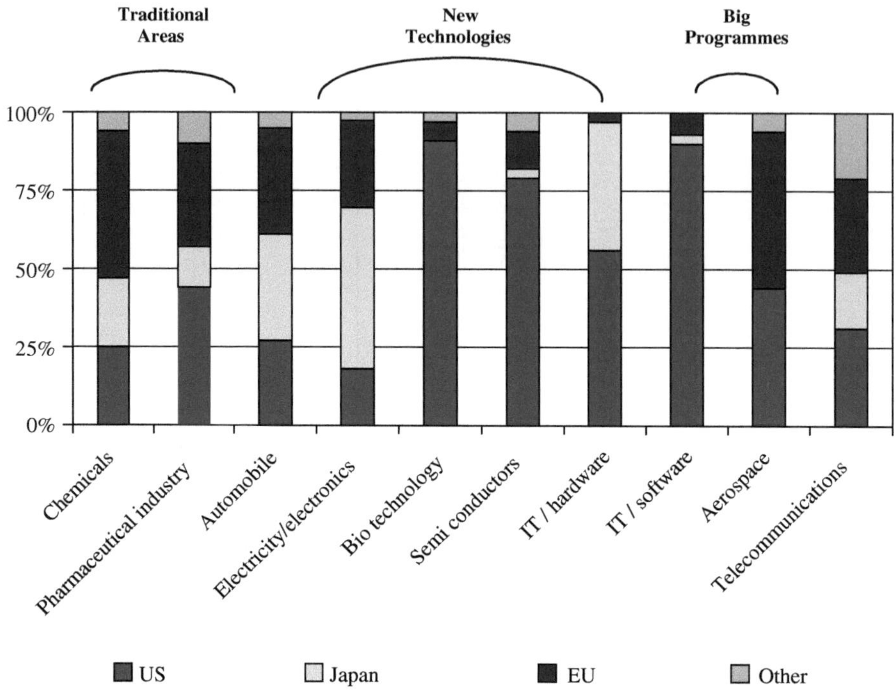

Fig. 2.6. Europe's position in new technologies – breakdown of R&D expenditures (%)
Source: Le Monde, 13/4/2004
Data: OECD

in R&D, but the situation is now reversed (Fig. 2.7). Much of this growth is due to the significant expansion of Chinese R&D efforts.

1.3. The need for more human resources

Europe also needs to invest more in researchers if it wants to reach the 3 per cent target for R&D intensity set at Barcelona. However, because of insufficient investment in the past, Europe has far fewer researchers in relation to its labour force than the United States and Japan (in 2003 EU-25: 5.4 per 1000, US: 9, Japan: 10.1 – see Fig. 2.8). Again, there is considerable variation between the Member States, with some (Finland, Sweden) boasting a higher proportion of researchers in their labour force than the United States, while others (Czech Republic, Latvia, Italy, Cyprus) have less than 3 per 1000.

It is worth noting too that some emerging economies with large populations, such as China and Russia currently (and no doubt India in the near future), have significant numbers of researchers in absolute terms, which are already approaching those of the EU and the United States (Fig. 2.9). Indeed China has more researchers in total than Japan. When it comes to growth, however, one sees the spectacular nature of the Chinese transformation: China managed to expand its stock of researchers by 7.4 per cent per year during the period 2000–2003 (Fig. 2.10).

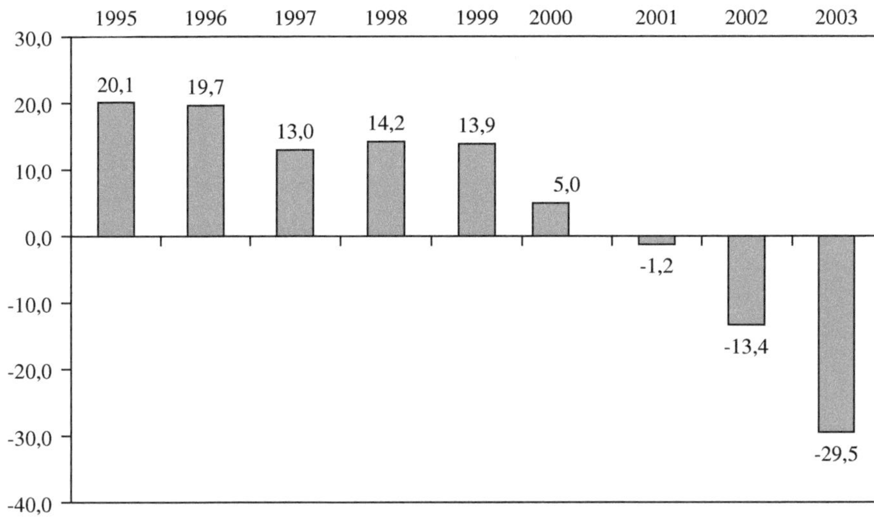

Fig. 2.7. An emerging R&D investment gap between Europe and Asia[1] ($ billion 2000, constant prices and purchasing power parities)
Source: DG Research
Data: OECD, MSTI

Note: (1) The data are for five selected Asian economies: Japan, South Korea, China, Taiwan and Singapore. (2) R&D spending has been converted into purchasing power parities (PPP) to facilitate international comparisons. It should be noted that China's expenditure in particular is much higher in PPP than in current US dollars. PPP are not easy to obtain for China, but arguably yield a more accurate measure of comparative price levels (see Schaaper (2004) for a fuller discussion of currency conversion rates for China).

In the face of these challenges, the EU needs to respond effectively in the way it cultivates and encourages its S&T human resources. To its credit, it is generating a higher proportion of graduates in science and engineering disciplines than the United States and Japan (in 2003, 24.2 per cent of new graduates in EU-25 were in Science and Engineering (S&E) disciplines versus 18.5 per cent in the United States and 23.1 per cent in Japan). The EU also produces more S&E doctorates per year in relative terms: 0.56 out of 1000 people aged 25–34 obtained a PhD in S&E in the EU, 0.41 in the United States and 0.25 in Japan. However, as will be seen below (Section 2.3.), a specific European problem is that its S&T graduates quite often either abandon their scientific careers or leave Europe because of the poor work opportunities for researchers in the EU. Europe thus needs to invest more to create the posts in laboratories, and to make employment in S&T more attractive, so that these graduates will pursue scientific careers, and do so in Europe. Creating more attractive conditions for foreign researchers, and foreign graduate students, to come and work in Europe will also help.

The ageing EU workforce is also a challenge: while the age distribution of human resources in S&T is not as acute as that for other employees, the age pyramid has

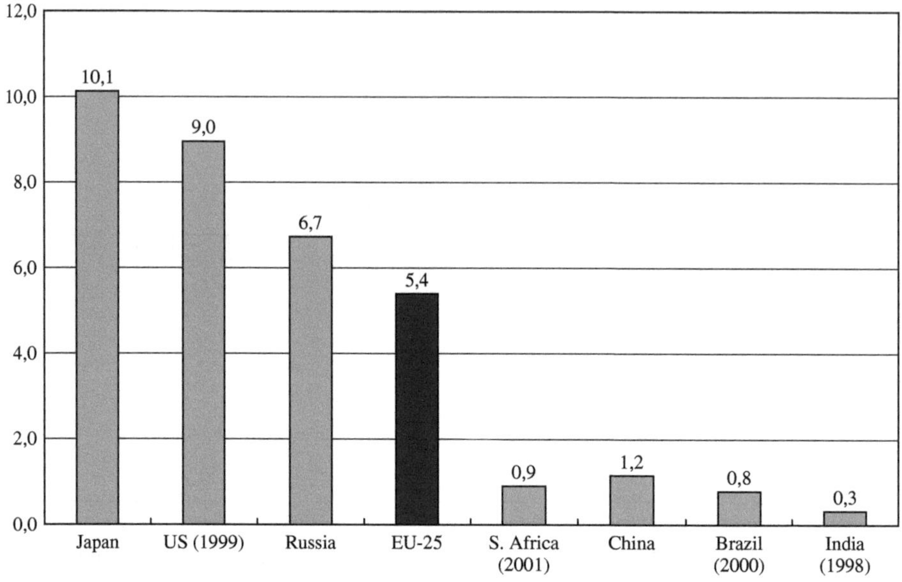

Fig. 2.8. Too few researchers in the EU workforce (number of researchers (full time equivalent) per 1000 labour force, 2003)
Source: DG Research
Data: OECD

a narrower base of younger than middle-aged people, indicating a possible future scarcity of workers in the EU in the field of S&T.[13]

There is of course another huge resource pool that should no longer remain under-exploited: women scientists. It must be ensured that both girls and boys choose to study scientific subjects at school. And later, better use must be made of the significant numbers of women who then graduate in S&E disciplines, but who either fail to enter scientific careers or find themselves stuck in the lower echelons of public and private research labs. Women are consistently under-represented as PhD graduates, as researchers – especially in the business enterprise sector – among senior university staff and as members of scientific boards. Only a third of researchers in higher education and government research institutions are women, and only 18 per cent of researchers in the business enterprise sector. Employment in high-tech sectors also remains dominated by men, who account for almost two-thirds of total employment in these sectors. Gender gaps in employment shares range from 25 per cent in Portugal to above 50 per cent in Greece, Belgium, the United Kingdom and the Netherlands.[14]

---

[13] Eurostat, *Ageing Work Force – How Old Are Europe's Human Resources in Science and Technology?*, Statistics in Focus, Science and Technology, 11/2006.

[14] European Commission, *Employment in Europe 2001 – Recent Trends and Prospects*, Brussels, 2001.

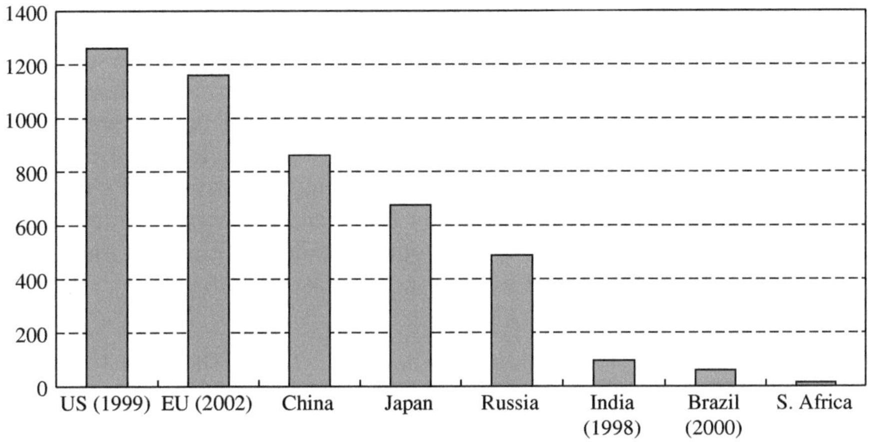

Fig. 2.9. Emerging economies have large potential S&T human resources (number of researchers (full time equivalent), per 1000 labour force, 2003)
Source: DG Research
Data: OECD

Unless steps are taken now to tackle these problems, a shortage of highly qualified S&T personnel in the next 10–15 years will pose a serious threat to the EU's innovative strength and productivity growth.

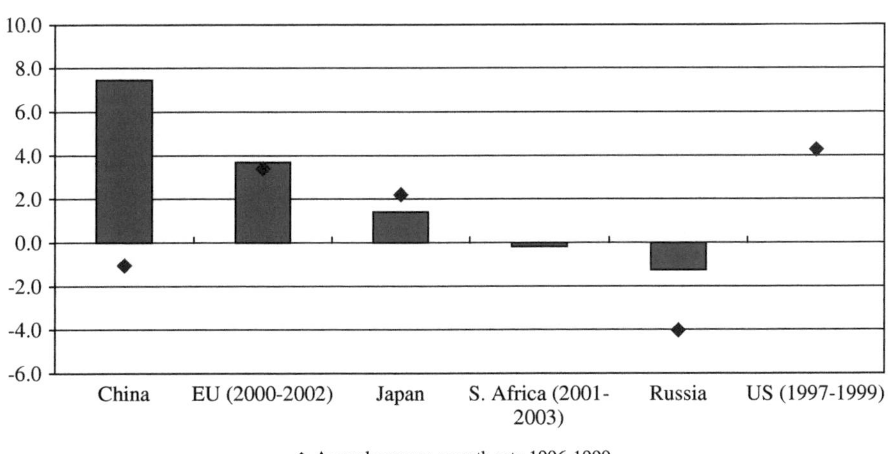

Fig. 2.10. China is on a rapid growth path (researchers (full time equivalent), annual average growth rate, 1996–99 and 2000–2003)
Source: DG Research
Data: OECD

1.4. When can government intervention help?

As has been seen in this section, Europe has for some time been under-investing in R&D. Yet much of the gap in spending compared with its competitors relates to R&D financed by industry. Below we will see that market failures prevent the private sector from investing in research at the socially optimum level. Governments have an important role to play here both by providing incentives and conditions which encourage more private investment in R&D, and by stepping in to support R&D in cases where business would not otherwise do so (Fig. 2.11 shows the importance of government support in Europe, Japan and the US).

Market failures lead to private under-investment in R&D The need for public support for research is generated by market failures, which prevent the private sector from investing in research at the socially optimum level.

A first market failure concerns uncertainty. At the start of a research project it is not at all sure whether the research efforts undertaken will actually result in new knowledge and innovation. Such uncertainty derives from technical complexity, time considerations and capital intensity.[15] This issue is particularly important in basic research and in emerging areas – entrepreneurial and dynamic research generally carries a high degree of scientific/technical risk, as multiple new directions in research are explored, before stable technological trajectories can be established.

Another market failure results from the fact that, even if the research initiative gives rise to new knowledge and innovation, it is not at all sure that the researcher or company that has undertaken the research efforts will be able to exclusively appropriate all the benefits deriving from it. Significant positive externalities exist in the form of important knowledge spillovers.[16] In other words, knowledge and innovation have some of the characteristics of a public good, that is, something that can be simultaneously consumed by everybody in a society. The consumption of a public good is non-rival (one person consuming it does not stop another person consuming it), non-excludable (if one person can consume it, it is impossible to stop another person from consuming it), and non-rejectable (people cannot choose not to consume it even if they want to). The combination of non-rivalry and non-excludability leads to private under-investment in research and justifies government intervention such as the provision of subsidies.

Companies may also be reluctant to invest in research out of fear that the new products they may come up with may make obsolete the products they are currently

[15] Gregory Tassey, Policy Issues for R&D Investment in a Knowledge-based Economy, In: *Journal of Technology Transfer*, Vol. 29, 2004, pp. 153–185.

[16] William J. Baumol, *The Free-Market Innovation Machine: Analyzing The Growth Miracle of Capitalism*, Princeton NJ and Oxford, 2002.

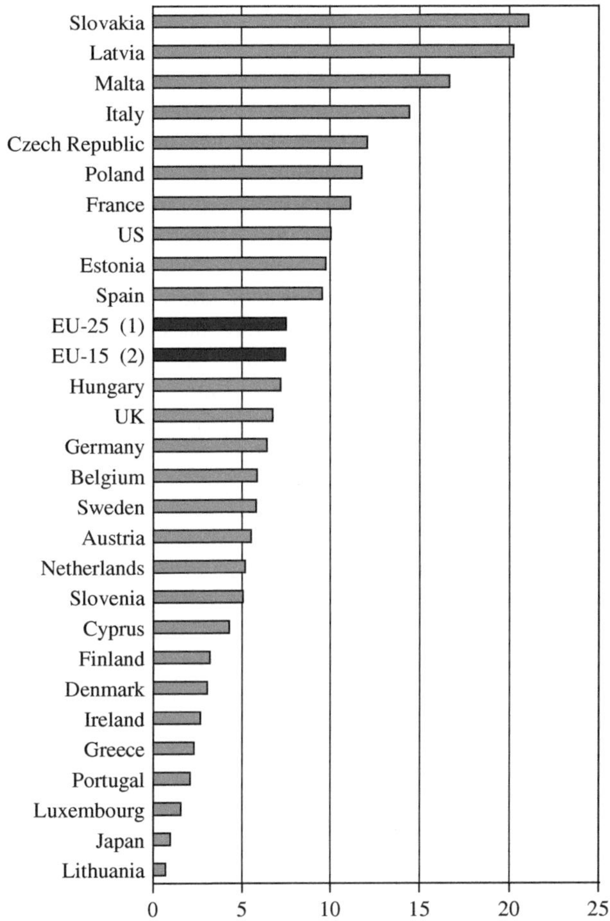

Fig. 2.11. Governments provide important support for business R&D (share of BERD financed by government (latest available year))
Source: DG RTD, Eurostat
Data: OECD, Eurostat
Notes: (1) AT: 1998; LU: 2000; DK, EL, IE, NL, PT, SE: 2001; DE, IT, US: 2003; (2) EU-25 does not include LU; (3) EU-15 does not include LU.

deriving substantial profits from.[17] The provision of public support may affect their calculation.

A good example of another severe market failure is provided by a recent NBER study, which examines why private pharmaceutical companies carry out very little

---

[17] Baumol, *Free-Market Innovation Machine*.

research on the development of vaccines for tropical diseases such as malaria, tuberculosis, and African strains of HIV. Michael Kremer argues that the reason is that the companies fear that were they to develop such products, governments would force prices down to a level that would not allow them to earn a satisfactory return.[18]

The need for public support of research also derives from the system nature of innovation, and from the importance to invest in human capital and networks to ensure the absorption of knowledge. The process of knowledge production is much more complex than the linear model suggests. There are many feedback effects between the various stages in the innovation process, which is best considered as a system, where institutional relationships and the flows of knowledge between actors in the system are of critical importance. The literature on national systems of innovation stresses that the innovative performance of a country, and thus its growth potential, depends upon the development of a balanced "system" of knowledge production and distribution.[19] In that context, coordination and institutional failures can occur and justify the intervention of the government. The role of government then is to invest in human capital, intensify relationships, and optimise the flows of knowledge.

Government intervention requires a careful design to avoid government failures
Once the decision has been made to intervene, this should be done in the best possible way. Government intervention can remedy some of the aforementioned market failures. But government intervention itself can also fail and social costs end up higher than social benefits. Examples include institutional inertia, a lack of reliable information (e.g. on the impact of policies), a lack of continuity and a long-term perspective, excessive red tape, bureaucratic rivalry, unintended side effects, and so on. Therefore, the benefits of solving market failures must exceed the costs of government failure.

From the above it is clear that, prior to public intervention, government must carefully design the way in which it will intervene. This will help to ensure that R&D support is provided in the most appropriate areas (with large spillover effects and where the private sector would not get involved on its own) and through the most effective instruments. A wide range of tools are available to optimise the research policy formulation process. These include the comparative analysis of S&T input and output indicators, foresight and technological assessment analysis, the benchmarking of national R&D policy actions and instruments, growth and competitiveness analysis, the consultation of stakeholders, the evaluation of the management and impact of past R&D programmes, and impact assessment and ex-ante evaluation.

---

[18] Michael Kremer, *Creating Markets for New Vaccines Part I: Rationale, Part II: Design Issues*, NBER Working Papers, 2000.

[19] Richard R. Nelson (Ed.), *National Innovation Systems: A Comparative Analysis*, New York and London, 1993.

First, a careful choice has to be made as to the area of intervention. The combination of finite resources and a multitude of new emerging research opportunities means that careful attention must be paid to funding priorities and requires wise choices by policy-makers. Is R&D support provided across the board, to all S&T areas, or is it concentrated in a smaller number of S&T priorities, and if so which ones? Is R&D support provided in research infrastructure, basic research, applied research, and human resources too, or is it concentrated in a single or just a few components of the innovation system, and if so which ones?

Second, the right choice has to be made as to the instrument of intervention. A wide range of possible instruments have emerged since the shift from the linear to the innovation systems paradigm.[20] Governments make use of a flexible and evolving toolkit of instruments adapting to the specific dynamics and composition of the innovation system and addressing existing bottlenecks in the system.[21] The Innovation Policy Trend Chart lists around 1340 instruments used in 28 countries.[22] These instruments can be classified according to different dimensions. Direct measures, for instance, are targeted at a specific scientific or technological theme, discipline or sector, while indirect measures refer to all sorts of schemes that sustain and reduce the cost of Research, Technology and Development (RTD) investment. Financial measures, such as tax credits or venture capital, give monetary support, while non-financial measures are aimed at improving the framework conditions: the legal framework, increasing attractiveness of science among boys and girls at all levels of schooling, an attractive environment for high-quality researchers and research careers, raising public awareness and understanding of S&T, and so on. Supply-side policy measures are intended to provide a transfer to firms of the resources and capabilities needed for innovation, while demand-side policy measures seek to increase the demand for innovative goods and hence increase the incentive for firms to perform R&D.[23] Paul Romer argues that, historically, policy has focussed too one-sidedly on stimulating R&D via tax credits or subsidies. However, if the supply of R&D resources – primarily technically trained people – adjusts only very slowly to increases in demand, then such policies will raise the wages of scientists and engineers without increasing research much. Effective government policies aimed at increasing the supply of scientists and engineers might counter these tendencies and ultimately increase the rate of commercial innovation and economic growth.

---

[20] Technopolis, *An International Review of Methods to Measure Relative Effectiveness of Technology Policy*, Amsterdam, 2001.

[21] Technopolis, *Policy Instruments for Sustainable Innovation*, Amsterdam, 2004. European Commission, *Third European Report on Science & Technology Indicators*, Luxembourg, 2003.

[22] European Commission, *Raising EU R&D Intensity – Improving the Effectiveness of Public Support Mechanisms for Private Sector Research and Development – Direct Measures*, Report to the European Commission by an Independent Expert Group, 2003.

[23] European Commission, *Raising EU R&D Intensity.*

Care should be taken not to introduce an excessive number of different instruments and to maintain a clear and understandable policy intervention logic. Consistency across instruments should also be ensured. Different instruments can work together to alleviate a particular problem, or they can counteract each other, sometimes in unanticipated ways. For instance, if fiscal incentives are too high, the attractiveness of grants and their impact are reduced. The policy mix chosen should also aim to maximise different kinds of benefits.[24] A first kind of benefit is that of input additionality, or whether resources provided to a firm are "additional", that is to say whether for every Euro provided in subsidy or other assistance, the firm spends at least one additional Euro on the target activity. If not, the public support is crowding out private funding. Another kind of benefit is that of output additionality, or "*the proportion of outputs which would not have been achieved without public support, measured in terms of patents, market share or profitability. A final kind of benefit is that of behavioural additionality, or change in the behaviour of a particular innovation actor as a result of public intervention. Behavioural additionality has been developed as an alternative criteria to measure the effectiveness of policy instruments in the context of their role in the innovation system and the specific objectives or targets groups they address, also referred to as "catalyst" effects of public support*".[25]

The effectiveness of the policy mix should be considered in the context of the innovation system and the pursued objectives. Evaluating the effectiveness of the different policy measures is important for several reasons. It helps to learn from the past; it helps to design or improve future policies, programmes, and initiatives; it provides periodic assessments of their intended and unintended effects, it justifies the continuation or cancellation of a policy/programme, and so on.

Policy support has to be provided at the most appropriate level, and consistency in support has to be ensured across all policy levels. In a world that is increasingly interlinked, government measures will generate effects that go beyond the sheer local, regional and national level. Multi-level governance means finding the most optimal combination of government intervention at all policy levels in order to create synergies which none of the policy actors will be able to achieve on their own. As will be seen in the next section, research policy can only be effective if a multi-level governance approach is applied in designing, implementing, and evaluating the success of the policy.

---

[24] Luke Georghiou *et al.*, *"Making the Difference" – The Evaluation of "Behavioural Additionality" of R&D Subsidies*, IWT Studies, Brussels, 2004.

[25] Mette Rye, Evaluating the Impact of Public Support on Commercial Research and Development Projects, In: *Evaluation*, Vol. 8, No. 2, 2002, pp. 227–248; *Institute for Fiscal Studies, Issues in the Design and Implementation of an R&D Tax Credit for UK Firms,* Briefing Note No. 15, London 2001; José García-Quevedo, Do Public Subsidies Complement Business R&D? A Meta-Analysis of the Econometric Evidence, In: *Kyklos*, Vol. 57, No. 1, pp. 87–102; OECD, *Government R&D Funding and Company Behaviour – Measuring Behavioural Additionality*, Pre-publication copy, Paris, 2006.

2. AN INEFFICIENT SYSTEM OF RESEARCH GOVERNANCE IN EUROPE

2.1. The R&D "Governance Gap"

Against the background of limited resources for R&D it has become even more important to ensure that scarce funds are spent as effectively as possible. However, the already negative effects of Europe's relatively low investment in research, as described above, are compounded by a number of structural deficits inherent in the European R&D system. These systemic weaknesses make Europe a less-attractive place for R&D investors and researchers, and produce a wasteful fragmentation of research efforts.

At the heart of the problem is the issue of research governance in the Europe. In particular, the question arises of how best to allocate policy competences and resources across the different organisational levels of public authority – local/regional, national and EU. As will be seen, the answer to this is complex. Most policy-makers accept that there are arguments for intervening at all of these levels, which leads to a further consideration: the need to ensure mutual consistency of policies across these various levels of governance.

In the EU there has been an increasing awareness of the need to better organise the multi-level governance systems for research in order to ensure greater complementarity of policies, to reduce fragmentation of funding and to avoid duplication of efforts. The debate on the development of a true European Research Area emerged from this growing perception of the need for better multi-level policy coordination in the EU.

While national policy remains the bedrock of European R&D policy, over the past twenty years a dual trend of regionalisation and Europeanisation of S&T policies can be detected. On the one hand, there has been a heightened recognition of the importance of developing policies at the local/regional level. Innovation-related actions have been popular because of the role of geographical proximity when it comes to spillovers and knowledge transfers. The success of areas like Silicon Valley and Cambridge has helped to convince governments of the need to create more of these innovative knowledge clusters.[26] At the level of research policy, there is also evidence of the increasing regionalisation. A number of factors have been proposed as possible drivers behind this trend: regional devolution, the growing importance of innovation, social problems of new technology, and the rising importance of supranational research and innovation systems.[27] The degree of regionalisation of research and innovation policies differs

---

[26] Philip Cooke, *Towards Regional Science Policy? The Rationale from Biosciences*, Presented to Conference on Rethinking Science Policy: Analytical Frameworks for Evidence-Based Policy, Science Policy Research Unit, University of Sussex, 21–23 March 2002.

[27] Centre for Urban and Regional Development Studies, University of Newcastle Upon Tyne, *Regional Science Policy and Instruments: An Overview of British Developments*, Background Paper Prepared for the "Strategy for Success" Scrutiny and Policy Development Panel, North East Assembly, 2004.

significantly across Member States. The German Länder have their own innovation programmes since the mid-1970s, while coordinating research and education policies with the federal level. In Belgium innovation policies are pursued at the sub-national level, while central government policies are mainly concerned with federal research organisations and international research programmes. In France and Portugal some decentralisation has occurred – for example, the establishment of the regional technology transfer centres in France – but with a different dynamic.[28]

On the other hand and as will be detailed in Chapter 4, there has been a significant growth in the scale and scope of EU level intervention. Since the 1st Framework Programme in 1984, European research policy has expanded in terms of its ambition and its budget. The objectives of EU research have progressively widened to encompass socio-economic goals such as improving competitiveness and employment, and more recently the re-structuring of the European research system.

However, despite these important developments, research and innovation policies continue to be pursued largely in parallel – at national, EU and regional levels – leading to what some have called a "governance gap" of poor integration and coordination between these different levels.[29] The division of responsibilities between these interdependent political layers is still evolving, but it is clear that there needs to be a better articulation and coordination between the different levels of governance. The *European Research Area* initiative was launched in 2000 to try and tackle these issues, but despite the progress doubtlessly achieved six years, later it cannot be claimed that there is a genuine European S&T policy.

Unlike the United States or Japan, European research still represents a jigsaw of national public systems. National activities, governed by 25 varying legislative, regulatory and financial structures, are still largely undertaken independently of one another. This governance gap leads to a series of weaknesses in the European research system: the fragmentation of research efforts (Section 2), the unattractiveness of Europe to R&D investment from abroad and to foreign researchers (Section 3), and uncompetitive framework conditions in Europe for carrying out research (Section 4).

2.2. *Fragmentation of R&D efforts*

The compartmentalisation of public research systems in Europe inevitably leads to an inefficient allocation of resources and a thinner spreading of research efforts than would be the case under a better coordinated system.

---

[28] Robert Kaiser and Heiko Prange, A New Concept of Deepening European Integration? – The European Research Area and the Emerging Role of Policy Coordination in a Multi-Level Governance System, In: *European Integration Online Papers*, Vol. 6, No. 18, 2002.

[29] Stefan Kuhlmann and Jakob Edler, Scenarios of Technology and Innovation Policies in Europe: Investigating Future Governance, In: *Technological Forecasting & Social Change*, Vol. 70, 2000, pp. 619–637.

As seen earlier in this section, relative to their GDP, some Member States spend a lot on R&D, while others invest at a much lower scale. Sweden and Finland, for example, spend over 3 per cent of GDP in research, whereas 8 Member States invest less than 1 per cent. However, even the best in the European class are outperformed at the international level. If one compares EU countries to the top-10 US states in terms of R&D intensity, then its best performer, Sweden, would rank only sixth. The EU's second best, Finland, would not reach the top 10. The same applies for the average annual R&D spending of a EU country: at just € 7 billion, this compares modestly with the Ford Motors company's annual investment of € 7.8 billion (Fig. 2.12).[30] Apart from the fragmentation of funding, the diversity among Member States induces a wasteful duplication of research efforts. At present, practically all Member States have their own national and regional research programmes. They lose valuable resources by setting similar priorities and investing in the same expensive facilities.

The example of basic research convincingly illustrates these issues. Its funding is dispersed across the Union, and consequently, many projects lack the necessary critical mass. The amount spent by Johns Hopkins University on basic research exceeds the individual efforts of 18 EU Member States, and is greater than the combined efforts of the 10 new Member States.[31] The prevalence of national rather than European-wide funding mechanisms leads to a lack of competition in Europe. Moreover, coordination of activities is limited due to the compartmentalisation of national programmes and support systems.

What can the Union do? At Community level, it already successfully helps to compensate for this insufficient coordination. This is mainly done via the Framework Programmes, which offer cooperation tools and set thematic priorities. For instance, funding of transnational collaborative research under the Framework Programmes has a positive impact on the coordination of research activities across the Union. However, the financial support the Community can offer today is limited. Community efforts still represent a 26th research policy, in addition to national efforts. Equipped with a budget of only around 6 per cent of national public R&D funding, they cannot be sufficiently dynamic to have a truly integrating effect on national policies.[32] At intergovernmental level, where there is no efficient

---

[30] Department of Trade and Industry, *2003 R&D Scoreboard*, London, 2004.

[31] The following figures illustrate this further: First, in the United States, federal funding for basic research increased by 26 per cent from 2001 to 2005 (i.e. 6 per cent per year). Federal funding for the National Science Foundation increased by 30 per cent to US$5.7 billion over the same period. Secondly, US federally funded nanotechnology and information technology R&D, both of which have the potential to bring about "radical" innovation, increased to US$ 1 billion and US$2 billion respectively.

[32] Court of Auditors, *Special Report No 1/2004 on the Management of Indirect RTD Actions Under the Fifth Framework Programme (FP5) for Research and Technological Development (1998 to 2002)*, Together with the Commission's Replies (Pursuant to Article 248(4) Second Subparagraph EC) (2004/C 99/01), 23 April 2004, Paragraph 5.

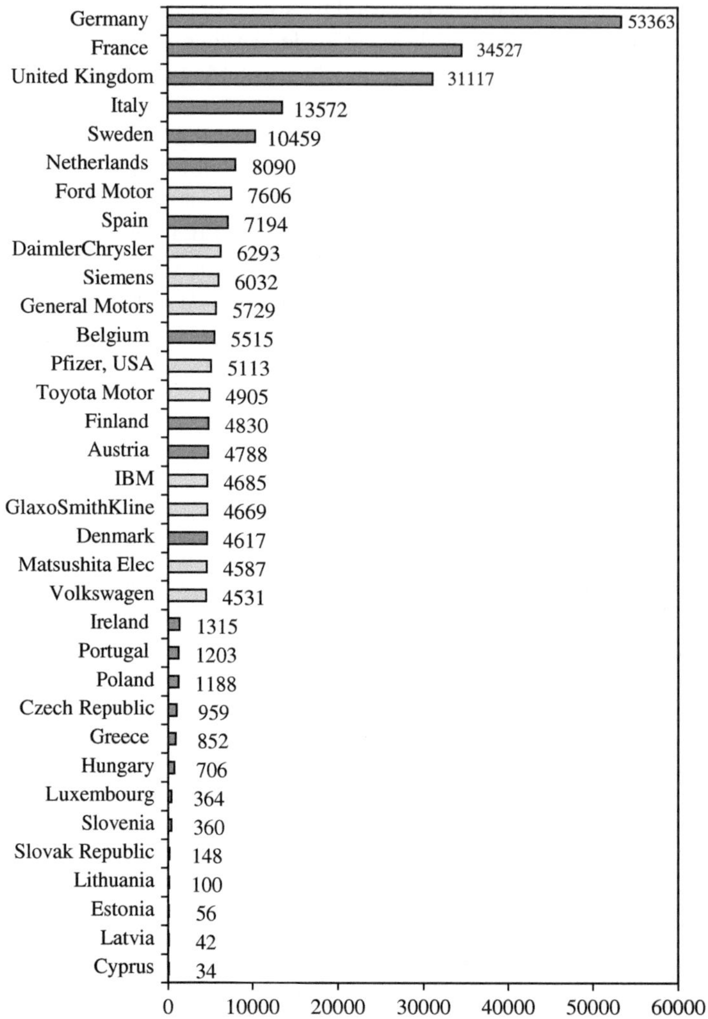

Fig. 2.12. Individual countries' R&D efforts are dispersed – total R&D expenditure of the 25 Member States compared with the top 10 R&D spending multinationals (2002, € billion)
Source: DG Research
Data: OECD, Eurostat, UK R&D Scoreboard
Notes: (1) LU: 2000; GR, IRL, BE, NL, SE, IT: 2001; (2) EU-25 was estimated by DG Research and does not include LU and MT; (3) MT is not included due to unavailability of data.

collaboration either, the Union can help by bringing together actors with the Commission acting as a catalyst. The Open Method of Coordination as defined by the Lisbon Council proves to have the potential to be a successful policy-making instrument in the research field. Its application through the Scientific and Technical

Research Committee (CREST) has resulted in a number of recommendations in key policies areas.[33] But further efforts are needed.

Acting alone is inefficient and prevents EU research efforts from achieving the necessary critical mass of human, technological and financial resources. Size does matter in the economic and technological performances of countries and systems, even more in the globalised economy.[34] The EU must therefore ask itself what are the costs and benefits of maintaining 25 separate research systems. In order to meet the American, Japanese and upcoming Indian and Chinese challenges, the choice is not to become a federal state with a central S&T policy. Nor is it to remain the jigsaw of 25 independent national policies. The EU needs to become a set of scientific states competing with one another while at the same time, within the context of ERA, collaborating effectively to joint advantage. Against this background, action at Community level is urgently needed. In order to overcome the fragmentation of research efforts, there is a need for greater coordination of policies and a substantial increase of the research budget at Community level.

The way the system is organised leads to inefficiencies which make Europe less attractive than it might be for R&D investors and researchers. Fragmentation of R&D funding across the Union, diversification of efforts and insufficient coordination of research activities are factors which, in the end, impact negatively on Europe's performance in S&T. In short, Europe needs to organise its research system better in order to create a true 'European Research Area'. Three aspects are particularly important: The European R&D system must further open up; its framework conditions must become more coherent and conducive to private investment; and, finally, Europe must better coordinate its national research efforts.

2.3. Difficulties in attracting R&D investment and researchers from abroad

The EU attracts most of the world's foreign direct investment.[35] Its macroeconomic stability, the quality of its labour force, and the size of its market (400 million consumers with substantial purchasing power) make the European economy attractive for foreign money. But Europe does not attract most of the world's foreign R&D. Quite on the contrary. Even private European firms themselves finance more R&D in the United States and Japan than they do in other countries in the EU. And United States and Japanese enterprises invest less in Europe than they get from it. In 2002, for example, the United States attracted more business R&D expenditure

---

[33] CREST (Scientific and Technical Research Committee) is an advisory body, whose function is to assist the Council and the Commission in performing the tasks incumbent on them in the sphere of RTD: the public research base and its links with industry, SMEs and research, Fiscal measures for research, IPR and research, Public research spending and policy mixes.

[34] Alberto Alesina and Enrico Spolaore, *The Size of Nations*, Cambridge MA, 2003.

[35] United Nations, *The World Investment Report 2003 – FDI Policies for Development: National and International Perspectives*, New York and Geneva, 2003; FT Business FDI, 05/01/2004 http://www.fdimagazine.com/news/fullstory.php/aid/546.

from EU companies than it allocated to the EU (€ 10 billion versus € 8 billion[36]). Moreover, between 1997 and 2002, R&D expenditures by EU firms in the Unites States increased by 54 per cent in real terms, while spending by US firms in the EU rose by just 38 per cent (Fig. 2.13).

The situation is similar for the flow of researchers: Europe does not succeed in retaining or attracting the best researchers (Fig. 2.14). An increasing number of highly skilled S&T personnel is lost by the Union to industrialised countries such as the United States, Canada and Australia. By 2001, for instance, more than 400,000 non-immigrants were admitted to the United States on a temporary basis came from the EU.[37] The more than 22,000 EU-born admitted to the United States in 2002 as temporary workers[38] mainly helped fill the demands for education, engineering and computer occupations. Unfortunately, Europe does not sufficiently succeed in

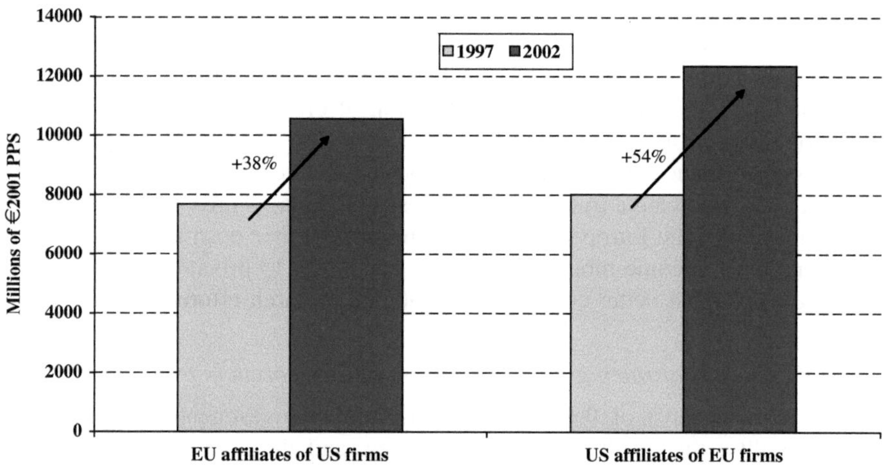

Fig. 2.13. The EU must do more to attract foreign R&D
Source: DG Research
Data: OECD (Activities of Foreign Affiliates Database)
Notes: (1) SK: 1996; FR: 1998; PT: 1999; PL: 2000; (2) HU: 1998; EL: 1999; CZ, FR, PL, SK, US: 2002; UK: 2003.

[36] These figures are expressed in constant 2001 prices and in purchasing power standards (PPS) to take account of inflation and purchasing power effects.

[37] Highly skilled workers on intra-company transfers and with H -1 B visa (a visa for workers with special skills such as IT). Results are from an excellent pioneering study in this area entitled "Brain Drain – Emigration Flows for Qualified Scientists" carried out under the leadership of Luc Soete and Wendy Hansen, 2003, downloadable from http://www.cordis.lu/indicators/publications.htm. The Canadian example analysed in this study illustrates the high education of EU immigrants: In 2002, most were in S&T occupations including management, natural sciences, social sciences and education.

[38] On H1-B visas (see reference above).

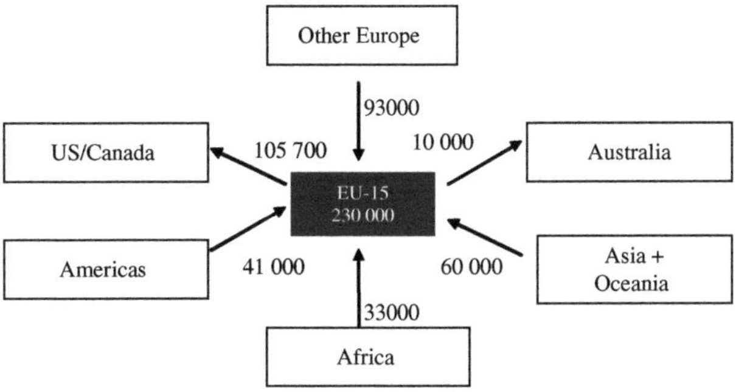

Fig. 2.14. Attractiveness of the EU for R&D personnel (2002)
Source: Third European Report on S&T Indicators, 2003

encouraging its own researchers to come back after a foreign sojourn: nearly 75 per cent of European PhD recipients plan to stay in the United States after their PhD,[39] whereas Europe appears to hold much less of an attraction notably to US researchers.[40]

When it comes to the academic brain drain, the United States is frequently cited as a primary destination for European academics. Part of this can be explained by the much higher levels of expenditure per third-level student (Fig. 2.15): in 2001 the average EU third-level student cost € 8600 to educate, compared with over € 20,000 in the United States.[41]

But it is not only the European researchers who are attracted by the United States. It is also the preferred destination for the best brains of China and India. Around two-thirds of highly skilled Chinese working in OECD countries are based in the United States, whereas the EU attracts fewer than 10 per cent.[42] Talented Indians are similarly attracted to the United States rather than Europe (Fig. 2.16).

In order to reverse the net outflow of researchers and R&D spending, the European research system must become more open. It has to dramatically improve Europe's attractiveness to third-country researchers and make it easier for them to come and work. Today, the EU overall does not offer particularly advantageous work conditions for researchers (such as salary and benefits, longer-term career perspectives and physical work environment). Its employment market and research careers are still mainly nationally determined. There is no transparent European-wide recruitment system. A number of Community initiatives, implemented with Member States and other stakeholders, have already been taken with regard to

---

[39] Representing an increase from 49 per cent in 1990 to 73 per cent in 1999. European Commission, *Third Report on S&T Indicators* 2003, p. 227.
[40] European Commission, *Third Report on S&T Indicators*, p. 232.
[41] Data are in euro purchasing power parities.
[42] Estimated figure from OECD data. Excludes Germany.

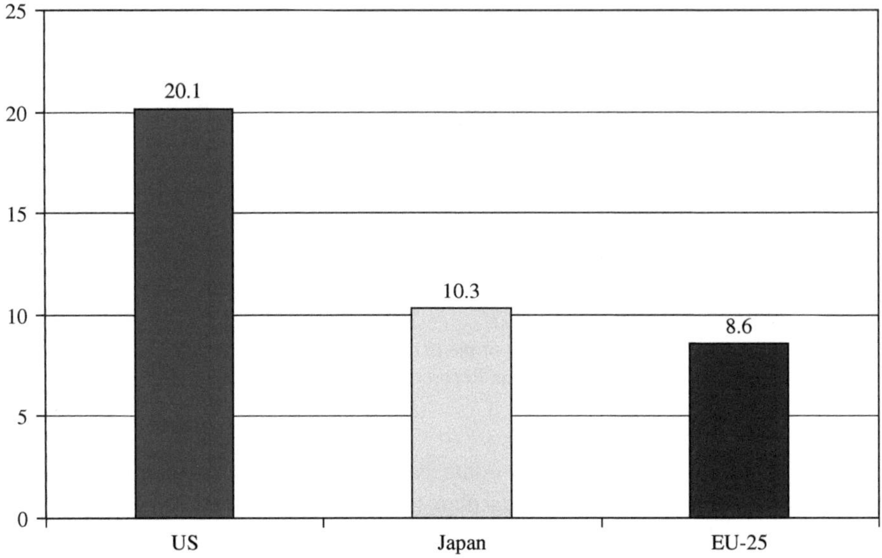

Fig. 2.15. The EU spends less per student – spending per third-level student in 2001 (in 000 Euro purchasing power standards)
Source: DG Education and Culture
Data: Eurostat (UOE data collection)
Notes: Total expenditure on public and private educational institutions per pupil/student in EUR PPS, ISCED level 5–6, 2001

researchers' careers, to assistance to researchers and their families[43] and with a view to reach out to and network European researchers abroad. These represent significant steps in the right direction, but the efforts towards the realisation of an open, competitive and transnational employment market need to be intensified. For example, recent US data reveals that among the top-10 supplying countries of specialised temporary workers (on H1-B visa) admitted in 2002, UK workers report a median income of $68,000, second only to the $70,000 reported by Canadians and well above the top-10 average of $53,000 they would earn in their home country. On the entry conditions of third-country researchers, efforts towards the adoption of European-wide immigration provisions must be actively pursued.[44] Combined with a substantial and targeted system of grants for researchers at different stages of

[43] Such as the initiative for a *European Researchers' Charter* and the *Code of Conduct* for the *Recruitment of Researchers*. The *European Researchers' Mobility Portal* (mobility assistance, vacant jobs) and the *ERA MORE Network* of Mobility Centres were launched in 2003.
[44] See European Commission, Communication from the Commission to the Council and to the European Parliament on the Presentation of a Proposal for a Directive and two Proposals for Recommendations on the Admission of Third-country Nationals to Carry out Scientific Research in the European Community, COM(2004) 178, 16 March 2004.

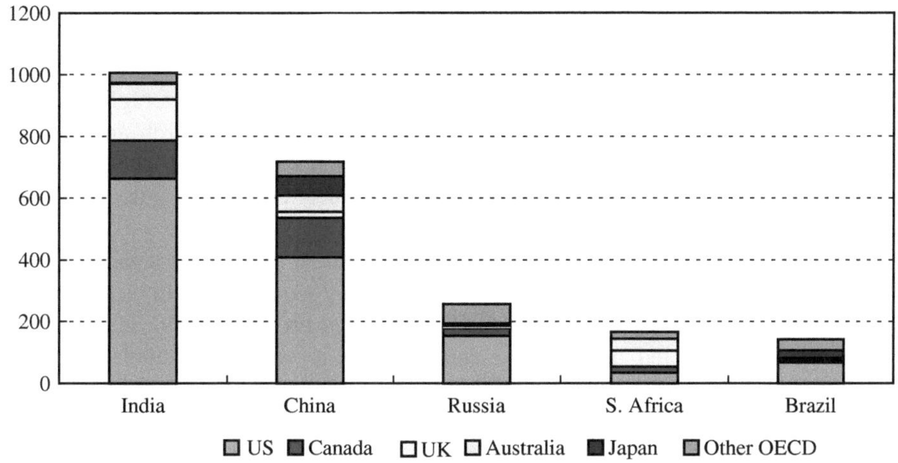

Fig. 2.16. The EU attracts far fewer highly skilled people from abroad than the US –foreign born, highly skilled people in the OECD, by country of birth and country of residence (1000s)
Source: DG Research
Data: OECD

their careers, with built in transnational, international and intersectoral mobility the European actions in human resources should help to remedy the current situation and improve Europe's attractiveness for the best researchers from within and outside Europe.

Efforts are also needed to attract more foreign graduate students in S&T disciplines. Injecting top talent in this way is crucial for the economy, for the education system, and for technological leadership in the modern world. According to a recent study in the United States,[45] for every 100 international students who receive science or engineering PhDs from American universities, the nation gains 62 future patent applications. International students have also founded many of America's most innovative firms, such as Sun Microsystems and Intel. Moreover, about one-third of America's engineering professors are foreign-born.

2.4. The need for better framework conditions for R&D

The same rules and conditions for carrying out research do not yet apply throughout the EU. This disparity is a major obstacle to the production and exploitation of knowledge, and is dissuasive to foreign investors.

[45] Gnanaraj Chellaraj *et al.*, *The Contribution of Skilled Immigration and International Graduate Students toUSInnovation*, World Bank Policy Research Working Paper 3588, May, 2005.

Harmonised and attractive framework conditions – such as tools to protect intellectual property, an environment with research and innovation-friendly regulations and competition rules, supportive financial markets and a favourable fiscal environment – can play a key role in stimulating R&D efforts and leveraging foreign R&D investment. Innovation depends not only on scientific research and technological development, but also on the conditions which facilitate the transformation of knowledge into innovation and marketable products and which guarantee legal certainty.

In Europe, the fragmentation of the European patent systems is a factor which inhibits innovation and the successful exploitation of good research. In many sectors, firms would invest more in R&D if their intellectual property could be better and more cheaply protected and if they could expect to reach sufficient returns to balance the risk inherent in such activities. The absence of a standard Community patent to cover all the European territory and provide affordable legal certainty to investors is a major disincentive especially for small and medium-sized enterprises (SMEs) and the academic sector (see box).[46]

Box 2.1. Reducing the Cost of Patent Protection with the Community Patent

The European Commission has proposed the creation of a Community Patent to give inventors the option of obtaining a single patent legally valid throughout the European Union. The proposal would significantly lessen the burden on business and encourage innovation by making it cheaper to obtain a patent and by providing a clear legal framework in case of dispute. The Community Patent will help to significantly decrease costs after grant, compared with national patents, in relation to translations and maintenance:

- The applicant will be responsible for obtaining translations of the claims of the patent into all EU official languages and filing these at the EPO except for any languages where states have renounced the translation requirement. A simple comparison suggests these costs should be less than the existing translation costs for an average European patent (estimated € 4845 compared to € 10,200).
- The annual renewal fees for maintenance of the patent will be payable in a single sum to the EPO, rather than to individual national offices as required for national patents. They will amount to not more than the combined annual renewal fees for an average European patent covering, say, eight states, and

[46] However, the legal framework alone is not sufficient, so long as those involved in the production of knowledge do not have the know-how to manage IPR properly. EPO's studies reveal that Europeans use patent information facilities less than Americans. EPO concludes that "*Europe would...do well to introduce policies which would lead to greater awareness and utilisation of the patent system, thereby establishing a stronger patent culture in Europe.*" (Letter from EPO Vice President Desantes to Research Commissioner Michel of October 2004).

could be less. The exact level will have to be decided nearer the time that Community patents will become available.

Taking both these aspects into account, calculations show that the Community patent will cost no more than a European patent covering five states. This would represent a cost saving to around two-thirds (68 per cent) of all applicants for European patents – at least 30,000 granted patents each year. For full EU coverage the Community patent would cost only 50 per cent of a European patent for the 25 individual Member States – a saving of at least € 30,000 per patent over 10 years.[47]

The Community Patent is one of the most actively debated dossiers in the European institutions. The first proposal was made by the Commission in a Green Paper already back in 1997. Even though the Lisbon Council called for a Community Patent to be available by the end of 2001, the proposal has been blocked by the Council of Ministers since 2000. The main concern has been the translation of patents into all or a number of EU languages.

With regard to fiscal conditions too, Europe is far from having a harmonised system. Taxation and incentive structures vary across the EU. For instance, national fiscal incentives for private investment to encourage business to invest in R&D are still different across Europe, although they are now reaching substantial levels (around 12.5 per cent of public spending on research in the Netherlands, 16 per cent in Austria and 42 per cent in Latvia).[48] This diversity can give rise to unhealthy tax competition.

In order to ensure that framework conditions are attractive for private investment in research in Europe, improvements are needed in the areas of product market regulation and standardisation, competition rules, financial markets, the fiscal environment and S&T human resources. New initiatives are underway both at Community and national levels. In the area of industrial property, mainly an area of Member State competence, the application of the Open Method of Coordination has resulted in mutual learning and will lead to the development of EU guidelines. In the area of competition rules, an exclusive Community competence, the framework for R&D State Aid expires at the end of 2006, and a revision of the rules aimed at adapting to changes in the research environment is currently being examined. In the area of product regulation, a shared competence, the impact of the introduction of new legislation is being assessed both by the Commission and by the Council. Also the setting of clear policy objectives, priorities and targets can provide for legal and political certainty to allow the private sector to invest into longer term oriented research.

---

[47] Source: http://europa.eu.int/comm/internal_market/en/indprop/patent/docs/2003–03-patent-costs_en.pdf

[48] Provisional figures of the R&D Investment Snapshot, European Commission, Brussels, September 2004.

What these examples show is the importance of improving and, where necessary, harmonising the framework conditions for research by means of greater coordination between Member States. They also serve to highlight a further key issue which Europe must tackle, namely the need to coordinate research policy with other related policies.

3. THE PROBLEM OF COORDINATING POLICIES

Research and development on its own cannot deliver solutions to all society's problems. Even in those areas where research might reasonably be expected to have an important role, its impact will depend on the extent to which other complementary policies are being pursued in parallel. For example, simply boosting investment in R&D will not suffice to raise Europe's economic growth rate, increase its competitiveness or create jobs if it is not combined with macro-economic stability and with horizontal coordination among structural policies (education, competition, energy, environment, internal market, transport, etc.). For example, raising Europe's expenditure on R&D can only be successful if more trained scientists and engineers become available through a targeted education policy, otherwise the result will simply be an increase in researchers' salaries. Similarly, the system should provide more opportunities for new entrants, greater mobility of employees within and across firms, and more retraining. Policies also need to be developed for attracting the best students and skilled workers from abroad, which will depend upon initiatives in migration policy. Taxation policies can be employed as well in order to stimulate R&D investment. Competition policies can help to drive down the costs of technologies, while open policies on trade and direct investment can let in new knowledge and ideas.

This systemic policy view underlies the conception of the Lisbon process,[49] which set the goal for the EU to become the most competitive knowledge-based economy in the world by 2010, to create more and better jobs and to achieve sustainable growth and social cohesion. The Lisbon strategy was based on the understanding that these goals could only be attained by launching a coherent and coordinated set of policy initiatives across a number of different fields – fiscal, macro-economic, employment, environment, innovation, education and training, and, of course, at its core, research.

Notwithstanding the problems the EU has encountered so far in implementing this ambitious agenda, Lisbon was a milestone for two reasons. First, it set research at the heart of a European policy for achieving growth, employment and other key societal objectives, and thus acknowledged the central role which will be played by knowledge in the economy and society of the future. Secondly, and crucially, Lisbon recognised that while research was an essential pillar of EU strategy, it could only be effective if it was coordinated closely with a range of other complementary policies.

---

[49] A more-detailed discussion about research and the Lisbon strategy will be presented in Chapter 3.

In 2005, half-way to the deadline for the achievement Lisbon strategy, the Commission looked at the progress achieved so far. A report by an external expert group (the "Kok report"[50]), contributing to this mid-term review, painted a gloomy picture: It blamed slow progress on an overloaded agenda, poor coordination and conflicting priorities. In particular, the report emphasised that "policies pulling in contradictory directions must be realigned so that instead they are mutually reinforcing". These problems were addressed by the Commission in the Renewed Lisbon Strategy launched in 2005. Among other things, it set out proposals for Member States to appoint a Mr or Mrs Lisbon at government level charged with coordinating the different elements of the strategy and presenting the Lisbon programme, along with the use of National Lisbon Programmes for growth and jobs which would become the major reporting tool on economic and employment measures.[51]

The difficulties encountered in implementing the Lisbon Strategy reflect a widespread problem associated with managing multiple-policy agendas. Coordinating research policy with the multitude of other complementary policies poses a real problem for governments and public authorities across the globe. A recent OECD level study concluded that the increasing need for more coherent innovation policy agendas spanning ministerial boundaries and including many other policy areas creates deep tensions which will have to be tackled, and pointed to the need for better governance systems to deal with these issues.[52]

Part of the problem is that the inter-linked system policy-makers are trying to influence is enormously complex, and there is no "unifying theory" of public policy to provide precise guidance about how best to go about it. The multi-pronged policy approach adopted by more and more countries is a reflection of the increasing complexity of the challenges facing modern society, and consequently of the solutions necessary to achieve them. It is also the result of a more-refined understanding of the systems and processes that policies are seeking to influence. This is notably the case for research and innovation policies, which have evolved in recent years against the background of a more nuanced view of national and regional innovation systems, which emphasises the role of institutions, linkages, geography and systemic failures as an underlying rationale for government intervention.[53]

Of course there are also institutional problems. Different policy areas come under the responsibility of different ministries or public authorities, which leads to the classic problems of institutional coordination. Policy-makers are also called upon to "square the circle" of various, often competing, interest groups such as businesses,

---

[50] *Facing the Challenge. The Lisbon Strategy for Growth and Development*, Report from the High Level Group Chaired by Wim Kok, Brussels, November 2004.

[51] European Commission, *Working Together for Growth and Jobs, A New Start for the Lisbon Strategy*, Communication to the Spring European Council 2005, COM(2005) 24, 3 February 2005.

[52] OECD, *Governance of Innovation Systems – Volume 1: Synthesis report* (Paris, 2005).

[53] See for example Lundvall, B-A. (Ed.), *National Systems oflnnovation: Towards a Theory oflnnovation and Interactive Learning* (London, 1992).

employees, trade unions, the public sector, the elderly, city dwellers, home owners, the unemployed, and so on. Meanwhile, the objectives of different policy areas may prove challenging to reconcile with each other: for example achieving growth while protecting the environment, or maintaining budgetary stability while boosting public investment in education and research.

Tensions can arise too from differences in emphasis among various lines of policy. Typically this is reflected in the regular budgetary conflicts between "competing" policy areas. At EU level, this was apparent during the debate on the Financial Perspectives for 2007–2013 when policy fields jostled for position, and the rift was exposed between those favouring a continuation of the Common Agricultural Policy and those preferring a re-balancing of the budget towards policies for the knowledge economy (see Chapter 7). At national level too tough decisions have to be taken, and the difficulties encountered by Member States in raising their public spending on R&D point towards the inherent problems of shifting finances from other important policy areas. Hilliard and Green cite an example from Ireland of different policy emphasis across government ministries: "While the Department of Enterprise, Trade and Employment is committed to the implementation of the Lisbon strategy, the Department of Finance is equally firmly committed to the goals of the Stability and Growth Pact and the associated Broad Economic Policy Guidelines (BEPG) in keeping tight control of public finances and debt".[54]

Such differences often reflect competing models or frameworks used by policy players to view economic and societal change. There are those who are convinced that the main emphasis for policy in Europe must be on macro-economic stability, deregulation and labour market reform. Others stress more the role to be played by structural policies such as R&D, innovation and education in moving Europe towards the knowledge economy. Others still take a strongly ecological viewpoint stressing sustainability. While such approaches are not mutually exclusive, it is important to appreciate that policy coordination does not always take place in a setting of shared understanding and vision. For example, in the case of Norway, Remøe reports that "there are deep tensions within the Ministry of Trade and Industry, basically between the division for economic policy, whose foundation is the neo-classical approach to economic policy, and the division for R&D and innovation, whose perspective is more in line with the innovation systems approach and evolutionary economics".[55] The same author also cites the case of the Netherlands, where the Ministry of Economic Affairs was overhauled to better support the "National Systems of Innovation" policy approach.

If the EU is to develop effective policies for research and innovation in the future, there needs to be an improvement in the systems of governance to promote more coordinated and joined-up policy-making and implementation. Most countries have

---

[54] Rachel Hilliard and Roy Green, Governance and Institutional Change in Ireland, In: *Governance of Innovation Systems: Vol. 2. Case Studies in Innovation Policy*, Paris, 2005.

[55] Svend Otto Remøe, S.O., Governing Fragmentation: The Case of Norway, In: *Governance of Innovation Systems*.

a complex set of institutions and arrangements with their own historical rationale. In an increasingly dynamic and inter-linked policy setting, governments are finding that these governance structures need to be renewed to avoid gaps and duplications and to enhance policy delivery. This is especially true for research and innovation policies which are transversal and systemic in nature. Moreover, when it comes to policy at EU level, there is another layer of complexity since its ambitious targets require coordination not only across different policy areas, but also, as seen in the previous section, between different levels of governance.

4. S&T AND THE PUBLIC: THE TRIPLE PARADOX

Even before touching on any paradox, we need to tackle this most fundamental question: Why does it matter what the public thinks?

First, S&T makes up a substantial component of public intervention and indeed of government spending – at EU as well as national level – so it is a matter of course that all sections of the public should be on board. Furthermore, even besides the role of public authorities in this regard, S&T is an integral part of daily life as well as at the heart of the most salient and persistent societal issues of the day.

Secondly, to make the issue immediately clear in the perspective of a chapter on the weaknesses of Europe's S&T system, genetically modified organisms and stem cells are examples where public distrust can be seen as a weakness affecting European S&T performance, and thus its growth and competitiveness. But that might be too simplistic a cut. Notably because the weakness might just as well consist in construing public recalcitrance as a failure, or in neglecting more sophisticated avenues of research and growth which can take the public – with all its wants and concerns – on board.

Thirdly and finally for now, the overarching issue: S&T is at the heart of the democratic state. In effect S&T is at the heart of the notion of *legitimacy* as the relation between the public and the state. This powerful finding – dating back to Max Weber and even Aristotle – has been compellingly documented in the work of James C. Scott[56] as well as Yaron Ezrahi.[57] His is a characterisation of the democratic state as ceaselessly seeking to legitimate itself through scientific and technological performances (e.g. large-scale projects, "modernisations", institutionalisation of scientific expert advice).

Thus the public's expectations as to S&T is closely tied with the public's expectations as to the state – and are of paramount importance. In contrast, in the classic conception of the role of the public as to S&T (attuned to the obsolete "linear model" or the "diffusion model"), the public is to be tamed or placated for roll-out to be optimised. Quite simply, in this canonical perspective, the public is a necessary

[56] James C. Scott, *Seeing Like a State: How Certain Schemes to Improve the Human Condition Have Failed*, New Haven, 1998.

[57] Yaron Ezrahi, *The Descent of Icarus: Science and the Transformation of Contemporary Democracy*, Cambridge, MA, 1990.

liability – and public distrust is a weakness. Hereunder we will be moving beyond this perspective, although not without heeding its warning.

The analysis in this section will be deployed in four stages. First, it will set out the triple paradox of public attitude to S&T as a heuristic reading grid. Second, it will develop some of its features by drawing on opinion surveys – an imperfect means to evidence public opinion. Third, it will pursue this inquiry into public engagement by scrutinising issues of trust in decision-making. It will conclude with further perspectives on Modernity and the democratic state.

4.1. The triple paradox

The literatures on the relationship between S&T and the public, including the specific literature on the "public understanding of science", often emphasise the ambivalent or paradoxical nature of the public's attitude. Be it only as a heuristic grid, it is most original and illuminating to present this attitude – or rather this *engagement* of the public with S&T and the state which is not reducible to "perception" or "understanding" – as comprising three interrelated paradoxes (Fig. 2.17). First, whereas S&T is seen as bringing about harmful results and being a causal factor of problems for society, S&T is also seen as bringing about beneficial results and solutions to society's problems. As a corollary, whereas S&T is seen as eliciting expectations and heightening hopes, S&T is also seen as eliciting concerns (and as heightening fears and not delivering on the hopes). Secondly, whereas S&T is seen as a means to settle controversies, it is also seen as a means to open, stir, and sustain controversies. Examples abound where scientific experts – brought to the table to determine the fact of the matter and thusly bring the matter to a close – do actually enrich the issue at stake, fuel the dissensus, and amplify its reach even beyond the technical domain alone.[58] It is perhaps comforting that experts rarely go without counter-experts. Thirdly, and in ways that build upon the two previous dimensions, S&T constitutes a resource (for governments or private companies notably) to increase trust, but can also be a source of distrust; "all our experts agree that our technology is safe" is a pronouncement which, liable to play both ways, can serve to illustrate this third dimension. The following pages will at various points elaborate on underlying features of this triple paradox.

4.2. Don't you agree? Don't you know?

Capturing public opinion in a meaningful way is a difficult task. This extreme difficulty as well as the severe limitations of opinion surveys is acknowledged by opinion surveyors themselves. The perseverance of many democratic states across the world as to these endeavours demonstrates the importance attributed to public opinion (to agnising it, to measuring it, and to having it in a readily usable form). As a collateral, opinion surveys can shape and justify both the

---

[58] That is very well shown in Les Levidow and Susan Carr (Eds.), Precautionary expertise for EU Agbiotech Regulation, In: *Science & Public Policy* (Special Issue), Vol. 32, No. 4, 2005, pp. 257–332.

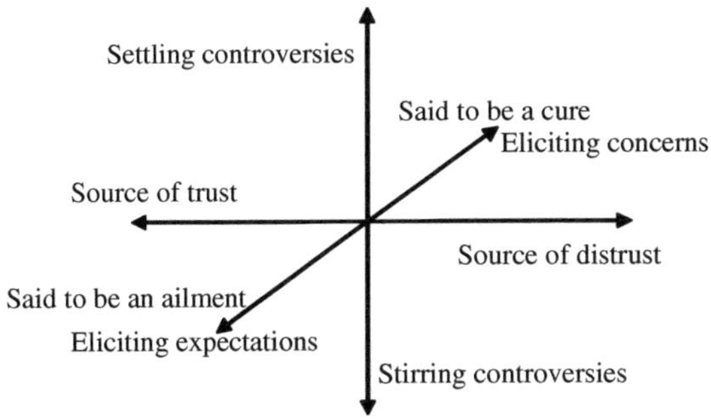

Fig. 2.17. The triple paradox of public attitude to S&T
Source: DG Research

political agenda and the research agenda, and indeed people's opinions themselves. Furthermore, results of opinion surveys are sometimes taken as "the public opinion".

The recent S&T Eurobarometer report of June 2005 shows a mixed picture about the relationship between science and society. On the one hand, S&T and scientists enjoy a high level of confidence. 88 per cent of Europeans agree that S&T progress will help to cure illnesses such as AIDS and cancer, 78 per cent agrees that S&T makes our lives healthier, easier and more comfortable and 73 per cent of European citizens are convinced that politicians should rely more on the advice of expert scientists. Despite these high expectations and the climate of confidence, the same survey shows that European citizens do not always have a positive perception of S&T. S&T developments raise important ethical issues (stem cell research, genetically modified crops, etc.) and the majority of those polled agreed that S&T changes our ways of life too fast, that many high-tech products are just gadgets and consider S&T to be responsible for most of the environmental problems we have today. Although they are divided on whether they consider themselves personally important in the decision-making process, they nevertheless express their wish to see more public involvement in decisions about S&T.[59]

In the United States, it is also a mixed picture that is elicited by the National Science Foundation's *Science and Engineering Indicators 2006* in regard to public views on S&T. Still largely premised on the "deficit model"[60] of the public understanding of science, the Report first notes the following:

---

[59] European Commission, *Eurobarometer 2005, Social Values, Science and Technology*, Luxembourg, 2005.

[60] See Meinolf Dierkes *et al.* (Eds.), *Between Understanding and Trust: The Public, Science and Technology*, Amsterdam, 2000 as well as Bruce V. Lewenstein, Science and the Media, In: Sheila Jasanoff *et al.* (Eds.), *Handbook of Science and Technology Studies*, London, 1995/2002, pp. 343–388, and Brian Wynne, May the Sheep Safely Graze? A Reflexive View of the Expert–Lay Knowledge Divide, in Scott Lash *et al.* (Eds.) *Risk, Environment and Modernity: Towards a New Ecology*, London, 1996, pp. 44–83.

Although Americans express strong support for science and technology (S&T), most are not very well informed about these subjects. Knowledge of basic scientific facts and concepts is necessary not only for an understanding of S&T-related issues but also for good citizenship. Most people do not think they are well informed about S&T. In fact, Americans generally know little about science – and science knowledge in the United States is not improving.

Americans have more positive attitudes about the benefits of S&T than are found in Europe, Russia, and Japan. In recent surveys,[61] 84 per cent of Americans, compared with 52 per cent of Europeans and 40 per cent of Japanese, agreed that "the benefits of scientific research outweigh any harmful results."

However, a sizeable segment of the US population has some reservations about S&T, including concerns about moral issues. For example, in 2004 surveys, more than half of the respondents agreed that "scientific research these days doesn't pay enough attention to the moral values of society", and that "scientific research has created as many problems for society as it has solutions". Contrast that with the 84-per cent figure above – this relates directly to the discussion of the paradoxical nature of the public's attitude to S&T. One can already note that the notion of "risk" has featured increasingly prominently in recent years as the leitmotiv of this quandary or balancing act between expected benefits, costs, and harms. Besides, in 2004, 83 per cent of Americans surveyed agreed that "even if it brings no immediate benefits, scientific research that advances the frontiers of knowledge is necessary and should be supported by the federal government".

In its concluding pages, the Public Attitudes chapter of the 2006 Report observes that, although the evidence about the public's interest in S&T is mixed, since 2002 more people have expressed confidence in the leadership of the scientific community than in any other profession except the military.

So – in the wake of the above caveats – what to make of the mixed results of these opinion surveys? Three observations thereon should be brought to the fore. First, opinion surveys construct an image of public opinion as much as they measure an extraneous reality. One of the messages that underpins this construction is that "the public should know better" (while another message is that "the public supports science and the government funding thereof", for instance). Secondly, there is not much that is compelling about the information provided as to that extraneous reality itself, and many a finding is contradicted by another. One can really talk of a "mixed picture".[62] Thirdly, what these exercises do compellingly exhibit, however,

[61] Allum, Nick *et al.*, *A Meta-Analysis of Science, Knowledge and Attitudes across Cultures*, Presentation at the Annual Meeting of the American Association for the Advancement of Science, Washington, D.C., February 2005 and National Institute of Science and Technology Policy, Ministry of Education, Culture, Sports, Science and Technology, *The 2001 Survey of Public Attitudes Toward and Understanding of Science & Technology in Japan*, Tokyo, 2002, both quoted in National Science Board, *Science and Engineering Indicators 2006*, Arlington, VA, 2006.

[62] One can argue that the most informative findings to come out of opinion surveys are elicited through time series, country/region comparisons, and indeed through unpacking the surveys themselves

is that the relations between society and science are permeated with issues of trust, leadership, and imaginaries – issues which the narrative of Progress has long tied together remarkably.

To close on a comparative note, while we left the US survey highlighting scientific (and military) leadership, the EU survey's coda highlights the public wish to see more public involvement in decisions about S&T. Though sharply contrasted, both thus point to trust and decision-making, which we have seen to be at the centre of the citizen–S&T–state triangle.

4.3. Risk, trust, and decision-making

In our societies, risk – collective and individual risks, risk production, distribution, imputation, (un)acceptability and regulation – is a pivotal concern, which finds ample translations not only in discourses but also in institutions.[63] In fact, many tangled and untractable issues related to S&T broadly constructed – such as climate change, endocrine disrupting chemicals, genetically modified organisms – are dealt with through institutional responses consisting in assessing, managing, and communicating (be it through downplaying) those issues as risks.

A crucial matter here is trust, as a relation between members of the public and those in a position of leadership, or the trustworthiness of the policy-makers and of the science-makers. In other words, these are issues of accountability and confidence, such as public confidence in public institutions, and these issues figure prominently in the national, European, and global agendas.

Lost trust It will thus be particularly useful to inquire into the diagnoses of "lost trust". Public mistrust, according to Ulrich Beck,[64] is a response to the failure of public institutions to control the risks, supplemented by a refusal to acknowledge the failures as well as a denial of responsibility. Another perspective is afforded by the influential work of Max Weber on "disenchantment" as the flipside of S&T developments, and on the "iron cage of the future" as the flipside of administrative and S&T developments. Meanwhile, with Maarten Hajer's[65] recourse to disenchantment, one is brought to view this response of the public as not only resulting from a disenchantment with regard to the consequences of previous S&T and state accomplishments, but also taking the form of a further disenchantment

---

(rather than merely their results). On the other hand, other forms of public engagement or stakeholders involvement can lead to more contextualised inputs – see Chapter 5.

[63] Dratwa, Jim, Social Learning at the European Commission and the Codex Alimentarius, in: Bob Reinalda and Bertjan Verbeek (Eds.), *Decision Making Within International Organizations*, London and New York, 2004.

[64] Ulrich Beck, *Risk Society: Towards a New Modernity*, London, 1992, and *World Risk Society*, Cambridge, 1999.

[65] Maarten.A. Hajer, *The Politics of environmental Discourse: Ecological Modernization and the Policy Process*, Oxford, 1995, and Ecological Modernisation as Cultural Politics, In: Scott Lash *et al.* (Eds.), *Risk, Environment and Modernity: Towards a New Ecology*, London 1996.

at the prospect of new S&T and state proposals. A vicious circle of sorts, or a dead end. A third perspective is advanced by Brian Wynne, preferring to talk of alienation to emphasise that the mistrust of the public has less to do with fear (or disenchantment) than with resentment. Those three perspectives are not mutually exclusive, rather focussing on different aspects of a (failed) relationship of trust. It should be noted that the fear would pertain to the issues or risks themselves as mediated by the risk (assessment, management, communication) institutions, whereas the resentment pertains to those very institutions.

Still, whereas it is a restoration of citizens and consumers *trust* that is sought by public authorities and risk institutions, it can be public *recalcitrance* that is elicited. In various ways, the public arises and resists. And controversies arise, too. Students of science, technology, and society have documented how scientific controversies allow to re-open the "black boxes of science made",[66] as well as how fruitful they can be for the advancement of scientific knowledge. Likewise, these public or societal controversies – where politics or outsiders might have seemed to irrupt inside science – can be learning processes in at least two important respects. First, they lead to the production and confrontation of new knowledge (from a variety of scientific disciplines as well as from "lay experts"). Secondly, they allow learning as to setups of expertise and decision-making by opening the "black boxes of decisions made", the black boxes of framings and implicit assumptions, thus bringing these to the fore and requiring a work of explication, of acknowledgement, of justification, and probably of reworking.

As we remarked above, scientific expertise is more and more contested – less believed in – and it is also more and more solicited, demanded. The point is more general: Science is more and more necessary to the socially binding delineation of truth, and at the same time less and less sufficient. It is necessarily thanks to scientific and technological creations, practices and practitioners that the invisible can be seen (the ozone layer and hole thereof, for instance) that some risks can be recognised. Science also proposes a hallowed ground, common if not neutral, a critical apparatus allowing the commensurability of claims. Better still, it can transmute into objective facts what consists largely of value conflicts (issues of social equity, ethical oppositions, or divergences on the common interest). But these feats of facts – those grand divides between facts and values, science and society, human and non-human, subjectivities and objectivity, grand divides which are the hallmark of Modernity – are increasingly difficult to achieve.[67] And it is

---

[66] See for example Michel Callon and Bruno Latour, Unscrewing the Big Leviathan: How Actors Macro-structure Reality and How Sociologists Help Them to Do so, in Karin Knorr-Cetina and Aaron V. Cicourel (Eds.), *Advances in Social Theory and Methodology: Towards an Integration of Micro and Macro Sociologies*, Routledge, 1981 and Bruno Latour, *Science In Action: How to Follow Scientists and Engineers Through Society*, Cambridge, MA, 1987.

[67] See Bruno Latour, *We Have Never Been Modern*, Cambridge, MA, 1993 and Aant Elzinga, The Growth of Science: Romantic and Technocratic Images [Nobel symposium "Progress in Science and its Social Conditions", 15–19 August 1983, Lidingö, Sweden], In: *Essays on Scientism, Romanticism and Social Realist Images of Science*, Göteborg, 1984.

also their distribution that is in question, and the necessary broadening of the base of expertise, of the base of knowledge – as regards further scientific disciplines and further forms of knowledge. Helga Nowotny insists on "socially robust knowledge" in addition to the classic "reliable knowledge" (corresponding to the phrase "it works" uttered in the laboratory, where she asserts that the technical reliability was negotiated), where Silvio Funtowicz and Jerry Ravetz referred to "*extended* peer review" as an extension of the criteria of validation on which science relies.[68]

There is no "vases communicants", no neat fit between the cause and the cure, the poison and the antidote (or as Ulrich Beck would have it the *Gifte* and *Gegengifte*). In fact the very first of our introductory paradoxes largely rests on the mistaken notion of S&T as a homogeneous whole, as a hypostasis. But it can thus be untied by acknowledging the diversity of S&T practices – for example by acknowledging that those S&T that played some part in the prevalence of asbestos or the hole in the ozone layer are not the same as those S&T that strive to identify, prevent, mitigate, or remedy such hazards or ill-effects.

Cautionary tale We are not in a position to judge whether these uninvited public reactions – in relation to risk, trust, and decision-making – are beneficial or detrimental on the whole, but they certainly can be significant and illuminating.

In this respect, one can usefully compare the reaction to agricultural biotechnologies (and more particularly genetically modified organisms, GMOs, intended for use as feed and food) in the European Union and in the United States. The issue is complex, ever-evolving, and predates the 1974–1975 Asilomar moratorium.[69] Still, to get to the gist, it is often convincingly argued that in the last 10–15 years the EU lost its technological edge to the United States largely because of differences in their respective home markets (with the scepticism of the European public contrasted with the placidity of the US public, regardless of further determinations as to differing attitudes to risk, but taking in the major part played by the retail sector in addition to the devisers and growers of GMOs).

The history of agricultural biotechnology is still too fresh and undetermined, but it can also be argued that the European public's recalcitrance stemmed from the realisation that those GMOs that were devised – initially at least – provided little benefits to cultivators and consumers. In the case of other new technologies, as indicated in Chapter 1, such a rather more precautionary and demanding stance has actually led to more robust – socially but also commercially robust – innovations. The involvement of various actors or voices or knowledge-claims – translated in the political domain as "participatory democracy" or "stakeholders consultation" – does

[68] Jerome R. Ravetz and Silvio O. Funtowicz, A New Scientific Methodology for Global environmental Issues, In: Costanza, Robert (Ed.), *Ecological Economics: The Science and Management of Sustainability*, New York, 1991.

[69] Through which, in the emerging field of genetic engineering, researchers from various parts of the world agreed to exercise self-restraint on certain experiments with recombinant DNA (until 1977).

also bear fruit in the domain of S&T innovation, as the most successful companies are acutely aware. Still, for better or worse, the EU–US contrast remains.

The wider implications of this divergency come in full view when one considers the two highest-promise S&T sectors – that were biotech and ICT – in the context of the elaboration of the Lisbon Strategy. becoming "the most dynamic and competitive knowledge-based economy in the world" had a conspicuous international-comparative perspective, and the European misfiring of the biotech boom (in contrast to its US development) is part of the reading grid with respect to ICT at the heart of Lisbon. It is precisely on further aspects of the policy context in which the Lisbon Strategy tenets – and the 7th Framework Programme – were conceived that Chapter 3 will concentrate.

Meanwhile, to conclude this section, much can also be derived from another, related, comparison: between the reaction to GMOs developed for agricultural use and the reaction to GMOs developed for pharmaceutical use, in Europe. While the former have been greeted with scepticism or antagonism, the latter have been generally well received, eliciting a quieter response, marked by more enthusiasm than concern.[70] Besides the issue of *trust* (and the story of GM food as a story of lost trust), this indicates that neither is *risk* the only determinant of public attitude to science. Also particularly important are thus the trade-offs between the various risks and the various benefits, as well as the *distribution* of those risks and benefits.

4.4. Coda: long is the way, and hard (on progress and modernity/ies)

United States President Franklin D. Roosevelt famously advocated a "New Deal" of his own to tackle the worldwide economic crisis which followed the Black Thursday stock market crash of 1929. His discourse resonates strikingly with the predicament we are facing today.

> The millions who are in want will not stand by silently forever while the things to satisfy their needs are within easy reach. We need enthusiasm, imagination and the ability to face facts, even unpleasant ones, bravely. We need to correct, by drastic means if necessary, the faults in our economic system from which we now suffer. We need the courage of the young. Yours is not the task of making your way in the world, but the task of remaking the world which you will find before you. May every one of us be granted the courage, the faith and the vision to give the best that is in us to that remaking!
> (Oglethorpe University Commencement Address, 22 May 1932)

But that is not all. The same Franklin Roosevelt also famously commissioned what would become Dr Vannevar Bush's *Science The Endless Frontier* report.[71]

Making social and political progress through science and technology. Such was the project of the Enlightenment, of Modernity. It was drawn upon to great effect by

---

[70] The literature on this contrast is abundant. For a recent assessment of Europeans' attitudes to biotechnology, see George Gaskell *et al.*, *Europeans and Biotechnology in 2005: Patterns and Trends*, Eurobarometer 64.3. A Report to the European Commission's Directorate-General for Research, 2006.

[71] Vannevar Bush, *Science The Endless Frontier*, A Report to the President, Washington, DC, 1945.

Roosevelt (albeit with a terrible war kicking in as an intermediary variable). And it is still a powerful source of inspiration for us today. Yet the order seems taller now, as we have grown even more sophisticated, demanding, disenchanted perhaps.

The matter here is obviously not to bring canon fodder to the so-called science wars pitting the "postmodern constructivist" strawman against its "modern rationalist positivist" counterpart. The matter is not, therefore, to relativise the scientific endeavour or the quest for the objectivity of hard facts into nothingness. Nor is it to ignore the key contributions of the social sciences to the understanding of S&T practices themselves. The matter is not less, but more. Scientific robustness is indispensable, but not sufficient.

So, yes it may be that, in solving some age-old problems, S&T has created the possibility for new problems to emerge. Yet even to address these new problems we can hardly do without S&T. But we can – and rightfully do – concern ourselves with the consequences of the solutions we devise. It is possible to draw lessons – and strength – from successes as well as from setbacks. Notably if one can henceforth open and pursue avenues of research which take the public, with all its wants and concerns, on board. Could that be why Europe is at the leading edge in environmental technologies, transport technologies, and some areas of biotechnologies?

CONCLUSION

Citizens and governments increasingly view science and technology as a panacea for all their problems. However, if Europe wants to be able to meet theses expectations more effectively, it needs to tackle a number of weaknesses of its S&T system. This chapter has analysed some of the key factors which currently prevent Europe from achieving its full S&T potential: Not only does it need to spend more on research, but also it needs to deploy these scarce resources more efficiently and to greater effect through more cooperation and coordination of S&T policies in the EU. This will require further improvement in the EU system of governance for research policy, as well as closer coordination between research policy and other related policies. Furthermore, the diverse relations between citizens and S&T must be recognised and taken into account if Europe is to advance in certain key fields.

These weaknesses are not altogether new or sudden. However, the growing perception that something needed to be done emerged from a new policy context. This will be shown in the next chapter.

CHAPTER 3

OPENING THE WINDOW OF OPPORTUNITY: THE NEW POLICY CONTEXT

Europe is faced with a substantial number of important – both long-standing and emerging – challenges of an economic, social, and environmental nature. S&T could play a significant role in meeting these challenges but is prevented from doing so because of important European S&T weaknesses. It is against this background that the 7th Framework Programme was conceived.

But there was another key factor influencing the design of the 7th Framework Programme: the new policy context. Around the turn of the century, several factors gave rise to a more acute awareness among European policy-makers of Europe's societal and S&T challenges, and to the formulation of the Lisbon Agenda. The Lisbon European Council of March 2000 and its aftermath created a new European research policy context, within which it was possible to pursue a Framework Programme much more ambitious in scale and scope than any past Framework Programme. This will be the topic of this chapter.

This chapter is organised as follows. Section 1 will demonstrate how at Lisbon, European policy-makers recognised unambiguously that to overcome its societal challenges, Europe would have to turn itself into a knowledge-based society. At the same time, it was acknowledged that putting knowledge at the centre would require more and better investment in the knowledge triangle of research, education, and innovation. This will be the topic of Section 2, which describes how the objective to invest more in research culminated in the 3 per cent objective, while the objective to invest better in research centred on the concept of the European Research Area. While Section 3 explains why impact assessment as a tool for better (research) policy-making emanates from this "Lisbon reform spirit", Section 4 paints a less-rosy picture when looking at the little progress made towards the achievement of these objectives after 2000. As will be shown in Section 5, policy-makers agreed that this was mainly due to weaknesses in the governance of the Lisbon strategy. Therefore, at their Summit in 2005, the European leaders set in motion a reform of the open method of coordination, and a general revaluation of the role of the Union in the achievement of the Lisbon objectives. Section 6 goes on to examine how this new policy context had important implications for the preparation of the 7th Framework Programme. In particular, the recognition of the relative effectiveness of action through the Union spilled over into the debate on Community funding programmes up for renewal, and opened the way for an ambitious new Framework Programme.

1. THE LISBON EUROPEAN COUNCIL: PUTTING KNOWLEDGE AT THE CENTRE

1.1. Taking stock of Europe's challenges

By the late 1990s, and for reasons discussed below, the various societal challenges analysed in Chapter 1 were felt more acutely by European policy-makers. It was therefore not without reason that the Presidency Conclusions of the European Council held in Lisbon in March 2000 began with taking stock of the Union's strengths and weaknesses. As the Council coincided with the peak of the Internet boom, the document could note with some confidence that "the Union is experiencing its best macro-economic outlook for a generation", and that "growth and job creation have resumed".[1] Yet it also observed a number of important weaknesses, mainly with regard to employment, the development of key economic sectors, and human capital formation: "More than 15 million Europeans are still out of work. The employment rate is too low and is characterised by insufficient participation in the labour market by women and older workers. Long-term structural unemployment and marked regional unemployment imbalances remain endemic in parts of the Union. The services sector is underdeveloped, particularly in the areas of telecommunications and the Internet. There is a widening skills gap, especially in information technology where increasing numbers of jobs remain unfilled".[2]

Almost as soon as the Lisbon European Council came to an end, in that same year 2000, the new economy boom collapsed. The economic situation worsened rapidly, and the need for faster economic growth in the EU was felt more urgently. The July 2003 "Sapir Report", for instance, identified growth as Europe's number one priority. Sapir and his team had undertaken a longer-term analysis of Europe's growth performance, and had found a steady decline of the average growth rate decade after decade.[3] The April 2004 "Strauss–Kahn Report" largely echoed the analysis of the "Sapir report", and identified five internal challenges to the European model (economic change, social change, the demographic challenge, environmental change, and democratic distrust), as well as two external ones (globalisation, new post-9/11 strategic reality).[4] Compared to previous reports, the November 2004 "Kok Report" did not come up with any new analysis or new recommendations, but it infused the debate with a sense of urgency. The independent assessment carried out within the context of the mid-term review of the Lisbon Strategy by a high-level group under the chairmanship of Wim Kok regretted the lack of progress made, but

---

[1] *Presidency Conclusions Lisbon European Council 23 and 24 March 2000.*

[2] *Presidency Conclusions Lisbon European Council 23 and 24 March 2000.*

[3] *An Agenda for a Growing Europe. Making the EU Economic System Deliver*, Report of an Independent High-Level Study Group Established on the Initiative of the President of the European Commission, Brussels, 2003.

[4] *Construire l'Europe Politique. 50 Propositions pour l'Europe de Demain*, Dominique Strauss-Kahn. Président de la Table Ronde "Un Projet Durable pour l'Europe de Demain" Établie à l'Initiative du Président de la Commission Européenne, Rapporteur Olivier Ferrand, April 2004.

positioned the Lisbon Strategy as "even more urgent today", since "the growth gap with North America and Asia has widened, while Europe must meet the combined challenges of low population growth and ageing".[5]

1.2. Recognising the need for Europe to turn itself into a knowledge-based society

The authors of the Lisbon Conclusions and of later high-profile policy reports shared the basic understanding that Europe to overcome its challenges would have to turn itself as soon as possible into a knowledge-based society.[6] That is why at Lisbon the Union set itself a new strategic goal to be accomplished by 2010, namely "to become the most competitive and dynamic knowledge-based economy in the world, capable of sustainable economic growth with more and better jobs and greater social cohesion".[7] The "Sapir Report" concluded in the same sense by ascribing the long-term decline of the European economic growth rate to Europe's failure to transform itself into an "innovation-based economy".[8] The "Kok Report" listed the "knowledge society" as the first of five policy areas requiring urgent action.[9] And the January 2006 "Aho Report" presented a strategy to create an "innovative Europe".[10]

The need for making the transition to the knowledge-based society was recognised not only at European level but also at Member State level. Numerous national reports were launched even from before the turn of the century calling for turning knowledge into the basis of the economy and society. To mention just a few examples, in Finland, the *Science and Technology Policy Council* published reports on the "innovative society" in 1993 and on the "knowledge-based society" in 1996.[11] In the United Kingdom, a key diagnostic report was published in 2003. Its core message was that the United Kingdom needed a new approach towards competitiveness. It had been successful in implementing market-based reforms such as deregulation, privatisation and competition, but these policies were now encountering diminishing returns. It needed a new approach focussed on improving skills, stimulating innovation, and fostering enterprise. Only by building such capacity, the United Kingdom would be able to move to the next stage of improving competitiveness, and achieve sustained higher levels of

---

[5] *Facing the Challenge. The Lisbon Strategy for Growth and Development*, Report from the High Level Group Chaired by Wim Kok, Brussels, 2004, p. 6.

[6] Luc Soete, The Challenges and the Potential of the Knowledge-based Economy in a Globalised World, In: Maria João Rodrigues (Ed.), *The New Knowledge Economy in Europe. A Strategy for International Competitiveness and Social Cohesion*, Cheltenham, 2002, pp. 28–53.

[7] *Presidency Conclusions Lisbon European Council 23 and 24 March 2000.*

[8] *An Agenda for a Growing Europe.*

[9] *Facing the Challenge.*

[10] *Creating an Innovative Europe*, Report of the Independent Expert Group on R&D and Innovation Appointed Following the Hampton Court Summit, Brussels, 2006.

[11] Science and Technology Policy Council, *Towards an Innovative Society: A Development Strategy for Finland*, Helsinki, 1993; Science and Technology Policy Council, *Finland: A Knowledge-based Society*, Helsinki, 1996.

prosperity.[12] In France, Daniel Cohen and Michèle Debonneuil, members of the Prime Minister's Economic Analysis Council, wrote in the year 2000 a report on the "new economy". They argued that an "industrial revolution" was taking place in the United States driven by Information and Communication Technologies (ICTs). The latter was gaining in importance because of R&D efforts, appropriate financing mechanisms, and the deregulation of old monopolies. France and Europe were lagging six to seven years behind and an important effort was needed for the diffusion and assimilation of new technologies, R&D, and production.[13] Five years later, Jean-Louis Beffa, presidential advisor, ascribed France's industrial weakness to among other things insufficient R&D, and argued in favour of supporting high-tech industries through large-scale industrial innovation programmes.[14] And in the Netherlands an innovation platform was launched in 2003, the mission of which was to strengthen the innovation potential to secure a leading role for the country in the European knowledge economy of 2010.[15]

1.3. Why knowledge took centre stage

Several factors could explain why around the turn of the century European policy-makers recognised the need for Europe to turn itself into a knowledge-based society.

Europe's knowledge weaknesses and the incomplete materialisation of new economy hopes The last decade or so of the twentieth century was marked by a significant acceleration in the pace of technological progress. Great steps forward were made in a number of key fields of technology including information and communication technologies (e.g. Internet, mobile communications), nanotechnology, and the life sciences (e.g. biotechnology, the human genome project). This rapid pace of technological progress boosted the expectations of policy-makers and society at large of the future potential of S&T, not in the least because associated with the arrival of the New Economy in the United States were a stellar economic growth performance and record-low unemployment rates.

Most of the technological progress appeared to come from outside Europe, however. Europe did benefit from the arrival of the New Economy, but only to a limited extent. It never succeeded in enjoying US levels of economic growth and

[12] Michael E. Porter and Christian H.M. Ketels, *UK Competitiveness: Moving to the Next Stage*, DTI Economics Paper No. 3, London, May 2003. See also subsequent more action-oriented reports: DTI, *Competing in the Global Economy – The Innovation Challenge*, DTI Economics Paper No. 7, London, November 2003. DTI, *Competing in the Global Economy: The Innovation Challenge*, Innovation Report, London, December 2003.

[13] Daniel Cohen and Michèle Debonneuil, *Nouvelle Économie*, Paris, 2000.

[14] Jean-Louis Beffa, *Pour une Nouvelle Politique Industrielle*, Paris, 2005.

[15] For Ireland, see *Building Ireland's Knowledge Economy. The Irish Action Plan for Promoting Investment in R&D to 2010*, Report to the Inter Departmental Committee on Science, Technology and Innovation, July 2004; For Estonia, see Marek Tiits *et al.*, *Competitiveness and Future Outlooks of the Estonian Economy – R&D and Innovation Policy Review,* Tallinn, 2003.

employment. The search for the cause of the growth, productivity, employment and technology gaps highlighted research, education, and innovation weaknesses (in addition to structural factors such as the fragmented European market, labour market rigidities, etc.). As a result, the awareness of fundamental European S&T weaknesses moved beyond the scientific community to a wider group of European policy-makers.[16]

More knowledge to cope with increased foreign competition Over the course of the last decade or so, major developing economies including, but not limited to, the so-called BRICs (Brazil, Russia, India and China) have opened up. Foreign direct investment has flowed into these economies, and foreign companies succeed in producing not just low-tech but also increasingly middle- and high-tech products. Because of trade liberalisation, they are able to export these products to advanced markets such as Europe. It is clear that this poses new competitive threats to developed economies such as the European one, forcing them to flee forward into ever higher-tech industries. At the same time, trade liberalisation also generates competitive threats from these countries to more traditional European industries (e.g. textiles), forcing them to upgrade too. Both factors push research to the forefront.

The knowledge society as a new major project for Europe Around the turn of the century, the Union could look back to the previous decade with a certain degree of pride. By the early 1990s, substantial progress had been made towards the establishment of a single European market. By the second half of the 1990s, a single currency was being introduced successfully under a carefully considered and seemingly unavoidable phase-by-phase and step-by-step approach. And the road towards the Union's most ambitious enlargement ever had been embarked upon. Within this context, the search was on for the next major project for Europe.

The realisation of the knowledge society presented itself as a suitable candidate. In this regard, Alyn Smith recently said that "research and the up-scaling of Europe's economy is Europe's new narrative. Europe used to be about coal and steel, then it was about agriculture and fisheries. For the future, Europe is going to be about research and development, and where Europe adds value in coordinating the two".[17]

In a way, this constituted a logical sequence to what came before. After the Union's large-scale enlargement, the search was for a project that could contribute to the consolidation and deepening of the Union. Given that the focus in the recent past had been mainly on putting in place certain non-dynamic economic fundamentals

---

[16] Such weaknesses included the mostly private sector R&D investment and expenditure gap with the United States and Japan; the European paradox, according to which Europe is strong in science but fails to quickly, effectively and efficiently turn new knowledge into new products, processes and services; the decreasing attractiveness of Europe as a place to carry out research; the lack of S&T mobility, both extra- and intra-European; and so on.

[17] Debate at the European Parliament's ITRE committee on 16 June 2006. Alyn Smith, Member of the European Parliament (Greens/EFA), was draftsman of the opinion of the European Parliament's Committee on Regional Development on the 7th Framework Programme proposal.

(single market, debt reduction, budget deficit reduction, single currency), the search was for a more dynamic policy. Given that previous major European projects had failed to make much of a positive impact on Europe's long-standing low growth and high unemployment problems, the search was for a project able to resolve this policy deadlock. Because of the macro-economic interdependence created by the introduction of the single currency, the search was for a project that could in the short to medium term contribute to the Union's economic convergence. Now that the introduction of the single currency had removed from the macro-economic toolbox currency devaluations and interest rate manipulations, the search was for a project that could nevertheless promote competitiveness. The realisation of the knowledge society was the only candidate project fitting the bill, and it was not too difficult to agree on this. It was a new project, having a "clean slate", and was untested. It was also non-controversial, holding the promise of a win-win situation for everyone, unlike rapid market integration, the introduction of the euro, or enlargement, where there always seemed to be winners and losers.

1.4. Investing more and better in the knowledge triangle

At both EU and Member State levels, it was immediately understood that making the transition to the knowledge-based society would require putting research, education, and innovation – the so-called knowledge triangle – at the centre, and that direct support programmes and suitable regulatory action would have to make more and better targeted resources flow into research, education and innovation. Thus the Lisbon strategy included major headings such as "Establishing a European area of research and innovation"; "Education and training for living and working in the knowledge society"; and "Creating a friendly environment for starting up and developing innovative businesses, especially SMEs".[18]

The call for investing more and better in knowledge triangle issues was repeated in the aforementioned high-profile policy documents. The "Sapir Report" recommended, first, to make the Single Market more dynamic and, second, to boost investment in knowledge.[19] The "Strauss-Kahn Report" put "investing in knowledge by emphasizing research and higher education" as a priority before the revitalisation of the internal market. The report contained 50 proposals for "Constructing Tomorrow's Europe", which impressively underlined the importance attached to research, innovation and education.[20] The "Kok Report" recommended

---

[18] *Presidency Conclusions Lisbon European Council 23 and 24 March 2000.*

[19] *An Agenda for a Growing Europe.*

[20] The first four recommendations were: (1) making research *the* budgetary priority for the union so that budgetary expenditure on public research would represent 0.25 per cent of community Gross Domestic Product, and progressively increase so that research would become the union's most important policy; (2) creating a European agency for science and research; (3) stimulating private research by fixing through a European directive a minimum tax break for enterprises investing in R&D; and (4) investing in higher education by establishing a network of European university centres of excellence, and fixing as an objective that 50 per cent of the European population would obtain a higher education degree[1] (*Construire l'Europe Politique*).

increasing Europe's attractiveness for researchers and scientists, making R&D a top priority, and promoting the use of ICTs.[21] The "Aho Report" recommended a paradigm shift going well beyond the narrow domain of R&D and innovation policy. A "Pact for Research and Innovation" would be needed entailing a combination of a market for innovative goods and services, focussed resources, new financial structures and mobility of people, money and organisations. The report concluded that "Europe and its citizens should realise that their way of life is under threat but also that the path to prosperity through research and innovation is open if large scale action is taken now by their leaders before it is too late".[22]

2. INVESTING MORE AND BETTER IN RESEARCH

2.1. Investing more in research

At Lisbon, and two years later at Barcelona, a number of concrete objectives were formulated in relation to research (Table 3.1). They concerned both more and better investment in research.

The objective to invest more in research found its clearest expression in the so-called "*3 Per cent Objective*". In 2002, the Barcelona European Council agreed that "overall spending on R&D and innovation in the Union should be increased with the aim of approaching 3 per cent of GDP by 2010. Two-thirds of this new investment should come from the private sector."[23] In formulating the 3 per cent objective, the Barcelona European Council built upon the Commission Communication *Towards a*

Table 3.1. Lisbon strategy targets/objectives in the field of research

Increase R&D spending to 3% of GDP by 2010. The proportion financed by business should rise to two-thirds of that total (target set at Barcelona).
Network national and joint research programmes on a voluntary basis around freely chosen objectives and develop an open method of coordination for national research policies.
Improve the environment for private research investment, R&D partnerships and high-technology start-ups.
Harness new and frontier technologies, notably biotechnology and environmental technologies.
Introduce a cost-effective Community patent.
Remove obstacles to the mobility of researchers, attract and retain high-quality research talent in Europe.
Roll out a world-class research communications infrastructure.

Source: European Commission, *Commission Staff Working Document in Support of the Report from the Commission to the Spring European Council, 22–23 March 2005, on the Lisbon Strategy of Economic, Social and environmental Renewal, SEC*(2005) 160, 28 January 2005, p. 17.

[21] *Facing the Challenge.*
[22] *Creating an Innovative Europe.*
[23] *Presidency Conclusions Barcelona European Council 15 and 16 March 2002.*

European Research Area (2000) and the Lisbon Conclusions (2000). The Communication noted that "more than ever, investing in research and technological development offers the most promise for the future. In Europe, however, the situation concerning research is worrying. ... Europe might not successfully achieve the transition to a knowledge-based economy".[24] The first reason listed was that the average research effort in the Union was much lower than in the United States or Japan, and that the gap appeared to be increasing.[25] In response, the Communication called among other things for "more dynamic private investment".[26] The Lisbon European Council Presidency Conclusions on the other hand called for the necessary steps to be taken to "improve the environment for private research investment, R&D partnerships and high technology start-ups".[27]

From the beginning there was agreement among policy-makers that increasing R&D expenditure would require the use of research, other knowledge triangle policies (education and innovation), and still other policies (e.g. competition, market, macro-economic), and thus horizontal policy coordination was essential (Fig. 3.1). In this regard, the Commission Communication *Towards a European Research Area* called for promoting more dynamic private investment by better use of instruments of indirect support to research, the development of effective tools to protect intellectual property, and the encouragement of the creation of companies and risk capital investment.[28] The Lisbon Conclusions called for improving the environment for private research investment, R&D partnership and high-technology start-ups by using tax policies, venture capital and EIB support, and emphasised the need for macro-economic policies to be supportive of knowledge investment: "As well as preserving macro-economic stability and stimulating growth and employment, macro-economic policies should foster the transition towards a knowledge-based economy, which implies an enhanced role for structural policies".[29] And the Commission Communication *More Research for Europe – Towards 3 per cent of GDP* distinguished between more effective use of public financing for business R&D on the one hand and more attractive framework conditions on the other hand.[30]

---

[24] European Commission, *Towards a European Research Area*, Communication from the Commission to the Council, the European Parliament, the Economic and Social Committee and the Committee of the Regions, COM(2000) 6 final, 18 January 2000, p. 4.

[25] European Commission, *Towards a European Research Area*, p. 4.

[26] European Commission, *Towards a European Research Area*, pp. 13–14.

[27] *Presidency Conclusions Lisbon European Council 23 and 24 March 2000.*

[28] European Commission, *Towards a European Research Area*, pp. 13–14.

[29] *Presidency Conclusions Lisbon European Council 23 and 24 March 2000.*

[30] European Commission, *More Research for Europe – Towards 3% of GDP*, Communication from the Commission, COM(2002) 499 final, 11 September 2002. Under the first heading, it called for the use of direct support measures, but also of fiscal incentives, guarantee mechanisms, and public support for risk capital. Under the second heading, it called for sufficient and high quality human resources, a strong public research base with improved industry links, entrepreneurship for, and through, R&D, effective adaptation and use of intellectual property rights systems, research- and innovation-friendly regulations, a competitive environment and supportive competition rules, supportive financial markets covering

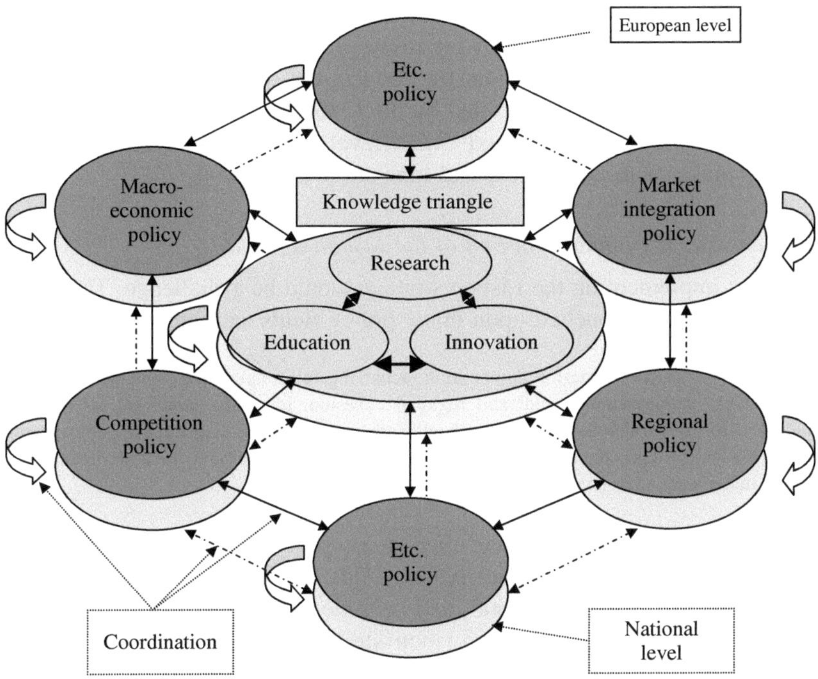

Fig. 3.1. The Lisbon strategy for getting to the knowledge-based society
Source: DG Research

2.2. Investing better in research

The objective of better investment in research concerned mainly the desire to overcome the fragmentation of European S&T through greater collaboration and coordination. Around the turn of the century, the fragmentation of European S&T was widely perceived as a problem, and calls for greater collaboration and coordination were widespread. In 1998, for instance, Caracostas and Muldur noted that the foundations of a *European System of Research and Innovation* were in place, but that consolidation of this system would require three kinds of coordination: "between European, national and (increasingly) regional policies; between Community measures and intergovernmental European measures; and between science and technology policies and other public policies".[31] A few years later, Kuhlmann and Edler noted that in Europe innovation policy initiatives were pursued in parallel at the regional, national, and transnational levels, and that this layer structure left open a governance gap of poor integration and

the various stages of development of high-tech and other innovative companies, and macro-economic stability and favourable fiscal conditions.

[31] Paraskevas Caracostas and Ugur Muldur, *Society – The Endless Frontier*, Luxembourg, 1998, p. 128.

coordination.[32] In response, the Union formulated the concept of the *European Research Area*, a Europe-wide market for research, characterised by mobility for researchers and knowledge, and scientific cooperation, and by improved policy coordination. The concept was confirmed by the Lisbon Conclusions, which stated that "research activities at national and Union level must be better integrated and coordinated to make them as efficient and innovative as possible".[33]

2.3. Instruments for the implementation of the Lisbon agenda

It was clear that implementing the Lisbon strategy would be a challenge. The initial broader Lisbon objective touched upon many policy fields and goals:

> completing the Internal Market, fostering competition, achieving full employment, raising quality and productivity in work, strengthening social and regional cohesion, launching structural reforms and modernising education and training, pension and healthcare systems, respecting the environment, and developing knowledge through education and training, research, innovation and the wide use of ICT, based on sound macro-economic policies.[34]

And successive presidencies added even more goals.[35] In addition, the legal competence for action in the different policy fields varied. In some areas, it was exclusively European, in some national, and in some shared.[36]

That is why for implementing the Lisbon strategy the full range of existing policy instruments was mobilised. This included EC legislation proposed by the Commission and adopted by both the European Parliament and the Council, and programmes and funding granted by the Community (especially in areas of shared competence), as well as, for instance, action plans setting out how legislation, policy guidance and funding would be brought together to achieve policy goals.[37]

In addition, a specific instrument was proposed for coordination between Member States in areas falling mainly under national responsibility (where the Community Method could not (always) be applied), namely the "*Open Method of Coordination*".[38] Its main purpose was to spread best practice and achieve greater convergence towards the main EU goals. It was designed to help Member States develop their own policies and involved

---

[32] Stefan Kuhlmann and Jakob Edler, *Governance of Technology and Innovation Policies in Europe: Investigating Future Scenarios*, p. 8.

[33] *Presidency Conclusions Lisbon European Council 23 and 24 March 2000.*

[34] European Commission, *Commission Staff Working Document in Support of the Report from the Commission to the Spring European Council, 22–23 March 2005, on the Lisbon Strategy of Economic, Social and Environmental Renewal,* SEC(2005) 160, 28 January 2005, p. 45.

[35] European Commission, SEC(2005) 160, p. 45.

[36] European Commission, SEC(2005) 160, p. 45.

[37] European Commission, SEC(2005) 160, p. 46.

[38] Although not called by this name at the time, the "open method of coordination" was first applied in EU employment policy, as defined in the Amsterdam Treaty of 1997. It was officially named, defined and endorsed at the Lisbon Council for a broad realm of policies.

fixing guidelines for the Union combined with specific timetables for achieving the goals which they set in the short, medium and long terms; establishing, where appropriate, quantitative and qualitative indicators and benchmarks against the best in the world and tailored to the needs of different Member States and sectors as a means of comparing best practice; translating these European guidelines into national and regional policies by setting specific targets and adopting measures, taking into account national and regional differences; periodic monitoring, evaluation and peer review organised as mutual learning processes.[39]

The open method of coordination is inter-governmental and soft: inter-governmental because it relies mainly on Member States interacting with each other, and soft because a possible lack of progress would not result in sanctions.

2.4. Implementation in the field of research

Following the commitments taken in Lisbon and Barcelona, all available means were employed to achieve the objectives defined in the field of research. The Community Method was used to improve research framework conditions (e.g. the efforts towards the introduction of a Community Patent). The Framework Programme was used to promote greater S&T collaboration and coordination. Actions plans were launched, for instance in relation to the 3 per cent objective.[40] And the open method of coordination was applied in five areas: the 3 per cent objective; human resources and researchers' mobility; science and society; networking and mutual opening of national and joint RTD programmes; and infrastructures of European interest.[41]

All in all, a large share of the responsibility for action in the field of research was assigned to the Member States. To a certain extent, this was to be expected. Research is an area of shared competence, and the bulk of R&D expenditure is situated at national level. But the Lisbon European Council also put extra emphasis on the principles of decentralisation and subsidiarity, and called for Community action that would not upset the delicate agreement reached on Agenda 2000, the Union's financial framework for the period 2000–2006.[42] As a result, the Union acted mainly through a 6th Framework Programme not much larger than the 5th once account was taken of enlargement, and played the role of catalyst and facilitator within the context of the open method of coordination.[43]

---

[39] *Presidency Conclusions Lisbon European Council 23 and 24 March 2000.*

[40] European Commission, *Investing in Research: An Action Plan for Europe*, Communication from the Commission, COM(2003) 226 final/2, 4 June 2003.

[41] European Commission, SEC(2005) 160, pp. 27ff.

[42] *Presidency Conclusions Lisbon European Council 23 and 24 March 2000*: "A fully decentralised approach will be applied in line with the principle of subsidiarity in which the Union, the Member States, the regional and local levels, as well as the social partners and civil society, will be actively involved, using variable forms of partnership"; "and by adding its own contribution to this effort under existing Community policies while respecting Agenda 2000".

[43] *Presidency Conclusions Lisbon European Council 23 and 24 March 2000*: "A method of benchmarking best practices on managing change will be devised by the European Commission networking with different providers and users, namely the social partners, companies and NGOs". "The Union's role is to act as a catalyst in this process, by establishing an effective framework for mobilising all available resources for the transition to the knowledge-based economy."

3. EVIDENCE-BASED POLICY-MAKING

In parallel with the launching of the Lisbon Agenda and in the overall reform spirit which emanated from the Lisbon Summit, the Union at the turn of the century also worked towards a new way of making policy. Within the framework of its 2001 *European Strategy for Sustainable Development* and its 2002 *Better Regulation Action Plan*, the Commission took several concrete actions to increase the transparency and quality of its policy design. One of its more important actions was the revision and strengthening of the policy cycle through the introduction of ex-ante impact assessment, which integrates several sectoral assessments into one global instrument.[44] As of 2005, a formal impact assessment is required for all legislative initiatives included in the Commission's Work Programme.

In an impact assessment, the likely impacts of public interventions are systematically analysed. Although such an assessment does not necessarily lead to clear-cut conclusions and recommendations, it does provide an important input by informing decision-makers of the likely consequences of different policy options. It does this by answering a common set of questions, and by assessing the issues at stake and the objectives to be pursued. It then identifies the main options for achieving these objectives and analyses their likely economic, environmental and social impacts. The advantages and disadvantages of each option are analysed, as well as any synergies, trade-offs and risks.[45] Impact assessment in the field of research was applied for the first time to the Commission's proposal for the 7th Framework Programme.[46]

4. THE LACK OF PROGRESS SINCE THE YEAR 2000

Since the year 2000, the progress made towards the achievement of the Lisbon objectives has been limited. In this regard, the 2004 Lisbon agenda mid-term review – or so-called "Kok Report" – spoke of "disappointing delivery", and the fact that "halfway to 2010 the overall picture is very mixed and much needs to be done in order to prevent Lisbon from becoming a synonym for missed objectives and failed promises".[47] And a Commission Communication to the 2005 Spring

---

[44] Impact assessment replaces existing requirements for business impact assessment, gender impact assessment, environmental assessment, SME assessment, trade impact assessment, and so on. Moving away from this partial approach towards one integrated impact assessment helps policy-makers to assess trade-offs and compare different scenarios when shaping a particular proposal.

[45] Ex-ante impact assessment is critically linked to ex-post evaluation. To properly inform future actions, it should be based on a solid understanding of the effects of past and ongoing Framework Programmes. At the same time, the objectives and performance indicators defined in the impact assessment will guide future ex-post evaluation work.

[46] European Commission, *Main Report: Overall Summary – Impact Assessment and Ex-Ante Evaluation (COM(2005) 119 final)*, Commission Staff Working Paper, Annex to the Proposal for the Council and European Parliament Decisions on the 7th Framework Programme (EC and Euratom), SEC(2005) 430, 6 April 2005.

[47] *Facing the Challenge*, pp. 6 and 10.

Council reported that "today, there is general consensus that Europe is far from achieving the potential for change that the Lisbon strategy offers. . . . the reality is that not enough progress has been made".[48]

Limited progress was observed with regard to the Lisbon growth and employment objectives. Between 1999 and 2003, the European employment rate increased somewhat compared to the United States. But living standards hardly improved, the number of hours worked per worker scarcely increased, and hourly labour productivity actually fell[49] (Fig. 3.2).

But a lack of progress was also observed with regard to the objectives in the field of research as "overall research and development (R&D) expenditure in percentage of GDP increased marginally" (and now appears to be decreasing again)[50] (Fig. 3.3). In addition it was noted that "a final agreement on the Community patent remains elusive as the Council failed to reach final agreement on aspects of the translation regime", and that

> initiatives have been taken both at European level and in Member States to improve the mobility of researchers, to attract and retain high-quality research talent in Europe, to improve the environment for private research investment, R&D partnerships and high-technology start-ups, but these are too recent to register any significant effect yet.[51]

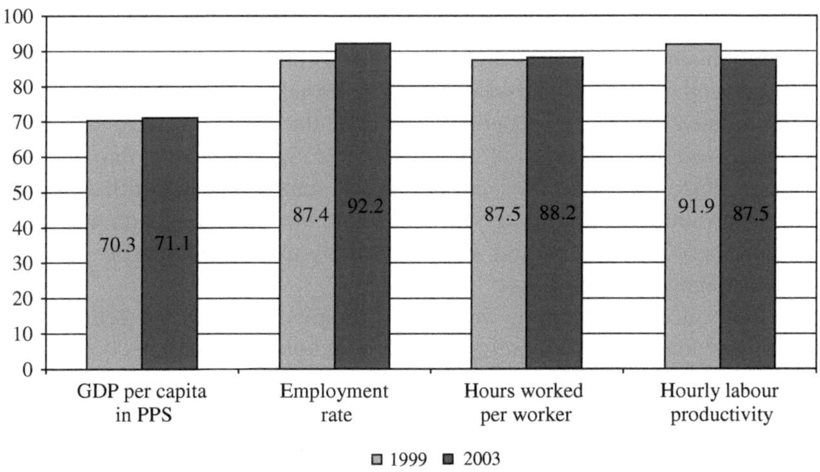

Fig. 3.2. Little progress made since Lisbon in economic terms (EU-15 Performance 1999–2003 (US=100))
Source: European Commission, SEC(2005) 160, p. 7.

---

[48] European Commission, *Growth and Jobs – A New Start for the Lisbon Strategy*, Luxembourg, 2005, p. 9.
[49] European Commission, SEC(2005) 160, p. 7.
[50] European Commission, SEC(2005) 160, pp. 2, 18 and 62.
[51] European Commission, SEC(2005) 160, pp. 3 and 20.

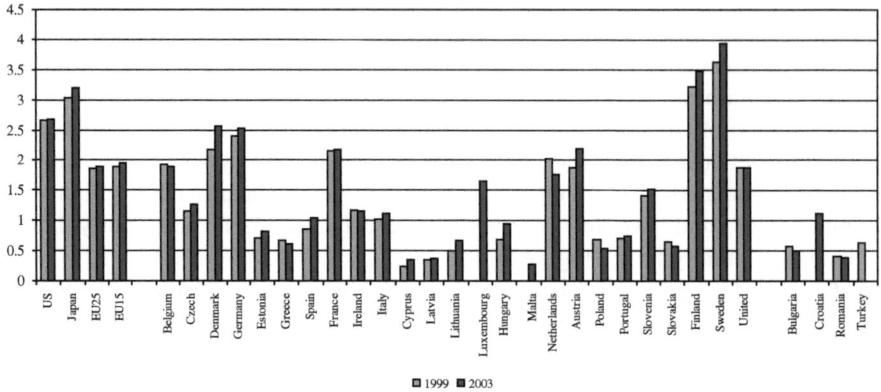

Fig. 3.3. Little progress made since Lisbon in the field of research (R&D intensity)
Source: DG Research
Data: Eurostat

5. EXPLAINING THE LACK OF PROGRESS

The overall consensus is that nothing was fundamentally wrong with the Lisbon agenda and objectives as such. Accordingly, the Commission noted in 2005 that "both the diagnosis and the remedies are not contested".[52] Still it is recognised now that the Lisbon agenda could have been formulated more carefully. For instance, "defined in a period of strong and dynamic growth, bringing favourable perspectives for both employment and macro-economic stability," the objectives were ambitious. The objectives were defined without taking Member State diversity into account. And too many objectives were pursued in parallel, some of them conflicting with each other in the short to medium term. In addition, the Lisbon agenda was never able to capture people's imagination, included highly unpopular reforms, and was perceived as an agenda for business.[53]

The two main factors explaining the limited progress towards the achievement of the Lisbon agenda were shortcomings concerning horizontal policy coordination (between different policy areas) and vertical policy coordination (between different levels of governance). First, little progress was made towards horizontal policy coordination (an issue also discussed in Section 3 of Chapter 2). This stemmed from the way in which horizontal policy coordination was understood at Lisbon. One of Lisbon's key propositions was that there should be coordination among knowledge triangle policies, and between knowledge triangle and other policies. Compared to the juxtaposition of different Community and national policies in the 1990s, this constituted a major step forward. But the basic spirit underlying Lisbon was that while there could be a slight refocussing of non-knowledge policies

[52] European Commission, *Growth and Jobs*, p. 9.
[53] European Commission, SEC(2005) 160, pp. 48–50.

in support of knowledge policies, there could be no fundamental rethinking of and mobilisation of those non-knowledge policies in support of the knowledge society. A good example is provided by the Stability and Growth pact, which was rigidly adhered to. As Philippe Aghion appropriately observed, "the only link between macro-policy and long-run growth that most policy-makers believe in, is that growth requires macroeconomic stability everything else remaining equal".[54] Yet the view that macro-economic policies could play a much more pro-active role is now gaining ground.[55] Another good example is provided by market integration, where the understanding is now gaining ground that "neutral" market integration does not suffice, and that there is a need to develop so-called lead markets.[56]

Not only was the original Lisbon understanding of horizontal policy coordination imperfect. The actual horizontal policy coordination mechanisms put in place were also rather weak, allowing at both Community and national level bureaucratic portfolio protection dynamics to take over, resulting in the continued juxtaposition of policies rather than real horizontal policy coordination.[57] Like those of other policy areas, the economics and finance policy-makers used "Lisbon" – the new scriptures affording diverse interpretations – to confirm and lock-in their prior orientations rather than to venture towards a new, more intermeshed, configuration.

The governance of the Lisbon strategy was identified as an important problem. In 2005, the Commission pointed out that "the governance of the Lisbon strategy needs radical improvement to make it more effective and more easily understood. Responsibilities have been muddled between the Union and its Member States. There are too many overlapping and bureaucratic reporting procedures and not enough political ownership." In addition, it noted that "we need to revamp the delivery process which has become too complicated and is poorly understood. It generates much paper, but little action. Responsibilities between the national and the European levels have become blurred".[58]

Within this context, a critical analysis was made in particular of the open method of coordination.[59] On the positive side, it was acknowledged that, for instance, it helped achieve a common understanding among Member States in relation to a particular policy area, that it was effective as far as the exchange of best practices was concerned, and that it was perhaps the best way for Member States and the Community to coordinate policies and efforts in areas where there was little or no Community competence. But the voluntary nature of the open method of coordination resulted in a lack of Member State commitment and weakened its

[54] Philippe Aghion and Peter Howitt, *Appropriate Growth Policy: A Unifying Framework*, 9 August 2005, p. 24.
[55] See Philippe Aghion's arguments in favour of a countercyclical macro-economic policy in Aghion and Howitt, *Appropriate Growth Policy: A Unifying Framework*.
[56] *Creating an Innovative Europe*.
[57] In this regard the 2005 OECD report on the governance of innovation systems constitutes interesting reading.
[58] European Commission, *Growth and Jobs*, pp. 12 and 31.
[59] European Commission, SEC(2005) 160, pp. 47 and 49.

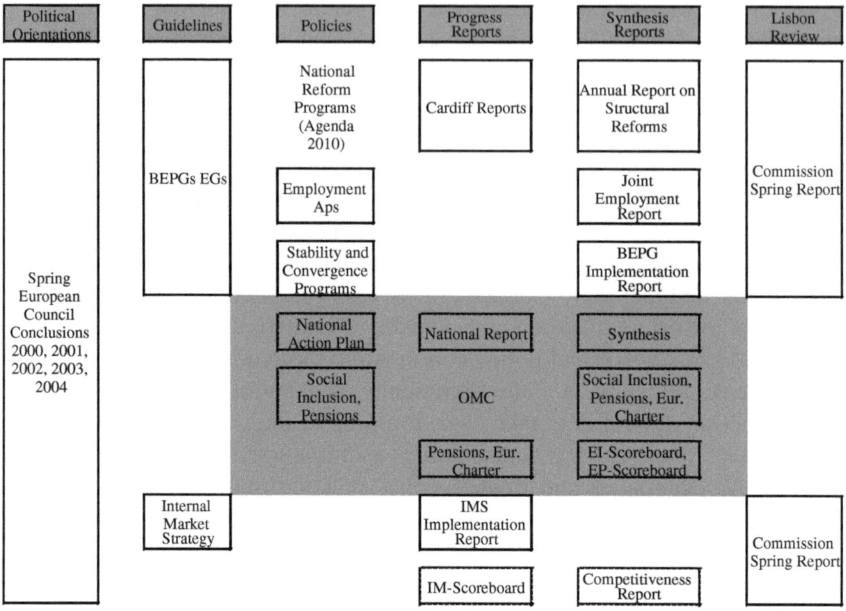

Fig. 3.4. The open method of coordination

Source: European Commission, *Delivering on Growth and Jobs: A New and Integrated Economic and Employment Co-ordination Cycle in the EU,* Companion Document to the Communication to the Spring European Council 2005 (COM(2005) 24) Working Together for Growth and Jobs, A New Start for the Lisbon Strategy, SEC(2005) 193, 3 February 2005, p. 6.

capacity to drive structural change. At the same time, there was not always sufficient control on how Member States translated agreed-upon goals into concrete national measures and on how coordination among different levels was ensured (Fig. 3.4).

The weaknesses of the open method of coordination were also recognised in the field of research. Commission services noted in 2005 that the "coordination of national research policies has had mixed success" and that "the experience to date suggests that Member States are willing to consider concertation and some degree of coordination between national policies, but that a higher level of involvement will be needed to commit the resources necessary to address sensitive issues requiring coordination between several Member States".[60]

6. A NEW POLICY CONTEXT

In response to the continued validity but excessive number of Lisbon objectives, and the perceived weaknesses of the Lisbon governance system, a number of reforms were instituted starting in 2005. The Lisbon agenda would focus on growth and jobs

[60] European Commission, SEC(2005) 160, p. 18.

by ensuring that Europe is a more attractive place to invest and work, "knowledge and innovation are the beating heart of European growth", and the policies are shaped allowing businesses to create more and better jobs. Broader support for and ownership of the Lisbon strategy would be built. The governance system would be simplified. And a new partnership between the Union and the Member States would be launched.[61]

Concrete actions included the drafting by the Commission of an overall Lisbon action programme, the integration of the broad economic policy and employment guidelines, the adoption of broadly agreed national action programmes for growth and jobs, the appointment of a "Mr." or "Ms. Lisbon" at national government level, and simplified reporting through a single Lisbon report at EU and at national levels on the progress made.[62]

Under the revised Lisbon strategy, knowledge remained at the centre, and the Union role was strengthened in line with high-profile calls in that direction: "The conclusion that imposes itself is therefore crystal clear: the inter-governmental method without commitment, which constitutes the basis of the Lisbon strategy, does not function correctly. We need an approach that is more restraining and Community-based".[63]

This revaluation of the possible role the Union could play resulted in a more proactive role for it, for instance, with regard to the national action programmes. But it also spilled over into the renewal of funding programmes such as the Framework Programme. It opened the way for a Framework Programme more ambitious in scale and scope than any past Framework Programme. It paved the way for "a new mindset in supporting and conducting research in Europe".[64]

CONCLUSION

The 1990s constituted a turbulent period for the Union. It took a major step forward towards complete market integration. In view of the introduction of a single currency, it achieved macro-economic stabilisation. It successfully introduced the euro. And it embarked upon its most ambitious enlargement ever. On the other hand, it was not able to deal conclusively with the problems of low economic growth, declining competitiveness, and high unemployment, while new challenges such as ageing were emerging on the horizon.

---

[61] European Commission, *Growth and Jobs*.

[62] European Commission, *Growth and Jobs*.

[63] Translation of French original "La conclusion qui s' impose est donc claire comme le jour: la méthode intergouvernementale sans engagement qui constitue la base de la stratégie de Lisbonne ne fonctionne pas correctement. Nous avons par contre besoin d'une approche bien plus contraignante et communautaire" (Discours du Premier Ministre Guy Verhofstadt prononcé devant le Parlement européen. Bruxelles, le 31 mai 2006).

[64] Intervention of Janez Potočnik at the European Parliaments' ITRE committee meeting of 13 June 2006.

The difficult take-off and, compared to the United States, limited impact of the new economy in Europe, as well as the new competitive threat of low-cost, high-tech developing countries, put the spotlight on knowledge, research, education, and innovation. The need for Europe to transform into a knowledge-based society was first made by academics, but the awareness spread to the community of European policy-makers.

The Lisbon European Council of March 2000 and its aftermath created a new European research policy context, within which it was possible to pursue a Framework Programme much more ambitious in scale and scope than any past Framework Programme. At Lisbon, European policy-makers recognised unambiguously that to overcome its societal challenges, Europe would have to turn itself into a knowledge-based society. At the same time, it was acknowledged that putting knowledge at the centre would require more and better investment in the knowledge triangle of research, education, and innovation. The objective to invest more in research centred on the 3 per cent objective, while the objective to invest better in research centred on the concept of the European Research Area. After 2000, however, little progress was made towards the achievement of these objectives. This was mainly due to weaknesses in the governance of the Lisbon strategy. This resulted in 2005 in a reform of the open method of coordination, and in a general revaluation of the role of the Union in the achievement of the Lisbon objectives. This recognition of the relative effectiveness of action through the Union spilled over into the debate on Community funding programmes up for renewal, and opened the way for an ambitious 7th Framework Programme.

Having taken stock of these "exogenous" factors – Europe's challenges and the role S&T can play in addressing them, the weaknesses of Europe's S&T system and, finally, the new policy context – the following part of the book will look at how the Framework Programme was actually conceived so as to respond to these issues. It will also be seen that the research community was supportive of boosting investment by the Union. However, the next chapter takes a look backward to the experiences gained from past Framework Programmes, and outlines how areas of improvement were taken up in the 7th Framework Programme proposal.

CHAPTER 4

BUILDING ON EXPERIENCE: LEARNING LESSONS FROM PAST FRAMEWORK PROGRAMMES

Building on past experience via ex-post evaluation is an increasingly important element in the design of research programmes and policies, notably at the European level.[1] Being aware of the strength and weaknesses of past Framework Programmes, knowing their impacts and realising in which fields potential impacts could not be achieved, is a precondition for the improvement of any new programme.

Based on the evidence of ex-post evaluations, this chapter looks back at past Framework Programmes and shows what lessons were helpful for the design of the 7th Framework Programme. The chapter starts with some methodological considerations linked to the evaluation of Framework Programmes (Section 1). Turning to the outcomes and impacts of Framework Programmes, the accumulated evidence convincingly shows that the Framework Programme has had a clear positive impact not only on participants' immediate competitive position, but also on their capabilities, behaviour, and ability to compete. It has improved Europe's scientific, technological, and innovative performance (Section 2). And it has strengthened Europe's human capital (Section 3) and integrated European research infrastructures (Section 4), and enhanced the coordination of Member State research policies and actions (Section 5). It has also generated wider economic, social, and environmental benefits (Section 6).

1. EVALUATING THE FRAMEWORK PROGRAMME IS NOT EASY[2]

1.1. Evaluating research projects is complex

Identifying the outputs and economic, social, and environmental impacts of individual research projects is complex. The first reason is the difficulty of attribution, which concerns the key question at the heart of every impact study: to what

[1] The Union, in response to its desire to regulate better, calls for transparency and accountability, as well as budget constraints, wants to know what impacts past policies and programmes have generated, and how it can increase those of future ones. The importance attached to learning lessons from the past and building on experience is reflected in the Commission's impact assessment guidelines.

[2] Over the years, an extensive and highly diverse body of Framework Programme ex-post evaluation literature has emerged. It consists of studies carried out for the European Commission focussing on the pan-European Framework Programme experience, and studies carried out for the Member States concentrating mainly on the programme's national impact. The spotlight of Commission studies was sometimes on the Framework Programme as a whole, sometimes on just one or a few Specific Programmes. Past Framework Programme evaluations were sometimes of an ad hoc and one-off nature, and sometimes carried out repeatedly within the context of the formal Framework Programme ex-post evaluation system

extent are measurable effects really the result of – that is causally attributable to – the project under review? [3] Usually it is not just the individual research project financed which has produced the solution to a particular problem or a certain impact. Project participants may have received funding from other sources as well. In the case of the Framework Programme, for instance, account has to be taken not only of purely national direct and indirect research funding, but also of other funding schemes supporting international collaboration such as bilateral agreements, COST, and EUREKA. Fig. 4.1 shows the evolution of European cooperative research funding schemes, which have grown from 6 per cent of total government expenditure on R&D in 1985 to 15.5 per cent in 2003.[4]

In addition, impacts may have resulted not only from intended effects directly related to programme goals but also from indirect and unintended effects going beyond programme goals. Next, factors other than research itself may have contributed to an impact.[5] Finally, attribution is hindered by the usually long period

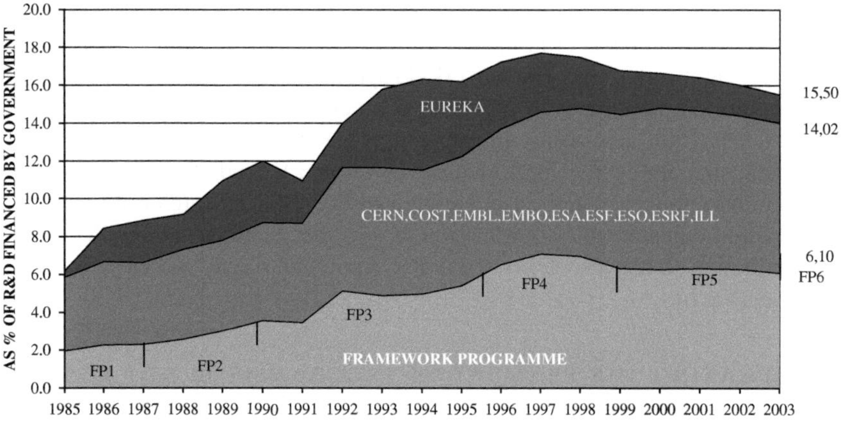

Fig. 4.1. The Framework Programme accounts for a major share of European Cooperative Research-Funding (Cooperative Research as a share of total government expenditure on R&D)
Source: DG Research
Note: Government Expenditure on R&D (GOVERD). Data are for EU-15 and EFTA-countries

(e.g. the so-called Five-Year Assessments). Some studies were based on survey results while other reports relied more on quantitative analyses of programme participation data. Lists of various FP evaluation studies can be found in e.g. PREST *et al.*, *Assessing the Socio-Economic Impacts of the Framework Programme*, June 2002, pp. 159–161; Erik Arnold, *What the Evaluation Record Tells Us about Framework Programme Performance*, May 2005, pp. 8–9.

[3] Joanneum Research Forschungsges.m.b.H *et al.*, *Evaluation of Austrian Participation in the 4th EU Framework Programme for Research, Technological Development and Demonstration*, 2001, p. 27.

[4] Estimates based on general budget data drawn from financial reports from different institutions.

[5] PREST *et al.*, *Socio-Economic Impacts*, pp. 85–86; Joanneum Research Forschungsges.m.b.H *et al.*, *Austrian Participation*, pp. 27–28.

of time between the funding and carrying out of research on the one hand, and the results of that research becoming visible on the other hand.[6]

The second reason for the complexity involved in identifying the outputs and impacts of individual research projects is the need to assess additionality or what it is that would have happened anyway even without public support.[7] Three important kinds of additionality are input additionality (did public support increase research and innovation inputs? Or did it just replace and crowd out inputs that would have been made anyway?); behavioural additionality (did researchers change their behaviour (importance attached to research and research collaboration, project management) as a result of receiving public support and participating in the project?); and output additionality (would the same outputs have been obtained without the policy action?).

1.2. The Framework Programme has evolved in terms of rationale, content, and budget

In addition to the methodological problems discussed in the previous section, puzzling together a consistent picture of past Framework Programmes and their outputs and impacts is hampered by the programme's main characteristics having evolved rather substantially over time. The Framework Programme constitutes a moving target, and this first of all in terms of its rationale.

The start of the Framework Programmes constituted a radical break with what came before. Until the late 1970s, European research policy was of an ad hoc nature. Because it was tied to particular sectors (agriculture, coal, nuclear energy, steel, etc.), it was also fragmented. Research policy was mainly a national affair, even though the Founding Treaties already provided the Community with a responsibility in the field.[8] The 1st Framework Programme (1984–1987), and the inclusion of a separate chapter on RTD in the Single European Act (1986), heralded the shift towards a legally solidly grounded integrated European research policy focussing on the competitiveness of European industry and the quality of life of European citizens.

This apparent unity of purpose has not precluded a certain evolution of the programme's rationale, however, accommodated by the fact that as some observers have noted the overall Framework Programme objective is extremely permissive.[9] The reason for this evolution is that each Framework Programme was prepared against a different background. The rationale for the 1st Framework Programme was

---

[6] Joanneum Research Forschungsges.m.b.H *et al.*, *Austrian Participation*, pp. 30–31.

[7] PREST *et al.*, *Socio-Economic Impacts*, pp. 105–115; Joanneum Research Forschungsges.m.b.H *et al.*, *Austrian Participation*, pp. 28–29.

[8] EC, Euratom, ECSC Treaties.

[9] Arnold, *Evaluation Record*, May 2005, pp. 3–4.

the perceived technology gap.[10] The 2nd Framework Programme (1987–1991) was intended to strengthen the research base of European industry in response to fierce Japanese competition. Developing information and communication technologies was high on the political agenda. The 3rd Framework Programme (1990–1994) was developed against the background of efforts to integrate the European market. The conceptualisation of the 4th Framework Programme (1994–1998) took place during the period of the Maastricht Treaty (1992) and the White Paper on Growth, Competitiveness and Employment (1993). The 5th Framework Programme (1998–2002) put increased emphasis on socio-economic values. And the 6th Framework Programme (2002–2006) was designed to help build the European Research Area (Table 4.1).[11] Together with its rationale, the Framework Programme's content has also evolved. The number and content of Specific Programmes has changed from one Framework Programme to the next. The number of thematic priorities has multiplied. And new instruments have been introduced.

Another indicator of substantial change has been the growth of the Framework Programme budget, which reached about €19 billion (at 2004 prices) for the four-year period 2002–2006 under the 6th Framework Programme, and may reach about €48 billion (at 2004 prices) for the seven-year period 2007–2013 under the 7th Framework Programme (Fig. 4.2). In 2006, research accounted for 4 per cent of

Table 4.1. Evolution of the Framework Programme

Periods	Main objective	Main priorities	New actions
Before (1975–1983)	Ad hoc approach	Energy oriented	–
FP1 (1984–1987)	Coordination of Community RTD actions	Energy and ICT oriented	Environment, international cooperation human capital and mobility
FP2 (1987–1991)	Information society	ICT oriented	Biotechnologies, marine resources, dissemination
FP3 (1990–1994)	Industrial competitiveness	Multiple priorities	–
FP4 (1994–1998)	Industrial competitiveness	Multiple priorities	Transport and social sciences
FP5 (1998–2002)	Innovation and social needs oriented	Multiple priorities	Nanotechnologies
FP6 (2002–2006)	Instrument for ERA	Multiple priorities	New instruments

Source: DG Research

[10] Stefano Breschi and Lucia Cusmano, *Unveiling the Texture of a European Research Area: Emergence of Oligarchic Networks Under EU Framework Programmes*, CESPRI Working Paper No. 130, July 2002, p. 5.

[11] European Commission, *Five-Year Assessment of the European Union Research Framework Programmes 1999-2003*, 15 December 2004, p. 4.

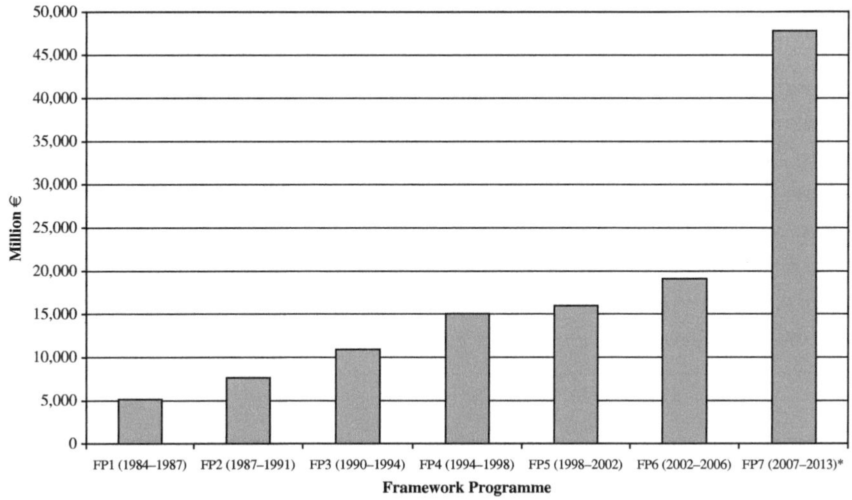

Fig. 4.2. The growing Framework Programme budget (evolution of the FP budget (€ million, 2004 Prices))
Source: DG Research
Note: *:Provisional

the EU budget (Fig. 4.3).[12] The Framework Programme also accounts for about 6 per cent of EU-15 non-military governmental RTD expenditure.[13] This share increases to almost 25 per cent when public support for research is more narrowly defined.[14]

1.3. Framework Programme data

Painting an adequate picture of the Framework Programme and its evolution over time has been hampered by a relative lack of data. Naturally the first purpose of databases holding information on Framework Programme applicants and participants has always been to facilitate proper Framework Programme contract and financial management, not the production of detailed statistics on Framework Programme application and participation patterns. Under the first few Framework

---

[12] European Commission, *General Budget of the European Union for the Financial Year 2006 – The Figures*, Luxembourg, 2006, pp. 13–15 (Sum of commitment appropriations for "Research" and for "Research Framework Programme" under "Enterprise", "Energy and transport", "Information society and media" and "Direct research").

[13] Court of Auditors, *Special Report No 1/2004 on the Management of Indirect RTD Actions under the Fifth Framework Programme (FP5) for Research and Technological Development (1998 to 2002), Together with the Commission's Replies (Pursuant to Article 248(4) Second Subparagraph EC) (2004/C 99/01)*, 23 April 2004, Paragraph 5.

[14] Court of Auditors, *Special Report No 1/2004*, Paragraph 5: "However, if institutional funding is deducted, the budget for FP5 indirect RTD actions amounts to approximately a quarter of total funding for publicly financed research projects in the European Union".

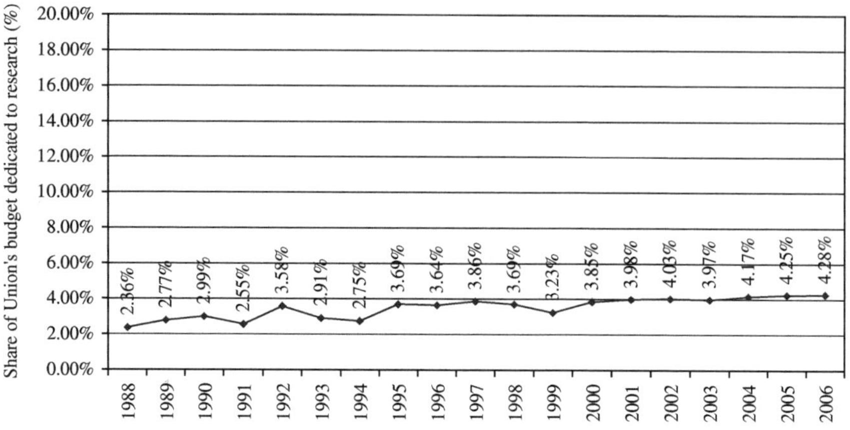

Fig. 4.3. Increasing share of the Union's budget dedicated to research
Source: DG Research

Programmes, different and not always compatible databases were in use managed by different Commission services in charge of a particular Framework Programme component, making it difficult to arrive at one integrated dataset for a single Framework Programme. As instruments and rules of participation evolved from one Framework Programme to the next, the labels attached in the databases to Framework Programme participants also changed, which makes it difficult to analyse, for instance, the evolution of the participant type structure (e.g. evolution of industrial participation). Some data on project outputs have been collected in the past via end-of-project reports, but these have tended to be incomplete, and this is not helped by the frequent perception that reporting such outputs represents a significant administrative burden on Framework Programme participants.

1.4. The Framework Programme monitoring and evaluation system

The European Commission has been involved in research programme evaluation since the late 1970s. Yet it was only in the early 1980s, when evaluation became a legislative requirement, that the institutionalisation of Commission research evaluation practices picked up pace.[15]

The ex-post evaluation system The Framework Programme ex-post evaluation system was introduced in the mid-nineties.[16] Its two main components are yearly monitoring exercises and five-yearly in-depth assessments (the so-called "Five-Year Assessments") carried out at overall and usually also specific programme

[15] Ken Guy *et al.*, *Strategic Options for the Evaluation of the R&D Programmes of the European Union*, Final Report Prepared for STOA, Brighton, 1999.
[16] European Commission, *Independent External Monitoring and Evaluation of Community Activities in the Area of Research and Technological Development*, Communication from the Commission to the Council and the European Parliament, COM(96) 220 final, 22 May 1996.

level. The annual monitoring exercise is intended to be rather light and enable a quick response to issues arising from ongoing programme implementation. The objective of the Five-Year Assessments, on the other hand, is to provide input for policy formulation and decision-making on the basis of feedback obtained from programme implementation.[17]

The current Framework Programme ex-post evaluation system has obvious strengths, for example its independence and legitimacy. However, the literature has also identified some important weaknesses which go beyond the issues plaguing all research evaluations (time lag, attribution, additionality, difficulties of measuring qualitative effects, etc.). According to some evaluation experts, the main problem affecting the current system relates to FP design: "The intervention logic that connects the high-level and operational goals of the FP is poorly articulated, making an overall evaluation of the FP difficult. The Framework needs more systematic planning, clearer objectives and a stronger link to an evidence base. This would ease evaluation and, arguably, improve FP performance".[18] This problem is of course not specific to EU programmes. The same difficulties apply when it comes to establishing quantifiable ex-ante objectives for programme evaluation at the national level.

Other (potential) problems have been identified mainly with regard to the available evidence base and the use of expert panels. A strong and timely evidence base constitutes the main tool for Five-year Assessment expert panels on which to base their assessment and recommendations. Yet a number of studies have highlighted weaknesses in the evidence base available.[19] Sometimes thematic ex-post evaluations have not (yet) been completed, or they focus on different issues, cover different periods, or have been carried out according to different methodologies. That means that panels must work with rather fragmented evidence, or with what is still possible to do in the very short run. Especially in recent years substantial efforts have been made to address this issue.

In Framework Programme ex-post evaluation, use is often made of external expert panels. As mentioned above, they contribute to independence and legitimacy. However, some observers have considered them to be time- and

---

[17] The careful timing of the Five-Year Assessments allows for the combination of an ex-post evaluation of the previous Framework Programme with a mid-term appraisal of the ongoing one to formulate recommendations for the next Framework Programme. To read more about the Framework Programme ex-post evaluation system, see, for instance, Luc Durieux and Gilbert Fayl, The Scheme Used for Evaluating the European Research and Technological Development Programmes, In: OECD (Eds.), *Policy Evaluation in Innovation and Technology – Towards Best Practices*, Paris, 1997, and Ken Guy and Wolfgang Polt, *Strategic Options for the Evaluation of the R&D Programmes of the European Union,* Plattform – Technologie – Evaluierung, No. 8, April 1999.

[18] Erik Arnold, *What The Evaluation Record Tells Us About Framework Programme Performance*, Technopolis, January 2005, p. ii.

[19] See Guy and Polt, *Strategic Options*, April 1999; European Commission, *Five-Year Assessment of the European Community RTD Framework Programmes – Report of the Independent Expert Panel Chaired by Viscount E. Davignon and the Commission's Comments on the Panel's Recommendations*, Luxembourg, 1997, and subsequent Five-Year Assessments; Arnold, *Evaluation Record*, January 2005.

resource-intensive.[20] Though strict rules apply, others have claimed that it is difficult to avoid completely conflicts of interest.[21] Some analysts have suggested that panel members may be pressurised into promoting unrelated agendas and specific interests. And with regard to the 2000 Five-Year Assessment, one scholar noted that "many of the recommendations drew not so much on an evaluation of past FP activities, but on the collective opinions and assessments of the panel member concerning the general structure, and organisation of RTD in Europe".[22] It should however be borne in mind that these problems are not unique to the Framework Programme, but exist in many other industrialised countries' R&D programmes as well.

The above observations on Framework Programme evaluation are all well known and well documented. Indeed, most of the remarks come from the Framework Programme evaluation process itself. But this is also one of the greatest strengths of the system, because transparent and constructive criticism is one of the main vehicles for promoting programme improvement. And indeed many important improvements to the Framework Programme in the past have emanated from comments made at the evaluation stage.

Topics, sources, and methodologies Above we have argued that, except for the recent introduction of impact assessment, the Commission's research evaluation system has remained relatively stable over the past decade. Greater change has characterised the evaluation studies carried out. New topics are being explored, studies are based on new kinds of sources, and use is made of innovative methodologies.

In past Framework Programme ex-post evaluations, substantial attention used to be paid to analysing participant characteristics (e.g. type of institutional actor, country of origin, region of origin, etc.) and R&D inputs. At the same time, much emphasis was put on counting project outputs in order to arrive at total and average (per project) numbers of publications, patents, and so on. This has not disappeared. But attempts are now made to profile programme participants in more innovative ways. This includes analysing their scientific (e.g. numbers of publications, numbers of citations, citation impact scores) and technological (e.g. numbers of patents) quality, the nature of their participation (one-time vs. repeat participation), the nature of their networking behaviour (stable vs. changing partnerships), and so on. From merely counting project outputs, the emphasis has also shifted to assessing

---

[20] Gustavo Fahrenkrog *et al.* (Eds.), *RTD-Evaluation Toolbox Assessing the Socio-Economic Impact of RTD-Policies*, JRC-IPTS and Joanneum Research, Strata Project HPV 1 CT 1999 – 00005, August 2002.

[21] Luke Georghiou, Assessing The Framework Programmes – A Meta-Evaluation, In: *Evaluation*, Vol. 1, No. 2, 1995 (Referred to in Arnold, *Evaluation Record*, January 2005). Panel members cannot have been FP contractors, or members of or experts to any FP Programme Committee during the preceding five years. The organisation for which the panel member works can, however, continue to participate in the FP. So does the panel member after the conclusion of his/her mandate.

[22] PREST *et al.*, *Socio-Economic Impacts*, p. 180.

Framework Programme aggregate impacts on Europe's scientific and technological performance, and on research capacity, or on the European economy and society.

The study of such more challenging topics can be embarked upon because of the mining of new data sources. Framework Programme ex-post evaluations used to be based mainly on surveys (interviews or written questionnaires) and end-of-project reports. However, attempts are now made, for instance, to create cleaned and consolidated participant databases at the individual scientist level rather than just the institutional level, and to link them with bibliometric and patent databases. The use of bibliometric and patent data almost by definition also entails the use of new methodological approaches and techniques. Within this context, mention should be made of the use of econometric models to estimate the impact of the Framework Programme on the European economy, as was done in the impact assessment on the Commission's proposal for the Seventh Framework Programme.[23]

The way forward Against this background, several steps are envisaged to improve the Framework Programme's evaluation system. A clearer formulation of the intervention logic in the 7th Framework Programme proposal with clear and measurable objectives and the monitoring of their progress will facilitate the ex-post evaluation of the 7th Framework Programme. A comprehensive effort will be made under the next Framework Programme to ensure that the ex-post evaluation will be based on a wide range of completed, focussed and methodologically standardised ex-post evaluation studies.[24]

Continued efforts will also be made to explore new topics, sources, and methodologies. There is, for instance, a great need to assess better whether project outputs/impacts were the result just of a research project being carried out, or of a research project being carried out at European level. Important questions also remain regarding the optimal mix of partners in a project, or the optimal size of networks. Also, what constitutes a project of critical mass? Does it relate to the number of partners? To the level of funding? How does critical mass interrelate with flexibility, with cohesion, with excellence? Are repeated Framework Programme participation and the stability of networks across Programme calls and even across different Framework Programmes good or bad, and when do they become excessive? To answer these questions, further efforts will have to be made to consolidate and link databases. New methodologies will also need to be explored, especially those being able to shed a light on questions of causality. For instance, does participation in the Framework Programme increase the quality of an individual scientist or institution,

---

[23] http://ec.europa.eu/research/future/index_en.cfm

[24] To read more, see also European Commission, *Building the Europe of Knowledge*, Proposal for a Decision of the European Parliament and of the Council Concerning the Seventh Framework Programme of the European Community for Research, Technological Development and Demonstration Activities (2007 to 2013) – Proposal for a Council Decision Concerning the Seventh Framework Programme of the European Atomic Energy Community (Euratom) for Nuclear Research and Training Activities (2007 to 2011), COM(2005) 119 final, 6 April 2005.

or does it attract already excellent scientists and institutions? As much as possible, methodological development in the field will be supported by including relevant research topics in the Framework Programme's work programmes. It will also be supported through Commission networking. Networking is first of all required in-house, so that methodologies may be standardised. But external networking is also needed so as to achieve complementarity between European Commission and national level Framework Programme evaluations.[25] Those networks also allow sharing and comparing hard evidence of the impact of research policies at regional, national and EU levels and will help identify what is done best at each level, and how to design the S&T governance model in the most efficient and effective way.

2. COLLABORATIVE RESEARCH AND ITS IMPACTS

Collaborative research constitutes by far the largest component of the Framework Programme, accounting for 70 per cent of the budget of the 6th Framework Programme. Collaborative research projects are implemented by transnational consortia composed of firms, universities, and research institutes. Under successive Framework Programmes, collaborative research projects have been organised under broad thematic priorities (e.g. Energy, ICT, etc.) covering a wide range of S&T disciplines. Growth in the collaborative research budget has been accompanied by increases in the number of collaborative research projects and participations.[26] Between the 2nd and 5th^{h} Framework Programmes, the number of projects increased by about 140 per cent from 2 779 to 6 712, while the number of participations grew by about 240 per cent from 13 000 to 44 000.

2.1. The size of collaborative research projects

The added value of European funded collaborative research projects is among other things to be found in the fact that they allow for pooling financial resources and knowledge across national frontiers. This allows research projects to attain the appropriate size or "critical mass" needed to achieve scientific breakthroughs. In other words, it helps break down the wasteful fragmentation of research efforts across Europe.[27] It is therefore interesting to analyse how the scale of Framework

---

[25] Within this context, DG Research set up the European RTD Evaluation Network in 1997 to enhance cooperation between the national RTD evaluation units/agencies or agencies concerned with evaluation and the relevant Commission Services. To read more about the RTD Evaluation Network, see http://www.cordis.lu/fp5/monitoring/rtd_evalnet.htm

[26] A participation is counted for each participant involved in an FP project. The same participant may be involved in several projects and therefore have several participations.

[27] See for example European Commission, *Five-Year Assessment*, 1997, p. 23: "The [European] scientific community's added value lies in it being a networked pool of talent that can compete internationally at a level beyond the capability of an individual Member State. Hence a European critical mass can be established in areas where no one Member State can separately mount a major effort"; Arnold, *Evaluation Record*, May 2005, p. i.

Programme research projects has evolved over time in terms of number of participations and funding. The average number of participations per project increased from 4.7 in the 2nd Framework Programme to 6.5 in the 5th Framework Programme, while the average Commission funding per project increased from €1.2 million to €1.4 million. At the same time, average EU funding per participation decreased from €256 000 to €196 000. The 6th Framework Programme appears to have initiated a trend towards decisively larger projects with more participations per project, and with higher funding per project and per participation (Table 4.2).

So the trend appears to be towards larger projects. But a better understanding has to be gained of what "critical mass" means, and of how it can be achieved. For example, does it refer to the number of participants in and the size of the funding allocated to a collaborative research project? If that is the case, what do we know about the ideal number of participants and the ideal size of funding? Is the relation between project scale and project achievement a simple linear one, or is there a minimum efficient project scale above which there is a disproportionately higher level of achievement?[28] Is it possible to make general statements about it, or is it

Table 4.2. Towards critical mass – the changing features of shared-cost research actions across Framework Programmes

Indicators	FP2-EU-12	FP3-EU-15	FP4-EU-15	FP5-EU-15	FP5-EU-25	FP6-EU-25
	Definitive data	Definitive data	Partial data	Definitive data	Definitive data	Interim data (first calls)
N° of projects	2 779	3 292	2 949	6 709	6 712	1 188
N° of participations (000)	13	18	21	41	44	16
Average n° of participations per project	4.7	5.6	7.0	6.2	6.5	13.5
Average n° of different Member States per project	3.0	3.5	4.2	3.7	4.0	6.5
Average EU funding per project (€000)	1 202	1 218	1 160	1 405	1 405	4 322
Average EU funding per participation (€000)	256	218	165	200	196	320

Source: DG Research
Note: FP4 Data Partial (01.01.1994-31.12.1996); FP6 Data Partial (to 11.03.2005); Average EU funding per participation under FP6 calculated by summing the EC contributions of all projects and dividing by the number of participations

[28] Evidence Ltd *et al.*, *The Role of Selectivity and the Characteristics of Excellence*, Report to the Higher Education Funding Council for England: A Consultancy Study within the Fundamental Review of Research Policy and Funding, October 2000.

thematic priority or even topic specific? These are tricky concepts to analyse and quantify, but more research on such questions would be very valuable.[29]

2.2. The characteristics of collaborative research participants

Who participates in the Framework Programme and receives Framework Programme funding? What kind of institutions are they? Where, that is what country or region, do they come from? How often do they participate? The answers to these questions provide important insights into the nature of the research collaboration induced by the Framework Programme.

Participation pattern by type of institution Available data indicate that over time the pattern of participation in collaborative research projects by type of institution has become more balanced.[30] Whereas business enterprises initially accounted for the largest share of funding and participations, that is no longer true. Higher education institutes and research centres now also account for sizeable participation and funding shares (Figs 4.4 and 4.5).[31]

Several factors could explain the growth of the collaborative research participation and funding shares held by higher education institutes and research centres. The wider trend in OECD countries is to put renewed emphasis on funding basic research.[32] In line with this, the 6th Framework Programme included a more basic research-oriented activity called *New and Emerging Science and Technology* (NEST), as well as a *Networks of Excellence* instrument less attractive to business enterprises. It has been argued that work programmes have shifted away from the strong applications focus of the 5th Framework Programme, and have become more academic in tone.[33] Higher education institutes and research centres may be

---

[29] In this respect, see also DTI – Office of Science and Technology, *Targeted Review of Added Value Provided by International R&D Programmes,* May 2004, p. 8: "The literature deals poorly with the question of 'critical mass,' providing no ready-reckoner to check whether more or better capacity is necessary."

[30] It is not possible to trace in a definitive manner across Framework Programmes the evolution of the collaborative research participation structure by type of institution. In the first section of this chapter, we discussed how painting an adequate picture of the Framework Programme and its evolution over time has been hampered by a relative lack of data. One reason is that instruments and rules of participation have changed across Framework Programmes. Therefore, definitions of especially industrial types of participants have also changed. This makes it more difficult to analyse the evolution of the participant type structure. The result is that a certain margin of error is associated with the statistics reported below.

[31] See European Commission, *Second Report on S&T Indicators, 1997 – Report*, Luxembourg, 1997, pp. 520–522, for a discussion on the drop between FP2 and FP3 and the seeming stabilisation between FP3 and FP4 of the industrial share of participations, and of the drop between FP3 and FP4 of the industrial share of funding. In European Commission, *Five-Year Assessment 1999–2003*, 15 December 2004, p. 6, the assessment panel, commenting on FP6, states having encountered "significant evidence of industrial discontent and reports from different regions of industry dropping out in unusually high numbers from the Sixth Framework Programme", but does not find supporting statistics. Nevertheless it calls for the next FP to have a stronger industrial focus.

[32] Arnold, *Evaluation Record*, May 2005, p. 13.

[33] Arnold, *Evaluation Record*, May 2005, p. 13.

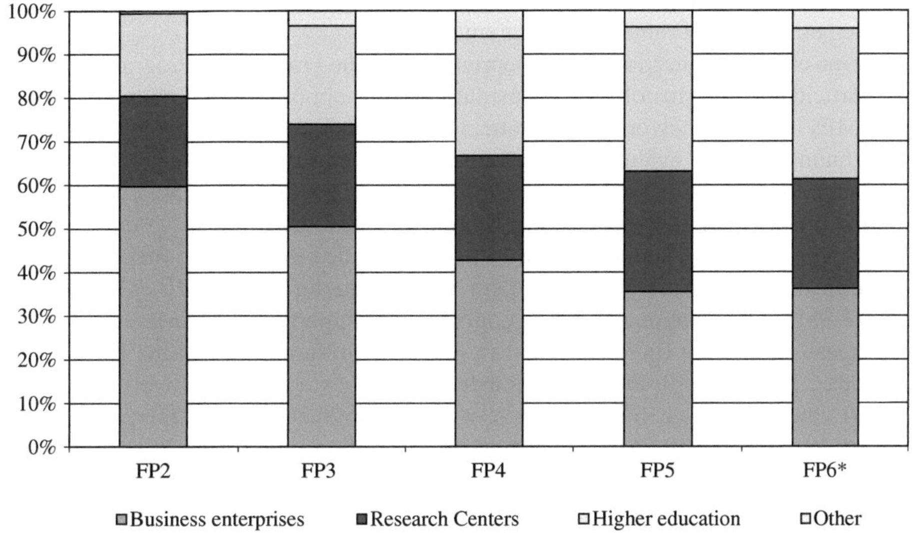

Fig. 4.4. How is FP funding shared between the main research actors? (% of FP funding received by type of participant (shared-cost actions only))
Source: DG Research
Note: *: FP6 data partial (to 11.03.2005)

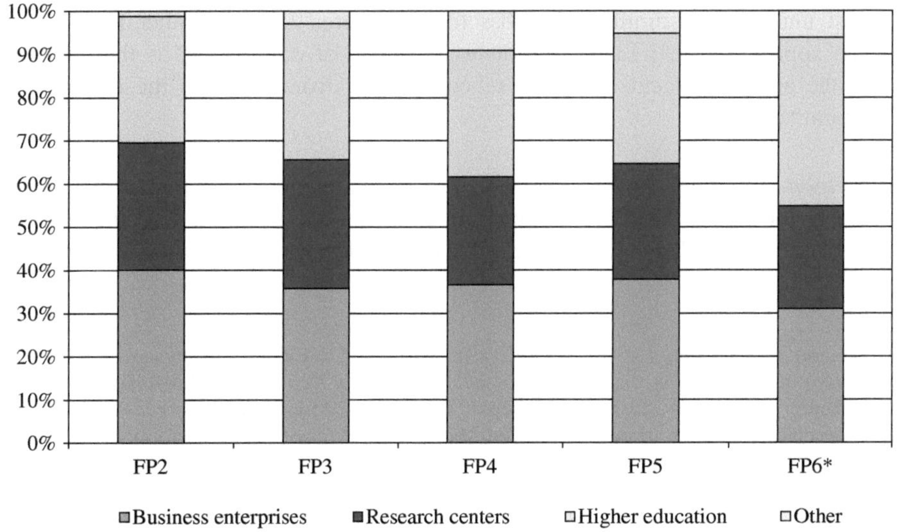

Fig. 4.5. How is participation spread across the main research Actors? (% of FP participations by type of participant (shared-cost actions only))
Source: DG Research
Note: *: FP6 data partial (to 11.03.2005)

less sensitive to the perceived increase over time in the cost of applying to and participating in the Framework Programme.[34]

Some observers question the appropriateness of the Framework Programme's mix of participants in terms of type of institution. The supposedly too low participation of SMEs in the Framework Programme, for instance, is a constant in the Framework Programme ex-post evaluation literature.[35] It has led to the establishment of SME participation targets of 5–15 per cent in the 4th Framework Programme (depending on the thematic area), and of 10 per cent in the 5th Framework Programme. In the 6th Framework Programme, at least 15 per cent of the budget of the first and second Specific Programmes was foreseen for research-performing SMEs. The concern about SME participation relates of course to the important role SMEs play in the European economy. There are between 8 and 12 million SMEs in the EU-15, and a further 2.5 million in the new Member States. They account for over 99 per cent of all enterprises, and for two-thirds of all employment in the enterprise sector. Some 30 per cent of SMEs – thus some 10 million enterprises in the EU-15 – regularly develop, apply, or acquire technology. Less than 3 per cent is involved in leading-edge research, however.[36]

Judging the appropriate level of SME or broader business enterprise participation in FP-funded collaborative research is difficult. Theory does not offer much insight beyond the one that it is quite natural for SMEs to be more irregular participants in the Framework Programme than large firms. They often lack a permanent R&D activity and staff. Given their scale, the relative cost of participation is higher. And there is no learning effect from regular participation. This argues against uncritically stimulating SMEs to participate.[37] On the other hand, there are no appropriate national or international points of comparison as the thematic coverage and instruments of these schemes differ from those of the Framework Programme.

Participation pattern by country of origin Framework Programme funding is awarded not on the basis of the nationalities of the applicants, but on the basis of their scientific excellence and the European added value of the research proposal.

[34] European Commission, *Five-Year Assessment*, 1997, p. 15; European Commission, *Five-Year Assessment of the European Union Research and Technological Development Programmes, 1995–99*, Luxembourg, 2001, pp. 18, 20, and 25; European Commission, *Five-Year Assessment Report Related to the Specific Programme: Quality of Life and Management of Living Resources Covering the Period 1995–1999*, p. 25; European Commission, *Report – Evaluation of the Effectiveness of the New Instruments of Framework Programme VI*, Report of a High-level Expert Panel Chaired by Professor Ramon Marimon, 21 June 2004; European Commission, *Five-Year Assessment 1999–2003*, 15 December 2004, pp. 9–12; European Commission, *Five-Year Assessment: 1999–2003 – Research and Technology Development in Information Society Technologies – Interim Panel Report*, June 2004, p. ii.

[35] Arnold, *Evaluation Record*, May 2005, pp. 27–28.

[36] EURAB, *EURAB Report On: "SMEs and ERA" – Report and Recommendations*, EURAB 04.028-Final, pp. 4–6.

[37] Arnold, *Evaluation Record*, May 2005, pp. 27–28.

Framework Programme funding is important to the Member States, however, especially the smaller ones, so that national stakeholders take an understandable interest in the amounts they receive.

The Framework Programme evaluation literature does not dwell much on the countries of origin of Framework Programme participants. It just observes that national Framework Programme participation shares are broadly in line with Member States' sizes and RTD capabilities (e.g. number of researchers).[38] This is confirmed by an empirical check. Fig. 4.6 displays for each EU-15 Member State (the highest ranking being 15 and the lowest 1) its ranking for three variables: the number of shared-cost action participations in the 5th Framework Programme, the shared-cost action funding received under the 5th Framework Programme, and the share of European GDP in 2000. It is clear that a Member State's Framework Programme participation is more or less in line with its share of European GDP. In other words, Member States accounting for large shares of European GDP participate most often in the Framework Programme and obtain most Framework Programme funding while the reverse is true for Member States accounting for small shares of European GDP. An important exception is Greece, which participates in the Framework Programme to a larger extent and obtains more funding

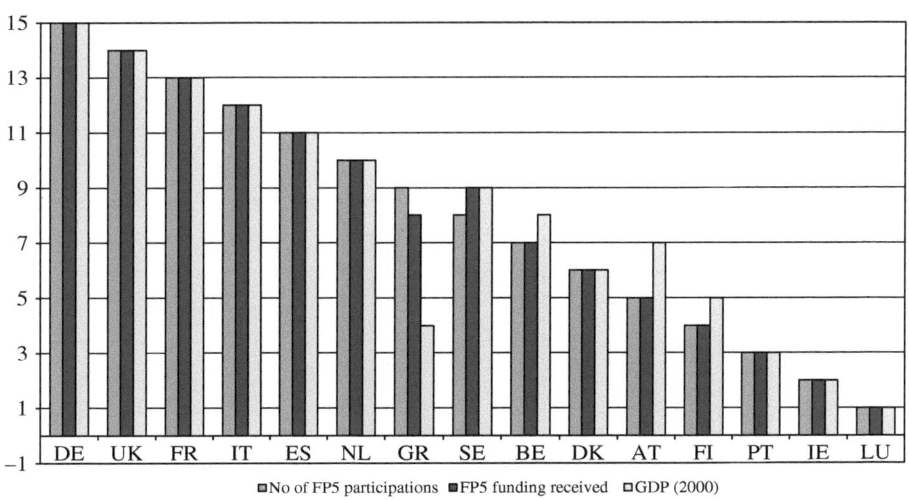

Fig. 4.6. Comparing Member States' FP participation with their share of European GDP (Member State rankings in terms of the number of FP5 shared-cost action participations, the amount of FP5 shared-cost action funding received, and the share of European GDP (2000))
Source: DG Research

[38] Arnold, *Evaluation Record*, May 2005, p. 16; European Commission, *Five-Year Assessment 1999–2003*, 15 December 2004, p. 4.

from the Framework Programme than one would expect on the basis of its share of European GDP. The reverse is true for countries like Belgium, Austria and Finland.

Similar results are obtained when the comparison is made not with the share of European GDP but with the number of researchers (Fig. 4.7).[39]

Almost completely opposite results are obtained when we rank Member States in terms of their intensity of Framework Programme participation, in other words in terms of how many Framework Programme participations and how much funding are obtained per unit of GDP, or per researcher (Figs 4.8 and 4.9). Then it becomes clear that smaller Member States participate more intensely in the Framework Programme than larger ones.

Fig. 4.10 shows that the relative importance of established Member States in terms of number of participations has not changed very much across Framework Programmes. New Member States, however, have usually been able to rapidly increase their number of participations across Framework Programmes (See e.g. Sweden, Austria and Finland in Fig. 4.10).

In addition to intra-European collaboration, the Framework Programme supports international collaboration beyond Europe's borders. The scale of this international collaboration has grown substantially across Framework Programmes. The number

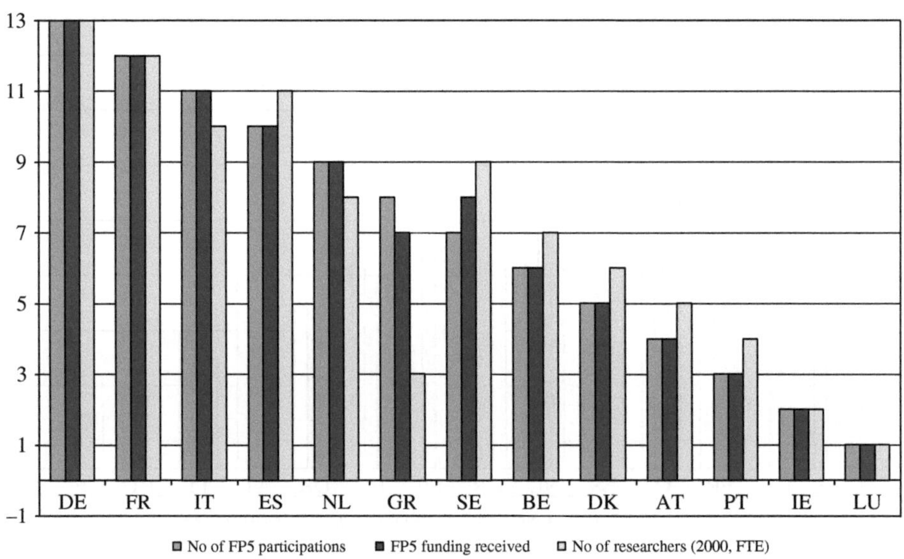

Fig. 4.7. Comparing Member States' FP participation with their number of researchers (Member State rankings in terms of the number of FP5 shared-cost action participations, the amount of FP5 shared-cost action funding received, and the number of researchers (2000, full time equivalent))
Source: DG Research

[39] See also European Commission, *Second Report on S&T Indicators 1997*, p. 548 (text) and 550 (graph): "The Member States' total participations in the last two Framework Programmes (1990–6) seem to correlate quite well with their human resources active in R&D, except in the cases of the three new Member States and those of Germany and Greece."

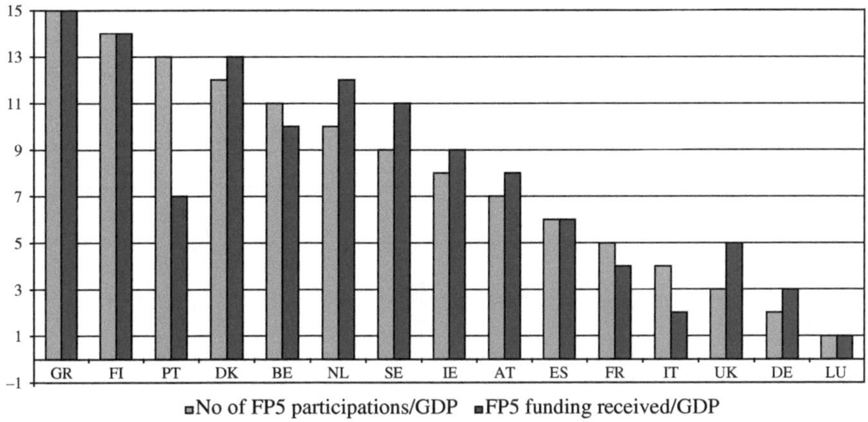

Fig. 4.8. FP participation intensities: FP participations and funding per unit of GDP (Member State rankings in terms of the number of FP5 shared-cost action participations and the amount of FP5 shared-cost action funding received per unit of GDP)
Source: DG Research

of non-European countries participating in the Framework Programme has increased from 30 in the 2nd Framework Programme to 140 in the 5th Framework Programme.

Participation pattern by region of origin The pattern of Framework Programme participation can be traced not just at national level, but also in Europe's regions. Some scholars argue that the regional level is the one at which the mechanisms of innovation and resulting economic benefits can be best understood.

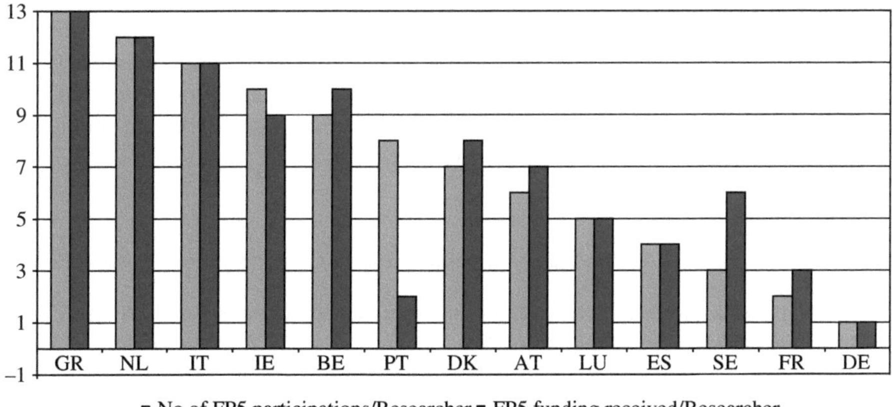

Fig. 4.9. FP participation intensities: FP participations and funding per researcher (Member State Rankings in Terms of the Number of FP5 Shared-cost Action Participations and the Amount of FP5 Shared-cost Action Funding Received per Researcher)
Source: DG Research

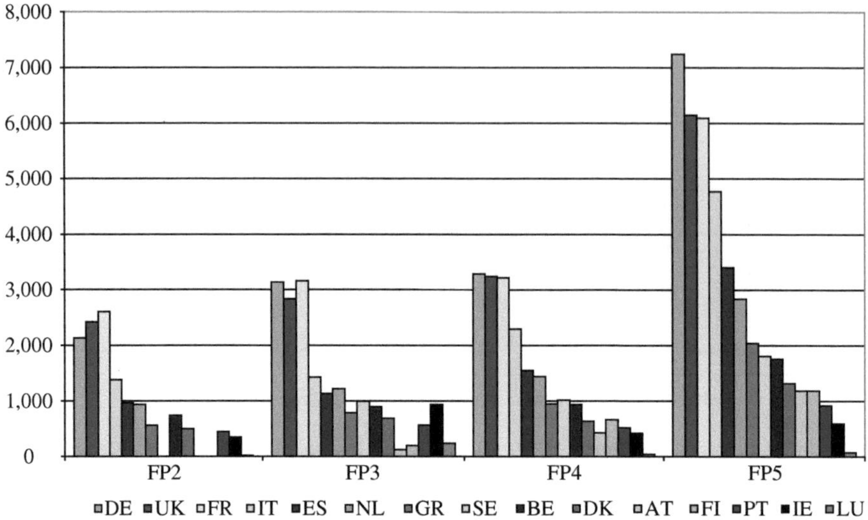

Fig. 4.10. The dynamics of participation (evolution of the number of shared-cost action participations by Member State across Framework Programmes)
Source: DG Research

A close look at the map of Europe reveals the great diversity in research capacity and innovation performance among its regions.[40] R&D "hot spots" generating above average high-tech employment rates exist alongside pockets of poverty and entrepreneurial deprivation.[41] R&D inputs – the presence of universities, higher S&T education enrolment, patterns of overall and particularly business R&D investment and expenditure, and so on – matter, but regional innovation performance also hinges strongly on the functioning of the region as an innovation *system*. Less favoured European regions are of special concern to policy-makers as they display a *regional innovation paradox*. This refers to the apparent contradiction between the comparatively greater need to spend on innovation in lagging regions and their relatively lower capacity to absorb public funds earmarked for the promotion of innovation compared to more advanced regions.[42]

---

[40] Andrés Rodríguez-Pose, *The Dynamics of Regional Growth in Europe: Social and Political Factors*, Oxford, 1998.

[41] Several typologies have been developed to demonstrate and better understand the diversity found. The report *Enlarging the European Research Area: Identifying Priorities for Regional Policy Focusing on Research and Technological Development in the New Member States and Candidate Countries* (FP6 contract n° COP6-CT-2004-00001) mentions the following typologies of regional innovation systems: Braczyk & Heidenreich (1998); Carrincazeaux & Lung (2003); Clarysse & Muldur (1997; 2001); Cooke (1998); Dunnewijk, Hollanders & Wintjes (2004); ECOTEC (2002); Muller *et al.* (2001); Tödtling & Trippl (2004).

[42] Christine Oughton *et al.*, The Regional Innovation Paradox: Innovation Policy and Industrial Policy, In: *Journal of Technology Transfer*, Vol. 27, No. 1, 2002, pp. 97–110.

A pilot study based on 15 years of Framework Programme data (1987–2002) has analysed in more detail the regional distribution of Framework Programme participation and funding.[43] There have been a number of interesting findings, in particular for Europe's less favoured regions. Successive FP programmes have created networks of research in less favoured regions which are at the base of the diffusion of knowledge spillovers and which undoubtedly have eased the diffusion of knowledge from the centre to the periphery. In terms of study findings, the study confirms firstly and not surprisingly that, in absolute terms, participations and funding are concentrated in the "usual suspects": London and the South-East in the United Kingdom; Île de France and Rhône-Alpes in France; the Stuttgart-Karlsruhe-Lower Bavaria axis in Germany; the Randstad in the Netherlands; the greater Brussels area in Belgium; and the North of Italy.[44] Secondly, within peripheral countries, European research activity is often, though not always, concentrated in those areas with the highest levels of GDP per capita and the greatest concentration of researchers and research activity (e.g. Helsinki, Lisbon, Madrid, Athens). Thirdly and most interestingly, a significant number of peripheral regions obtain a share of European research funds which is larger than one would expect on the basis of their overall R&D effort, the size of their economy, or their relative wealth (e.g. Scotland in the United Kingdom; Catalonia and Murcia in Spain; Steiermark in Austria). Put differently, the European Framework Programme offers less favoured regions an important alternative source of research funding. Given the fact that they often face a more limited choice of funding possibilities than more advanced regions, this study indicates that the Framework Programme plays an encouraging role in reducing the research investment gap in less favoured regions. From the viewpoint of the more advanced – in terms of research endowment and innovation performance – regions it can be said that successive Framework Programmes have supported excellent research without being detrimental to cohesion (Fig. 4.11).

The degree of concentration of collaborative research participation In the previous two sections, we have looked at the distribution of Framework Programme participations and funding across countries and regions. This section considers how participation is dispersed at the level of individual organisations. Looking across Framework Programmes, it appears that some institutions have been very effective in competing for Framework Programme funds. Geuna found on the basis of 1st, 2nd, and 3rd Framework Programme data that the distribution of Framework Programme university participations is skewed. Many institutions are present in the system only one or a few times, but a small number of universities achieve a high number of participations.[45] Breschi and Cusmano identified

---

[43] *Regional Analysis of Framework Programme Participants from 1987 (FP2) until 2002 (FP5)* (FP6 contract n° EVA6-CT-2005-000002).

[44] See also European Commission, *Second Report On S&T Indicators*, pp. 552–557.

[45] Aldo Geuna, *University Participation in Community Programmes: How Does the Selection Process Work?* (Prepared for the Conference "New Research Findings: The Economics of Scientific and Technological Research in Europe", University of Urbino, 24–25 February, 1995), February 1995, p. 25.

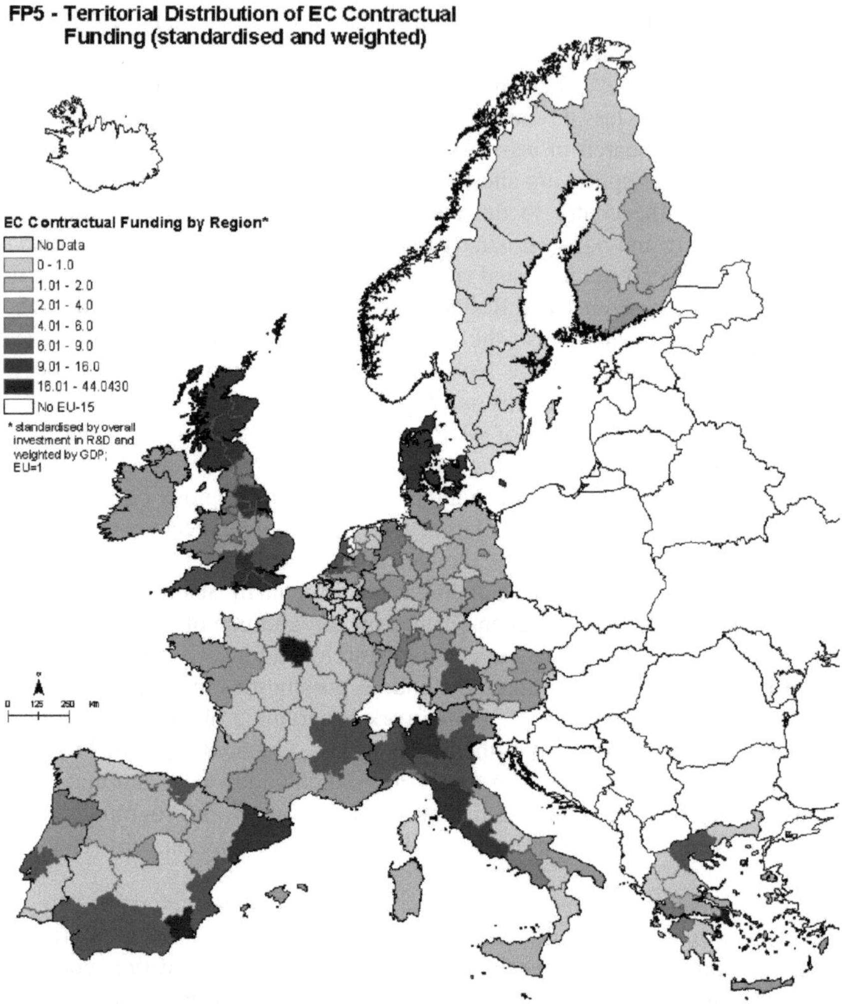

Fig. 4.11. A regional map of FP funding marrying scientific excellence and social cohesion
Source: DG Research.
Data: DG Research, data treatment by London School of Economics

on the basis of 3rd and 4th Framework Programme data a large "oligarchic core", whose centrality and connectivity strengthened over programmes.[46] The Five-Year Assessment 1999–2003 discovered a relatively small core of organisations participating multiple times and across several Framework Programmes and estimated that these organisations accounted for about 20 per cent of all participants.[47] A recent

[46] Breschi and Cusmano, *Unveiling the Texture*, p. 2.
[47] European Commission, *Five-Year Assessment 1999–2003*, 15 December 2004, p. 7.

summary of a series of Framework Programme ex-post evaluations comes to the same conclusion: many organisations' participation in Framework Programme is short-lived, but there is a core of frequent participants.[48] Finally, a recent network analysis of partial data on the 6th Framework Programme found skewed distributions of connectedness, with a central core of highly linked organisations and a periphery of less well connected entities.[49]

A certain degree of concentration of participation in a selective funding scheme based primarily on scientific excellence is inevitable, and a certain degree of continuity of participation is indeed to be welcomed. The literature identifies a number of factors that may promote a certain degree of concentration of participation in (internationally collaborative) research programmes. Large institutions are more likely to apply multiple times to research programmes as they have more staff, and to internationally collaborative research programmes as they are more likely to have prior experience with international research collaboration.[50] Large, well-known institutions are better placed than small, lesser-known ones to put together research consortia.[51] More than small institutions, large organisations have the institutional capacity to handle complex application and project management procedures, and to do so multiple times.[52] Research project proposals submitted by consortia involving a large, well-known institution with a well-established research reputation are more likely to get selected.[53] A lock-in effect then ensues. This results first of all from the fact that successful applicants can familiarise themselves even more with complex application and project management procedures, which remain stable to a certain extent over time. This facilitates later applications. Successful participation, on the basis of which scientific and technological outputs can be produced, further consolidates the participant's scientific reputation and reinforces the information signalling effect to both potential consortia participants and proposal evaluators on the occasion of a next application. In the literature, this phenomenon is referred to as the "Matthew effect" and cumulative advantage.[54]

It is not possible to judge now whether the degree of concentration of Framework Programme participation is appropriate. A study would be required of the evolution

---

[48] Arnold, *Evaluation Record*, May 2005, pp. 13–15.

[49] Caroline S. Wagner *et al.*, *ERAnets – Evaluation of NETworks of Collaboration among Participants in IST Research and Their Evolution to Collaborations in the European Research Area (ERA) – Final Report*, March 2005, pp. 17–18.

[50] "Selection effect".

[51] "Information signalling effect".

[52] "Selection effect".

[53] "Information signalling effect".

[54] For the need for prior international research collaboration/FP experience, see Arnold, *Evaluation Record*, May 2005, pp. 13–15. For information signalling, see Geuna, *University Participation*. For how complex FP application and project management procedures act as barriers to FP participation, see above. For the Matthew effect and cumulative advantage, see Robert K. Merton, The Matthew Effect in Science, In: *Science*, Vol. 159, No. 3810, 1968, pp. 56–63, and Robert K. Merton, The Matthew Effect in Science, II – Cumulative Advantage and the Symbolism of Intellectual Property, In: *ISIS*, 79, 1988, 79, pp. 606–623.

of the degree of concentration of participation over time. A comparison would have to be made with the degree of concentration of participation in Member State, intergovernmental, and foreign (collaborative) research schemes. And an assessment would have to be made of the relationship between programme participant and network characteristics on the one hand, and project outputs and impacts on the other hand.

In the meantime, the Commission is aware of the importance of not discouraging small innovative institutions from participating regularly in the Framework Programme. The Commission's repeated efforts on the one hand to streamline and simplify the Framework Programme bureaucracy, and on the other hand to provide measures to encourage applicants with particular needs, such as SMEs, are aimed at ensuring that this does not happen.

2.3. The characteristics of collaborative research networks

Collaborative research networks described Framework Programme collaborative research projects are implemented via cross-border, cross-sector, interdisciplinary Framework Programme networks. These networks embody the unique European added value of the Framework Programme, which is to bring together different types of high-quality participants from different countries and with complementary expertise in productive partnerships, thus overcoming the sectoral and national fragmentation of the European research system. Over the past two decades, the number of links established between participants in FP-supported collaborative research projects has increased substantially to reach 160 000 under the 5th Framework Programme (Fig. 4.12).[55]

Collaborative research project participants do not just establish collaborative links within their own sectors (businesses just collaborating with businesses, universities just with universities, etc.). FP-funded collaborative research networks succeed in bringing together and creating links between different types of institutions. Over time, the typical configuration of a Framework Programme project has evolved, as expected given the aforementioned changing overall Framework Programme participant structure.[56] At the heart of these networks, however, still lies science–industry collaboration. This is a feature greatly appreciated by Framework Programme shared-cost action participants.[57] It is also a key component of innovation systems.[58]

The average number of different Member States per project has increased from Framework Programme to Framework Programme. Starting from an average of 3 different Member States per project in the 2nd Framework Programme, interim data for the 6th Framework Programme show an average of 6.5 (Table 4.2).

[55] For the number of collaborative links created under FP3 and FP4, see also European Commission, *Second Report On S&T Indicators*, pp. 559–560.

[56] European Commission, *Second Report on S&T Indicators 1997*, pp. 520–541.

[57] European Commission, *Five-Year Assessment 1995–99*, 2001, p. 17.

[58] See for example recent OECD reports on *Benchmarking Industry-Science Relationships* (2002), *Turning Science Into Business – Patenting and Licensing At Public Research Organisations* (2003), and *Governance of Public Research* (2003).

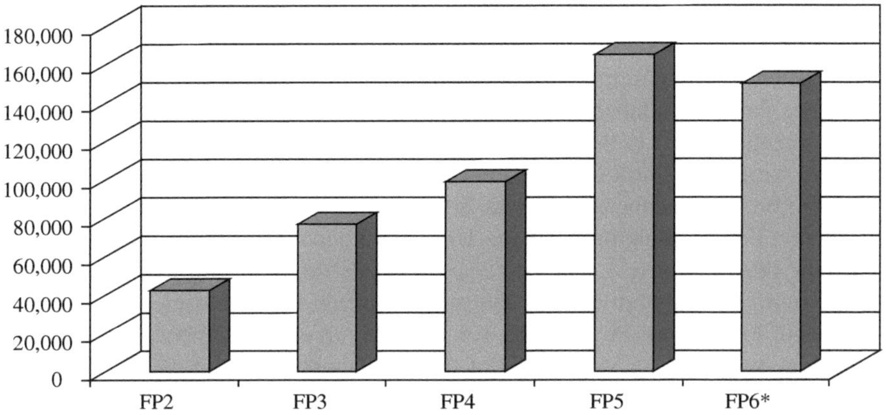

Fig. 4.12. Collaborative links (national+international) established through FP funded shared-cost actions
Source: DG Research.
Note: *: Partial data (to 11.03.2005)

Over time, the configuration of intra-European transnational collaboration has also changed (Figs 4.13–4.15). The number of privileged collaboration partners has increased across Framework Programmes. No less than eight Member States are now situated at the core (Austria, Belgium, Greece, Ireland, Netherlands, Portugal, Spain, and Sweden). The groups of "intermediate" and "peripheral" countries have each become more integrated. Countries in these two groups increasingly collaborate amongst themselves thus reducing dependency on the core and diversifying collaboration patterns. Under the 5th Framework Programme, this was reflected in the emergence of two distinct sub-networks in the group of "intermediate" countries –Southern European (Greece, Italy, Spain, and Portugal) and Northern European (Netherlands, Denmark, Sweden, and Finland). Finally, a strong cohesion effect is evident. In the 6th Framework Programme, formerly "peripheral" countries – Austria, Ireland, and Luxemburg – joined the two inner circles of collaboration. In summary these three trends highlight a shift from projects dominated by just a few core countries and relying on geographical proximity to projects with a more balanced national representation and detaching themselves from traditional ties. This illustrates the integrating effect of European collaborative research.

Box 4.1. Methodology for presenting the global configuration of Community collaborations

Figs 4.13–4.15 show the changes across Framework Programmes in the positioning of the Member States in Community collaboration networks. For each Framework Programme, a cluster analysis was performed on the correlation matrix of all Member State collaborative links. The results show the degree to which different countries have collaborated with each other (indicated by the percentages) and the relative importance of countries as preferred partner

(indicated by the different zones). They do not provide any indication of the importance of each country in terms of the absolute number of participations in the Framework Programme or the absolute number of links created with other countries through collaborative research projects.
The percentages next to the arrows show the "collaborative links" of a particular country with the countries situated in the light grey zone. In the 5th Framework Programme, for instance, Italy had 50.45 per cent of its collaborations with Germany, France, Belgium, and the United Kingdom (and thus 49.55 per cent with the other countries). The percentages inside the light grey zone show the proportion of collaborative links that exist between the countries of that zone. In the 6th Framework Programme, for instance, all eight Member States in the light grey zone had 34.56 per cent of their collaborations with each other (and thus 65.44 per cent of their collaborations with other countries).
The central group – (countries situated in the light grey zone): The countries situated in the light grey circle are, first of all, very densely collaborating with each other. At the same time, these countries are also involved in collaborative links with the countries situated in the dark grey and black zones. This is indicated by the percentages in the figures.
The intermediate group – (countries situated in the dark grey zone): These countries are linked to the central group to an extent shown by the percentage figures. It emerges from the cluster-analysis that these intermediate countries also form one or more highly connected sub-networks, encased in checked or dotted zones. Countries that are part of these sub-networks have dense collaborative links with each other (within the dotted and checked cluster, respectively).
The peripheral group – (countries in the black zone): These countries are linked to the central group to an extent shown by the percentage figures and are not members of a particularly densely connected sub-network.

Modern mainstream innovation theory recognises the importance of networks bringing together different types of actors (e.g. science and industry) in regional, national, and supranational systems of innovation. However, an important trade-off exists. Some authors question the value of uncritically promoting transnational collaboration at a time when, they would argue, competitive advantage is mainly determined by advances made by excellent single-country research teams.[59] Even if we accept the value of networking, important questions remain to be addressed. One question concerns the optimal size of networks, as some have suggested

[59] See Giovanni Dosi *et al.*, *Evaluating and Comparing the Innovation Performance of the United States and the European Union*, Expert Report Prepared for the TrendChart Policy Workshop 2005, p. 3: "We suggest that effective European catching up would require much less emphasis on various types of "networking", "interactions with the local environment", "attention to user need" — current obsessions of European and national policy makers – and, conversely, much more on policy measures aimed to both strengthen "frontier" research and, at the opposite end, strengthen European corporate actors."

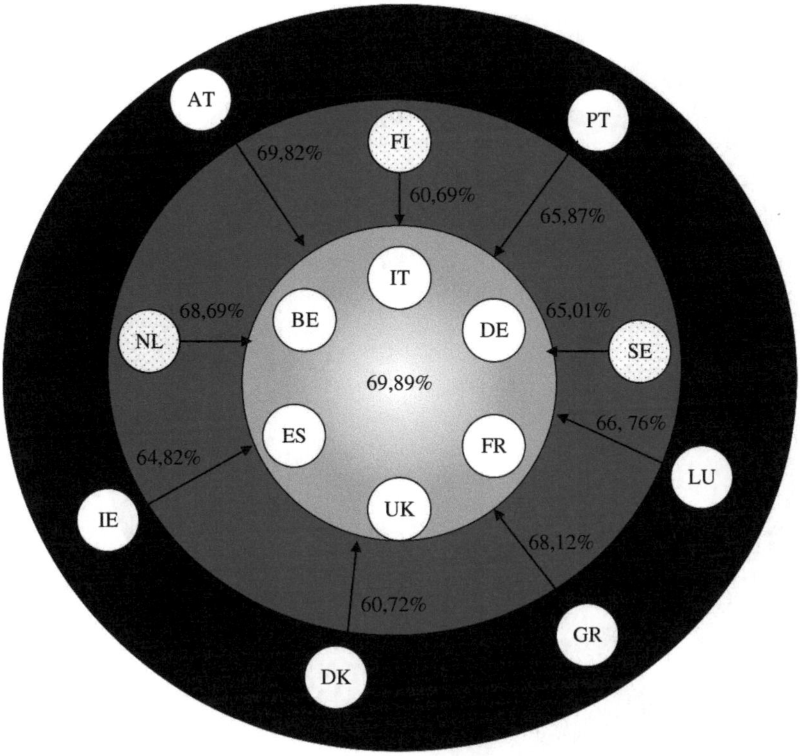

Fig. 4.13. Evolution of the configuration of intra-European trans-national collaboration – Fourth Framework Programme
Source: DG Research

that the large networks promoted under the 6th Framework Programme may be less effective than smaller ones.[60] Another question concerns the optimal mix of participant types and of domestic/international partners. A first step towards answering these kinds of questions should consist of analysing in a more in-depth manner the configuration and performance of networks by Framework Programme, Specific Programme, thematic priority, and call. In tracing and assessing networks, innovative (bibliometric, patent) methodological techniques should be used.

The degree of stability of collaborative research networks FP-funded collaborative research networks are relatively stable across calls, and even across Framework

[60] Arnold, *Evaluation Record*, May 2005, p. 17: "Two earlier evaluations suggest that larger networks such as those promoted by FP6 may be less effective than smaller networks. The FP4 Impact Biomed2 assessment states that scientific quality in the programme was negatively related to the number of partners in a project, and that the quality of the larger network projects was problematic. The FP5 Impact Growth study also found that larger networks (averaging 16 partners) generated limited impacts, compared with smaller ones."

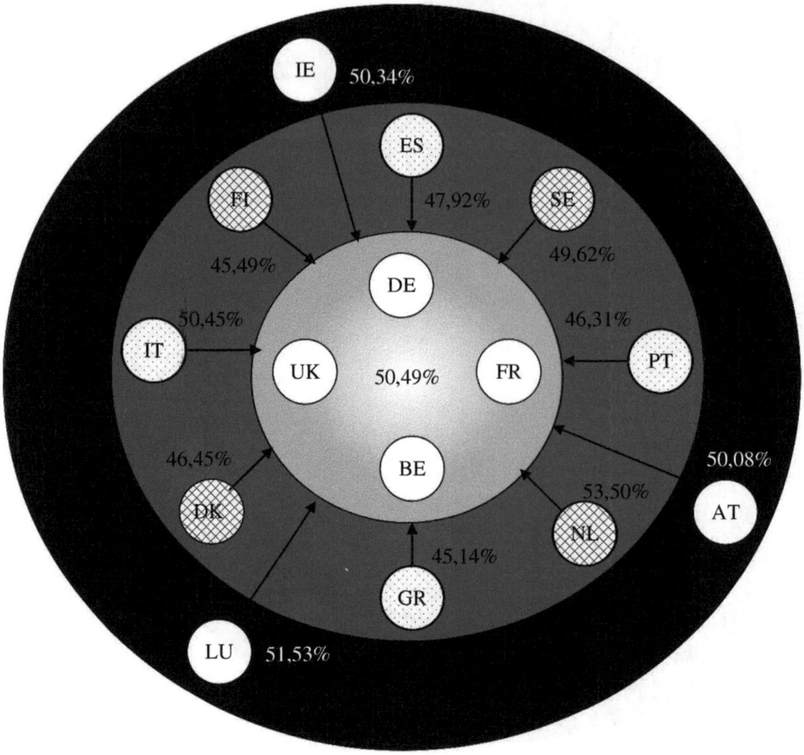

Fig. 4.14. Evolution of the configuration of intra-European trans-national collaboration – Fifth Framework Programme
Source: DG Research

Programmes. The Five-Year Assessment 1995–1999 found evidence to suggest that many of the links formed between academic and industrial researchers persisted beyond initial projects.[61] An analysis by Breschi and Cusmano showed that the centrality and connectivity of the "oligarchic core" of organisations at the heart of the Framework Programme has increased over time.[62] The Five-Year Assessment panel 1999–2003 took the view that the Framework Programmes have played an important part in the formation and reinforcement of inter-organisational networks.[63] A recent summary of Framework Programme ex-post evaluations concluded that the Framework Programme does not generate wholly new R&D networks, but causes network extension. In other words, the basic networks remain stable but they are able to absorb new participants.[64]

---

[61] European Commission, *Five-Year Assessment 1995–99*, 2001, p. 18.
[62] Breschi and Cusmano, *Unveiling the Texture*, p. 2.
[63] European Commission, *Five-Year Assessment of the European Union Research Framework Programmes 1999–2003. Executive Summary*, p. 2.
[64] Arnold, *Evaluation Record*, May 2005, pp. 13–15.

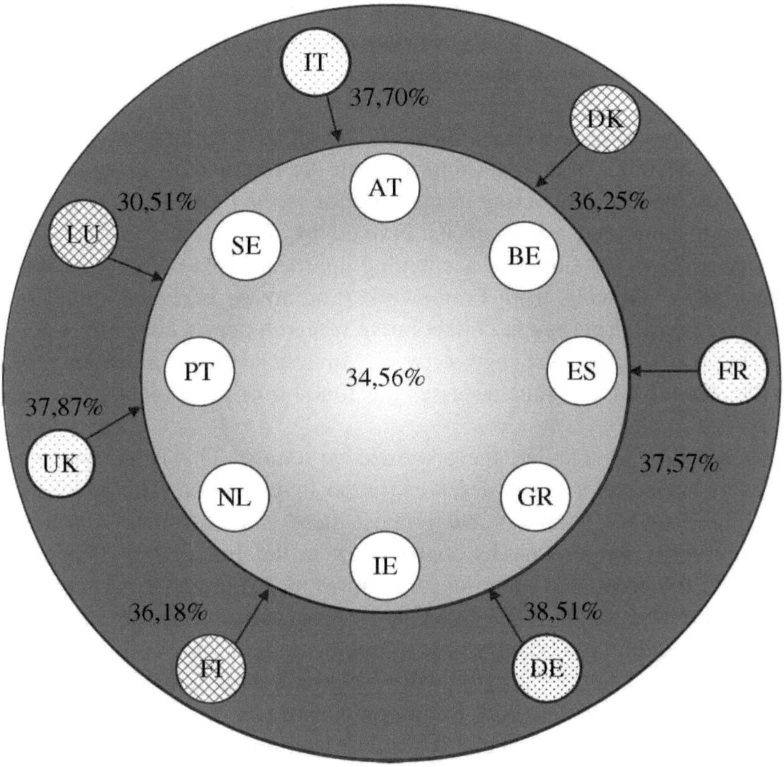

Fig. 4.15. Evolution of the configuration of intra-European trans-national collaboration – Sixth Framework Programme (data as of end of April 2004)
Source: DG Research

Is network stability beneficial? Certainly one of the aims of Framework Programme support is to tackle the fragmentation of the EU research system by encouraging the formation of transnational consortia. If the networks thus created are durable, then, one could argue, a real contribution has been made to the long-term restructuring of research in Europe. Continuity over successive Framework Programmes can allow complex research fields to advance within a stable cooperative structure. However, there are also some important caveats. It may be that, once established, certain networks would have continued without further Framework Programme funds. In such instances, there would be no "behavioural additionality" from repeated Framework Programme support (although there may still be input or output additionality). At the same time, a potential danger of too much stability is that it may make it correspondingly difficult for new or more peripheral institutions to enter the game.[65] This in turn may have negative

[65] Wagner *et al.*, *ERAnets*, p. 25.

implications for the innovativeness of research undertaken while raising issues in relation to competition. Finding the right balance between stability and dynamism is therefore important and merits ongoing analysis.

From financial to knowledge returns – the multiplier effect of collaborative research via networks A Member State's net returns from the Framework Programme are traditionally calculated by subtracting its contribution to the Framework Programme budget from the combined value of its Framework Programme participations. However, such a restricted financial accounting approach does not capture all the "knowledge returns" flowing from Framework Programme participation. Participating in a Framework Programme collaborative research consortium offers access to an EU-wide knowledge exchange network. In other words, a single project participant benefits from and thus accesses the funding of all project participants combined.

The question is how to calculate these knowledge returns. This is an issue which so far has barely been explored and deserves further investigation. The preliminary, relatively straightforward approach we propose here is to estimate a country's knowledge returns by subtracting its contribution to the Framework Programme budget, not from the combined value of its Framework Programme participations, but from the combined value of the Framework Programme contracts in which it participates. For example, if a country contributes €1 to the Framework Programme budget, the combined value of its Framework Programme participations is €2, and the combined value of the Framework Programme contracts in which it participates is €10, then that country's net financial return is €1 while its net knowledge return is €9, and thus the multiplier effect is of a factor 9.

Calculations carried out by DG Research show that for each Euro invested in the 5th Framework Programme, countries' net knowledge return averaged €19.8 for the EU-25 (Fig. 4.16). Participation in the Framework Programme was thus a win–win situation for all parties involved: all countries enjoyed positive net knowledge returns under the 5th Framework Programme. However, the size of these returns tended to be inversely related to a country's number of Framework Programme participations. Countries with a smaller number of Framework Programme participations (e.g. smaller EU Member States, and the then Acceding Countries) enjoyed higher net knowledge returns than countries with a larger number of Framework Programme participations (e.g. the bigger Member States). This probably happened because a smaller number of Framework Programme participations translates into a pattern of widely dispersed single participations per project, while a larger number of Framework Programme participations translates into a pattern where regularly two or three participations from the same country can be found in the same project, which partially dampens the net knowledge return.

2.4. Themes, quality, and riskiness of collaborative research

The collaborative research component of the Framework Programme is organised around a number of thematic priorities which have grown in number under

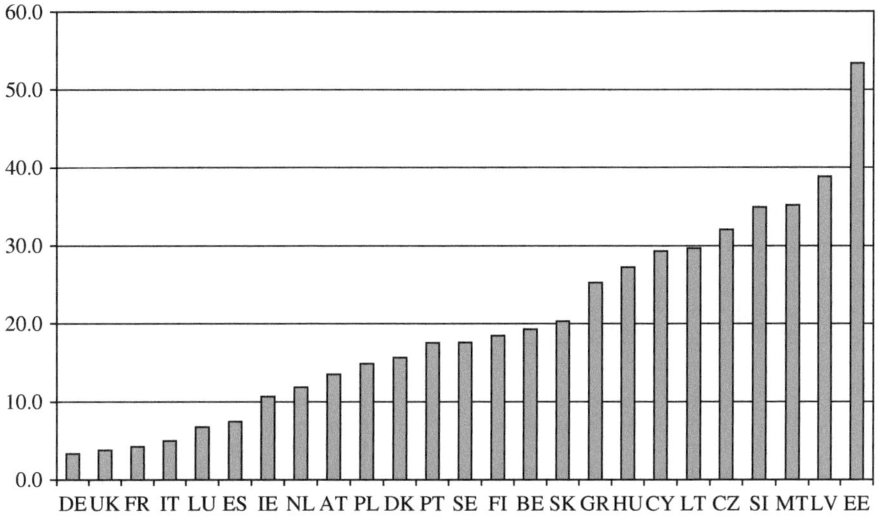

Fig. 4.16. Net knowledge return per €1 invested in the FP
Source: DG Research
Note: Malta not included; calculation: [(Value of shared cost contracts in which participating (no double counting))/Contribution to FP shared cost actions budget) – 1]; EU-15 contribution to FP5 shared cost actions budget calculated on the basis of their share in EU-15 GDP, Acceding Countries' contribution to FP5 shared cost actions budget calculated on the basis of then total contributions to FP5.

successive Framework Programmes (Fig. 4.17). The focus of the first two Framework Programmes, for instance, was clearly on energy and ICT, which together accounted for 75 per cent of funding under the 1st Framework Programme, and 65 per cent under the 2nd Framework Programme. On the other hand, the 6th Framework Programme was organised around seven thematic priorities.[66] And the 7th Framework Programme will centre around ten thematic priorities.[67]

The coverage of the Framework Programme is comprehensive, and does not appear to leave out any important thematic priority. Though in general the Framework Programme ex-post evaluation literature says surprisingly little on thematic coverage, the 1999–2003 Five-Year Assessment Panel could not identify any major thematic gaps in the parts of the Framework Programmes examined.[68]

[66] Life sciences, Genomics and Biotechnology for Health; Information Society Technologies; Nanotechnologies and nano-sciences, knowledge-based multifunctional materials, new production processes and devices; Aeronautics and Space; Food Quality and Safety; Sustainable Development, Global Change and Ecosystems; Citizens and Governance in a knowledge-based society.

[67] Health; Food, Agriculture and Biotechnology; Information and Communication Technologies; Nanosciences, Nanotechnologies, Materials and new Production Technologies; Energy; Environment (including Climate Change); Transport (including Aeronautics); Socio-economic Sciences and the Humanities; Security; and Space.

[68] European Commission, *Five-Year Assessment 1999–2003*, 15 December 2004, p. 5.

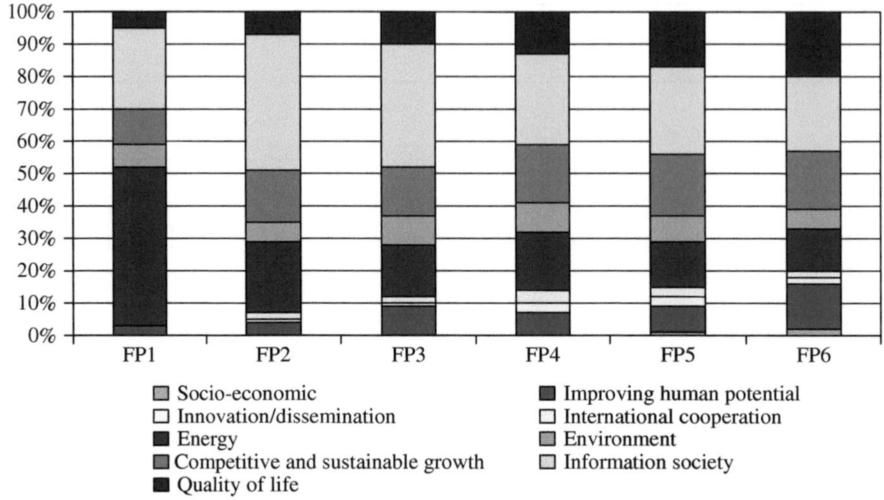

Fig. 4.17. Evolution of FP priorities
Source: DG Research[69]

The Framework Programme thematic priorities also correspond more or less to those of major US and Japanese research support programmes.[70]

Within this context, it is legitimate to ask whether the Framework Programme's increasingly comprehensive coverage may have led to a degree of fragmentation in the sense that Framework Programme resources may be spread too thinly across too many priorities. This is always an important consideration because, to make a real difference, the level of support in a particular S&T area should be sufficient to achieve critical mass and a minimum efficient scale of research. However, there are no indications as yet in the Framework Programme evaluation literature or elsewhere that this is a major cause for concern.

A rigid predefined thematic priority structure can make it more difficult to accommodate important scientific advances and new needs which emerge during the life-time of the programme. For this reason, there have been a number of calls to introduce greater flexibility in the Framework Programme. The Davignon Five-Year Assessment, for instance, noted on the basis of the 3rd and 4th Framework

---

[69] European Commission, *Towards a European Research Area – Science, Technology and Innovation – Key Figures 2002*, Luxembourg, 2002, p. 22.

[70] See the US National Science Foundation Internet webpage for a listing of programme areas: Biology; Computer and information sciences; Crosscutting programs; Education; Engineering; Geosciences; International; Math, physical sciences; Polar research; Science statistics; Social, behavioural sciences. See Council for Science and Technology Policy Japan, *Japan's Science and Technology Basic Policy Report*, 27 December 2005, p. 17, for Japan's priorities under its Second S&T Basic Plan: life sciences; ICT; environmental sciences; nanotechnology and materials; energy; manufacturing technology; infrastructure; frontier (outer space and oceans).

Programme experience that adjustments to meet new needs, or to reflect new scientific advances, required a tortuous and time-consuming legal process.[71] In the recent ICT Five-Year Assessment, the Panel was of the opinion that, in a fast-moving area such as ICT research, the programme should be willing and able to respond rapidly – if justified – to changes in the economic, technological, or even policy context.[72] It cannot be denied, however, that such flexibility is gradually being built into the programme. The 6th Framework Programme, for instance, has been able to fund research on, for example, SARS and avian flu. Under the EU-supported SEPSDA (Sino-European Project on SARS Diagnostics and Antivirals), scientists from across Europe and China have spent the past years screening over 8 000 existing drugs to find prime candidates for combating SARS, and identified a number of promising anti-SARS compounds.[73] The European Commission recently also announced that it would be making a further €20 million available for research into avian and pandemic influenza. Relating to animal health, the relevant call for proposals would cover issues such as developing vaccines for avian species, improved diagnosis and early warning systems, and technology transfer to third countries. Relating to human health, it would cover issues such as clinical research on pandemic influenza vaccines, better understanding of the influenza virus, and strengthening support to surveillance.[74]

In terms of quality, FP-supported collaborative research has generally been found to be of a high standard.[75] Industrial participants have reported that Framework Programme projects are of relatively high scientific and technological complexity, while university participants have identified major benefits in the form of knowledge enhancement.[76] A first important factor underlying the high quality of FP-funded research is the well-functioning Framework Programme proposal assessment and project monitoring system.[77] For instance, it has been suggested that the strong correlation between the number of participations by country in the 4th Framework Programme and the number of researchers in that country provides some evidence that the proposal selection system is working as it should.[78] Another crucial factor is that Framework Programme participants themselves are generally of high quality. A recent study on the networks resulting from funding under the 6th Framework Programme in the field of ICT concluded that networks under this Framework Programme attract knowledge leaders from corporate and academic centres of excellence, evidenced by the rate at which relevant patent holders and institutions

---

[71] European Commission, *Five-Year Assessment*, 1997, p. 14.

[72] European Commission, *Interim Panel Report*, p. 28.

[73] EU-Chinese Co-operation Pinpoints Potential Treatments for SARS, In: *European Research Headlines*, 29 July 2005.

[74] *News Alert – Commission Releases New Funds for Avian Flu Research*, 17 January 2006.

[75] Arnold, *Evaluation Record*, May 2005, pp. I, 6, 12 and 29.

[76] European Commission, *Five-Year Assessment 1999–2003*, 15 December 2004, p. 7.

[77] European Commission, *Five-Year Assessment 1999–2003*, 15 December 2004, pp. 9–10; Arnold, *Evaluation Record*, May 2005, p. 12.

[78] Arnold, *Evaluation Record*, May 2005, p. 16.

participate.[79] A recent bibliometric study on participants in the 5th Framework Programme in the fields of the life sciences and nanotechnology convincingly demonstrated that the level of Framework Programme participants is generally above world level and that they reach citation impact scores higher than what could be expected on the basis of the journals in which they publish or the fields in which they are active.[80] A substantial number of Nobel Prize laureates in medicine, chemistry, and physics are involved in research projects and networks under the 6th Framework Programme, many of them oriented towards medical purposes.[81]

Although Framework Programme collaborative research is of high quality, it has been observed that it can often be too mainstream and insufficiently risky and "at the frontier". The Five-Year Assessment 1995–99 found that the Framework Programme should retain its overarching emphasis on excellence, but make extra efforts to ensure that "risky" projects are not excluded.[82] Efforts were made to address this in the NEST programme under the 6th Framework Programme, and while this was welcomed by the subsequent Five-Year Assessment panel, they felt it was still a too narrow approach. The panel therefore recommended that an ambition to fund more risky projects should be embedded in the Framework Programme objectives, instruments and implementation as well as in the assessment of achievements and impacts.[83] Under the 7th Framework Programme, risky frontier research will be supported through the planned European Research Council.

2.5. The additionality of collaborative research

Before assessing the various impacts generated by FP-funded collaborative research, it is important to establish whether or not Framework Programme collaborative research funding schemes have generated effects that are additional to what would have happened anyway in the absence of such funding schemes. In other words, in the absence of the Framework Programme collaborative research component, would projects have been carried out at all or would they have been carried out differently?

The consensus appears to be that Framework Programme collaborative research funding schemes are characterised by a high degree of additionality. Without Framework Programme funding, projects would in general not have been carried

---

[79] Wagner *et al.*, *ERAnets*, March 2005, pp. vii–viii.

[80] *Bibliometric Analysis for Participants Under the 5th European Framework Programme for Research and Technical Development (Contract Number EVA6-CT-2005-00001) – Final Report.*

[81] Nobel Excellence in EU Research Projects, In: *European Research Headlines*, 9 December 2005: Medicine: Erwin Neher (1991 Nobel laureate), Christianne Nüsslein-Volhard (1995), Rolf M. Zinkernagel (1996), Stanley B. Prusiner (1997) and Tim Hunt (2001); Chemistry: Harmut Michel (1988 Nobel laureate), John E. Walker (1997), Kurt Wüthrich (2002) and Aaron Ciechanover (2004); Physics: Theodor W. Hänsch (2005).

[82] European Commission, *Five-Year Assessment 1995–99*, 2001, p. 21.

[83] European Commission, *Five-Year Assessment 1999–2003*, 15 December 2004, p. 5.

out or they would have been carried out differently, often entailing substantial losses in terms of efficiency and effectiveness. Typical is the conclusion of a UK study on the EU Framework Programmes, which reported a broad agreement that Framework provides considerable added value.[84] It expands the funds available to national researchers over and above that which is available to them through national research funds alone. It provides UK participants with access to foreign researchers and research outputs in a way that national funds cannot. By pooling financial resources, it boosts investment in important research topics, and by pooling competencies it increases the likelihood of a breakthrough in a given area. Complex issues are resolved more quickly and more thoroughly as a result of larger projects and portfolios and multiple projects across successive Frameworks.[85]

Supportive evidence is wide-ranging and convincing. Almost 95 per cent of respondents to a survey of Norwegian participants in the 5th Framework Programme answered that EU-funding was very important for getting the project started, 90 per cent that it was important for the size of the project, and over 80 per cent that it was important for how fast the project could be carried through.[86] Among Irish participants in the 4th Framework Programme responding to a survey, 82 per cent would not have proceeded with their project without Framework Programme funding. Of the 18 per cent who would have continued, over 70 per cent would have done so with reduced objectives and reduced funds, while close to 40 per cent would have done so over a longer timescale, and with fewer partners.[87] An Austrian study found very high additionality of the EU Framework Programmes. Of those questioned, 70 per cent said they would not have continued their research project had it been rejected. Of the remaining 30 per cent who would have continued the project without EU subsidy, only 14 per cent said that they would have done so without any adjustment. Typical adjustments entailed a modified goal orientation, a reduced number of project partners, and a different project volume.[88] Of Finnish participants in the 5th Framework Programme, 70 per cent would not have carried out the project without EU funding, 40 per cent would have carried it out on a smaller scale, and 35 per cent would have carried it out more slowly.[89]

2.6. The impact of collaborative research

Evaluations of past Framework Programmes have shown that impacts are important and wide-ranging. The evidence is considerable that the Framework Programme

---

[84] DTI – Office of Science and Technology, *The Impact of the EU Framework Programmes in the UK*, July 2004, p. 3.

[85] DTI – Office of Science and Technology, *EU Framework Programmes*, p. 10.

[86] NIFU *et al.*, *Evaluation of Norway's Participation in the EU's 5th Framework Programme*, 2004, p. 100.

[87] Forfás, *The 4th Framework Programme in Ireland – An Evaluation of the Operation and Impacts in Ireland of the EU's Fourth Framework Programme for Research and Development*, 2001, p. 43.

[88] Joanneum Research Forschungsges.m.b.H *et al.*, *Austrian Participation*, p. 83.

[89] Marjo Uotila *et al.*, *Finnish Participation in the EU Fifth Framework Programme and Beyond*, 2004, p. 28.

helps to improve participants' research capabilities, and modifies their behaviour, in addition to generating significant scientific and technological outputs, and contributing to competitiveness and innovation by boosting productivity and encouraging the development of new products and processes.

Impact on capabilities, behaviour, and ability to compete FP-funded collaborative research projects strengthen participants' research capabilities by enhancing their knowledge base, and improving the skills of staff.[90] Behaviour is also permanently changed resulting in a better ability to compete. For example, participants able to achieve short-term networking goals are more likely to carry out research through networks in the future.[91] An analysis of the Third Community Innovation Survey (CIS-3) shows that Framework Programme participating enterprises are more likely to engage in innovation cooperation with other partners in the innovation system, such as other firms and universities (Fig. 4.18).

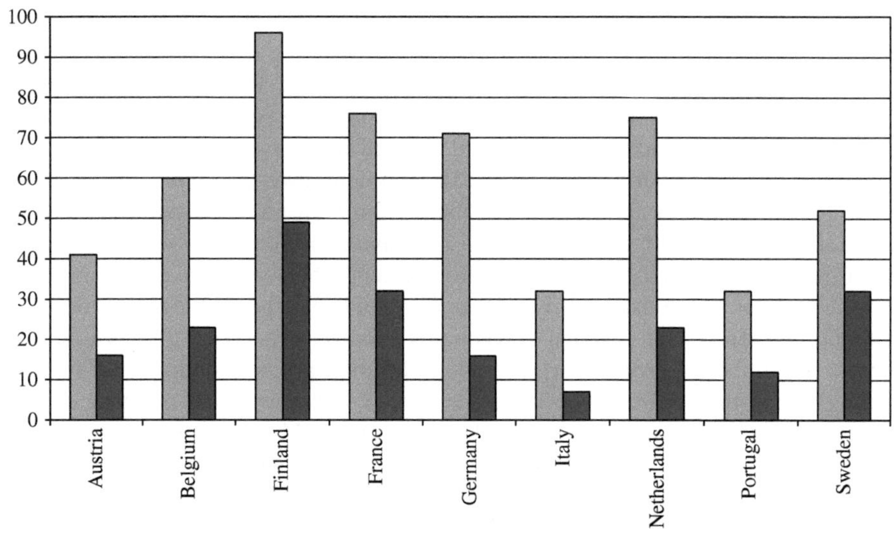

Fig. 4.18. FP participants are more likely to collaborate
Source: DG Research, Eurostat
Data: Eurostat
Note: Results here are for firms in the manufacturing sector

[90] Databank Consulting *et al.*, *IST Impact Study – Microelectronics & Microsystems, Health, Mobile Communications – Draft Final Report D.6*, 4 August 2004, p. 24; Arnold, *Evaluation Record*, May 2005, p. 20–21; European Commission,*Five-Year Assessment 1999-2003*, 15 December 2004, p. 7.
[91] Databank Consulting *et al.*, *IST Impact Study*, p. 24; Arnold, *Evaluation Record*, May 2005, pp. 20–21, 25.

Impact on scientific performance The extension of the knowledge base is to some extent codified and made explicit through scientific publications. Framework Programme projects generate many publications and thereby contribute directly to Europe's total output in terms of scientific publications, an important indicator of scientific performance.

The publication of peer-reviewed scientific work is an important goal for Framework Programme participants, especially those from universities and research institutes, now accounting for most Framework Programme participations and funding. For instance, 94 per cent of British public sector researchers participating in the 4th and 5th Framework Programmes rated refereed publications in journals and books as important outputs.[92] A similar survey in Austria found that around 80 per cent of respondents considered publications an important or very important goal of participation.[93] And a study in Ireland showed that Framework Programme participants from the public sector ranked publications as one of the ten most important goals.[94]

Framework Programme participants generally achieve their publication goals, and thus generate large numbers of scientific publications. For instance, 92 per cent of Finnish university participants in the 4th Framework Programme (and 91 per cent of such participants in the 5th Framework Programme) had achieved their publication goals.[95] And 85 per cent of 3rd/4th Framework Programme public sector participants declared that they had achieved their publication goals.[96] In a study of Irish participants in the 4th Framework Programme, 83 per cent of university/research institute respondents listed publications in refereed journals as important outputs.[97] Over 80 per cent of Austrian university participants in the 4th Framework Programme responding to a survey (and over 70 per cent of such research institute participants) had already reached their publication goals or expected to do so within the next three years.[98] A recent IST impact study obtained for publications a score for goal achievement of over 3 (important goal achieved as expected).[99] In the majority (79 per cent) of projects under the 4th and 5th Framework Programmes with British participation surveyed, at least one peer-reviewed publication was produced, while in around 10 per cent of those projects more than 20 such outputs were produced.[100] An analysis of impact variables resulting from research projects in the fisheries and aquaculture domain of the FAIR programme under the 4th

---

[92] DTI – Office of Science and Technology, *EU Framework Programmes*, p. 23.

[93] Joanneum Research Forschungsges.m.b.H *et al.*, *Austrian Participation*, p. 72.

[94] Forfás, *The 4th Framework Programme*, p. 35.

[95] Uotila *et al.*, *Finnish Participation*, p. 34.

[96] Decisia *et al.*, *Final Report – Assessment of the Impact of the Actions Completed Under the 3rd and 4th Community Framework Programmes for Research; Survey for the Five-Year Assessment of Community Research Activities (1999-2003)*, 2 March 2004, p. 22.

[97] Forfás, *The 4th Framework Programme*, p. 41.

[98] Joanneum Research Forschungsges.m.b.H *et al.*, *Austrian Participation*, p. 79.

[99] Databank Consulting *et al.*, *IST Impact Study*, p. 24.

[100] DTI – Office of Science and Technology, *EU Framework Programmes*, p. 22.

Framework Programme identified 711 publications in 219 peer-reviewed journals by the participants of 82 projects, an average of 8.7 peer-reviewed publications per project.[101] Participants in the BRITE-EURAM II programme under the 3rd Framework Programme reported 3 621 publications for 454 projects, an average of 8 publications per project.[102] Participants in the transport programme under the 4th Framework Programme reported 3 766 publications for 269 shared-cost projects, an average of 14 publications per project.[103]

The scientific publications resulting from FP-funded collaborative research projects are often intra-European international co-publications, which points once more to the integrating effect of these projects. Direct evidence for this comes from a recently completed bibliometric study on participants in the 5th Framework Programme in the life sciences and nanotechnology, which found that for those Framework Programme participants the growth in intra-European international co-publications was much larger than that in international co-publications with US authors or in the overall number of scientific publications.[104] This may explain the fact that also at the aggregate level the share of intra-European international co-publications in Europe's international co-publications has increased, while that of international co-publications with the United States has decreased.[105]

The scientific publications resulting from Framework Programme collaborative research are generally of high quality. It is well known that the citation impact score of international co-publications is generally higher than that of purely national publications.[106] Direct evidence, however, also comes from the aforementioned recently completed bibliometric study on participants in the 5th Framework Programme, which found that the citation impact scores of those Framework Programme participants were higher than could be expected on the basis of the journals in which they published, and that they scored above world level when allocated to specific fields.[107]

Impact on technological and innovative performance Framework Programme participation enhances the development and use of new tools and techniques; the

---

[101] Gesche Pluem, *Analysis of Impact Variables Resulting from Research Projects in the Fisheries and Aquaculture Domain of the European Commission's Fair Programme (1994-1998)*, 2003, pp. 12, 14.

[102] European Commission, *Five Year Assessment Report Related to the Specific Programme Competitive and Sustainable Growth Covering the Period 1995–1999*, June 2000, p. 25.

[103] European Commission, *Competitive and Sustainable Growth*, June 2000, p. 28.

[104] *Bibliometric Analysis.*

[105] Laurence Esterle and Ghislaine Filliatreau (Eds.), *Indicateurs de Sciences et de Technologies Édition 2004,* Rapport de l'Observatoire des Sciences et des Technologies, Paris, 2004, pp. 284–285.

[106] Wolfgang Glänzel, National Characteristics in International Scientific Co-Authorship Relations, In: *Scientometrics*, Vol. 51, No. 1, 2001, pp. 69–115; Isabell Gómez *et al.*, Collaboration Patterns of Spanish Scientific Publications in Different Research Areas and Disciplines, In: M.E.D. Koenig and A. Bookstein (Eds.), *Proceedings of the Biennial Conference of the International Society for Scientometrics and Informetrics*, Medford, NJ, 1995, pp. 187–196; Wolfgang Glänzel *et al.*, A Bibliometric Analysis of International Scientific Cooperation of the European Union (1985–1995), In: *Scientometrics*, Vol. 45, No. 2, 1999, pp. 185–202.

[107] *Bibliometric Analysis.*

design and testing of models and simulations; the production of prototypes, demonstrators, and pilots; and other forms of technological development.[108] One finds confirmation of a strong impact on the innovative performance of European firms in analyses of the Community Innovation Survey (Fig. 4.19).[109] The results show that firms participating in the Framework Programme, irrespective of their size, tend to be more innovative than those that do not participate. Although no causal links can be "proven" by these results, they nevertheless provide a strong indication that public funding for research strengthens innovation performance.[110]

FP-funded collaborative research projects also generate a large number of patents, once more exerting a positive influence on an important S&T indicator. A 2002 assessment of nearly 1 900 non-nuclear energy research and demonstration projects under the 4th Framework Programme found that they had resulted in about 400

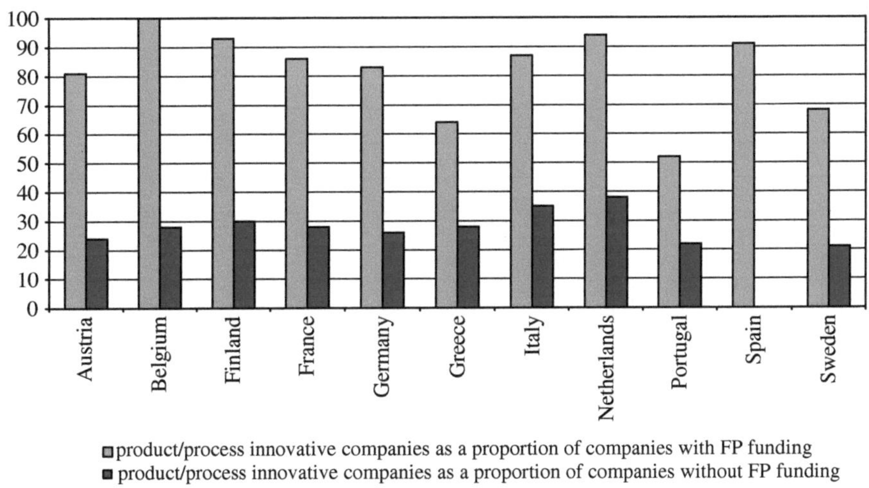

Fig. 4.19. FP participants are more likely to produce product/process innovations
Source: DG Research, Eurostat
Data: Eurostat
Note: Results here are for firms in the manufacturing sector

[108] Databank Consulting *et al.*, *IST Impact Study*, p. 24.

[109] In the CIS-3 survey, organisations are defined as "innovative" when they had either introduced a product or process innovation in the course of the past three years, or had engaged in innovation but had not (yet) completed or abandoned it. They were asked whether they had received EU funding in general and FP funding more specifically. This made it possible to look for a relation between the receipt of FP funding and other variables such as the degree of innovativeness, the likelihood to apply for a patent, the likelihood to hold a patent, and the likelihood to cooperate with other partners in the innovation process.

[110] Even when causality is assumed, it is not clear in which direction causality is operating: does the FP attract more innovative participants or do companies become more innovative as a result of FP participation? See also Chapter 2 of European Commission, *European Competitiveness Report 2004*, Luxembourg, 2004.

patent applications.[111] Corroborating evidence is again found in the Community Innovation Survey which shows that Framework Programme participating enterprises are more likely to apply for patents than non-participants (Fig. 4.20). In Germany, for example, FP-funded firms submit three times as many patent applications as non-participating firms. Again, no causal links can be "proven" by these results. Nevertheless they provide a strong indication that public funding for research strengthens innovation performance.

The micro-economic benefits of collaborative research The Framework Programme not only strengthens participants' mid- to long-term ability to compete, but also directly enhances their competitive position. A wide range of ex-post evaluation studies show that as a result of Framework Programme participation firms are able to realise increased turnover and profitability, enhanced productivity, improved market share, access to new markets, reorientation of the commercial strategy, enhanced competitive position, enhanced reputation and image, and reduced commercial risk.[112]

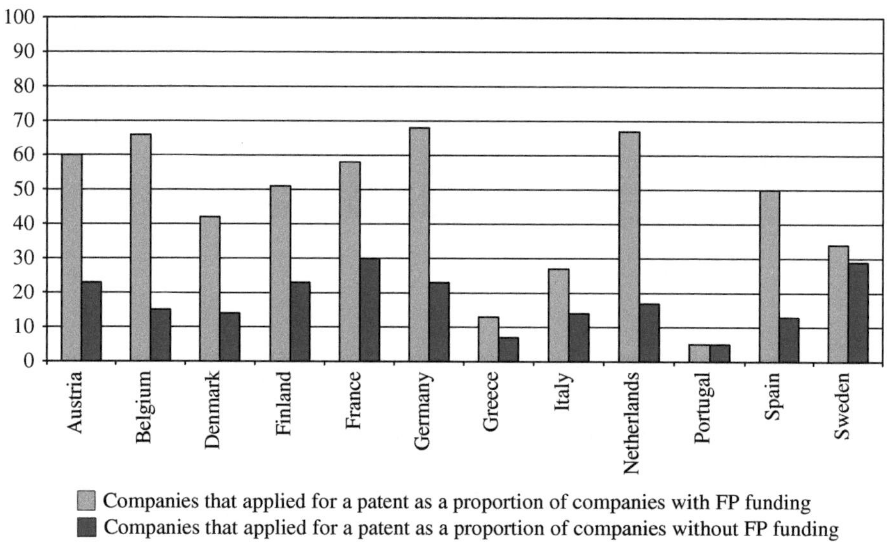

Fig. 4.20. FP participants are more likely to patent
Source: DG Research, Eurostat
Data: Eurostat
Note: Results here are for firms in the manufacturing sector

[111] European Commission, *Clean, Safe and Efficient Energy for Europe - Impact Assessment of Non-Nuclear Energy Projects Implemented Under the Fourth Framework Programme - Synthesis Report*, Luxembourg, 2003, p. 13.

[112] Databank Consulting *et al.*, *IST Impact Study*, p. 36; Arnold, *Evaluation Record*, May 2005, pp. 24–25.

3. THE IMPACT OF THE FRAMEWORK PROGRAMME ON HUMAN RESOURCES

One of the important effects of Framework Programme collaborative research projects is the enhancement of skills of participating research staff, which helps to strengthen research capabilities. Some Framework Programme actions also explicitly target the training, mobility, and career development of researchers, notably the "Marie Curie" actions. These commenced under the 3rd Framework Programme and now account for about 10 per cent of the 6th Framework Programme's budget. The purpose of the Framework Programme human resources schemes is to enable researchers to participate in top transnational teams and high-level projects, and benefit from training and knowledge sharing. This should ultimately lead to better R&D, while also having a positive impact on the attractiveness of the EU as a place to pursue a scientific career for European and third country researchers.

The Framework Programme human resources schemes are very much in demand. In the period 1994–2002, almost 12 000 researchers undertook international research training funded by the Marie Curie fellowships under the 4th and 5th Framework Programmes.[113] The 6th Framework Programme is characterised by high numbers of applications as well.[114] Widely appreciated and prestigious, it is useful to take a closer look at the characteristics of and impacts achieved through the Marie Curie fellowships.[115] Most fellows (79 per cent) applied to the scheme to gain international experience, and a substantial share of them (28 per cent) would not have gone abroad without a fellowship (which has a bearing on the question of additionality). In addition, the scheme enabled many fellows (50 per cent of post-graduate and about one-third of post-doctoral researchers) to gain their first international experience. Fellows mostly came from Spain (16 per cent), Italy (14 per cent), and Germany and France (13 per cent each). They mostly went to the United Kingdom (28 per cent), France (17 per cent), Germany (12 per cent), the Netherlands (9 per cent), and Spain and Italy (6 per cent each). This turned some countries (e.g. the United Kingdom, Denmark, Netherlands, Norway) into clear "host" countries (hosting more fellows than sending) and other ones (e.g. Slovak Republic, Iceland, Hungary) into clear "home" countries (sending more fellows than hosting), with the reminder reaching a rough balance. Most fellowships by far took place in an academic setting, though the 5th Framework Programme included a special fellowship programme for industry in which 8 per cent of respondents to a survey participated. About one-fifth of fellows were engaged in university–industry collaboration during their fellowship.

---

[113] Daphne van de Sande *et al.*, *Impact Assessment of the Marie Curie Fellowships Under the 4th and 5th Framework Programmes of Research and Technological Development of the EU (1994–2002)*, IMPAFEL 2 – Contract nr. IHP-D2-2003-01, June 2005, p. iv.

[114] European Commission, *Implementation Report 2004 on "A Mobility Strategy for the European Research Area" and "Researchers in the ERA: One Profession, Multiple Careers"*, Commission Staff Working Document, SEC(2005) 474, 6 April 2005, p. 33.

[115] The remainder of this section is based on van de Sande *et al.*, *Marie Curie Fellowships*.

Marie Curie fellows identified as most important impacts for themselves having gained international research experience, having had dedicated time to carry out research, and having developed research skills. Marie Curie fellowships generated a number of tangible research outputs. Scientific publications were produced, their number depending on the duration of the fellowship and the experience of the fellow. The longer they stayed, the more likely they were to publish. Of those on fellowships of up to 6 months (for 94 per cent at postgraduate level), 57 per cent produced at least one publication. Among fellows on fellowships of between 6 and 12 months (for 65 per cent at postgraduate level), the per centage rose to 77 per cent. And among fellows on fellowships of at least 12 months (for 92 per cent at postdoctoral level), it rose to 90 per cent. Looking only at postdoctoral fellows, the average publication rate was 3.1 publications per year for a fellowship at a university, 2.6 for one at a research centre, and 1.2 for one in industry. Of respondents who had already concluded their fellowship, 7 per cent reported that their research project had resulted in one or more patents (one-third owned by the fellow and two-thirds owned by the host organisation). Supervisors were of the opinion that in 43 per cent of cases, the work of the fellow had led to the creation of new technologies.

After their fellowship, a majority of fellows eventually returned to their home country, though planning a new mobility experience within five years. They generally returned to positions of more responsibility under more stable contracts. They took with them the networks established and reinforced through their fellowship. The scheme strengthened existing connections for 34 per cent of postgraduate researchers, 40 per cent of postdoctoral researchers, and 73 per cent of senior researchers. Over 70 per cent of fellows reported that their stay abroad generated new contacts influential for subsequent career progression, and 86 per cent maintained contacts with their Marie Curie host institution after the fellowship.

4. THE IMPACT OF THE FRAMEWORK PROGRAMME ON RESEARCH INFRASTRUCTURES

If Europe's research teams are to remain at the forefront of all fields of science and technology, then they will have to be supported by state-of-the-art infrastructures. The successive Framework Programmes have been active in this sense, promoting the development of a fabric of research infrastructures of the highest quality and performance in Europe, and their optimum use. The term "research infrastructures" refers to facilities and resources that provide essential services to the research community in both academic and/or industrial domains. Research infrastructures may be "single-sited" (single resource at a single location), "distributed" (a network of distributed resources, including infrastructures based on Grid-type architectures), or "virtual" (the service being provided electronically). Examples of research infrastructures include large-scale research installations, collections, special habitats, libraries, databases, integrated arrays of small research installations, high-capacity/high-speed communications networks (e.g. Géant), networks of computing facilities (e.g. Grids), and infrastructural centres of competence, which provide a service for the wider research community based on an assembly of techniques and know-how.

To develop a strategic approach for research infrastructures at European level, comprehensive and up-to-date information about the current pattern of existing infrastructures is essential. To this end, the Commission organised a *Survey of European Research Infrastructures* between December 2004 and January 2005. Existing research infrastructures of clear European dimension, as well as a certain number of infrastructures under construction, were invited to participate The first round of the survey resulted in feedback from 585 existing research infrastructures and 157 under construction, that is, in total 742 research infrastructures were reported.

The technical review panel for the access to research infrastructures action under the 5th Framework Programme concluded that it succeeded in providing researchers with access to unique or outstanding infrastructures in a broad range of disciplines, and thus contributed to the production of high quality research.[116] It also noted that the Framework Programme helped to develop interdisciplinary approaches between several areas as well as cooperation between European researchers, while giving a European dimension to several national facilities and, in several cases, organising more efficiently their networking. Through infrastructure RTD projects, it also contributed to improving the technical performance of several categories of infrastructures and thus to enhancing European scientific competitiveness in several fields.

A questionnaire to participants who received 5th Framework Programme funding for transnational access to infrastructures, carried out as part of this technical review, found that 88 per cent of respondents would have been unable to carry out their project at this research infrastructure without EU support. Of these, 71 per cent indicated that they would have been unable to pay for travel and subsistence, while 41 per cent could not have afforded the user fees.

5. THE FRAMEWORK PROGRAMME AND INTERGOVERNMENTAL COOPERATION AT THE PROGRAMME LEVEL

Although the vast bulk of public R&D spending in Europe is still carried out at national or regional level, these substantial resources remain fragmented and poorly coordinated. As was noted in the Communication on the European Research Area, "the European research effort as it stands today is no more than the simple addition of the efforts of the 15 [now 25] Member States and the Union. This fragmentation, isolation and compartmentalisation of national research efforts and systems and

[116] European Commission, *Enhancing Access to Research Infrastructures – Improving the Human Research Potential and the Socio-Economic Knowledge Base (1998–2002)*, Technical Review of IHP-ARI Contracts – Panel's Technical Review Report, January 2003.

the disparity of regulatory and administrative systems only serve to compound the impact of lower global investment in knowledge".[117]

Before the 6th Framework Programme, successive Framework Programmes focussed on fostering cooperation between research actors at "project" level by bringing together universities, research agencies, and companies. They were not used as vehicles for promoting the better coordination of national R&D programmes. This changed under the 6th Framework Programme when following the ERA initiative the first steps were taken to move cooperation forward to the next level by coordinating at programme level. The main tools proposed under the 6th Framework Programme for promoting the coordination of national and regional research at programme level were the ERA-NET scheme and the application of Article 169 of the EC Treaty.

In early 2002, EU research ministers recognised the importance of the mutual opening of national research programmes. CREST followed this impulse by launching five pilot actions for the mutual opening of national programmes in March 2002.[118] During the course of 2003, CREST came to the conclusion that the exchange of good practices between national programme managers was very useful, but that there was a need for a framework for further discussion. The ERA-NET scheme was considered to be the most suitable mechanism for pursuing the first stages of coordination.

The ERA-NET scheme was a highly innovative component of the 6th Framework Programme.[119] Its objective was to contribute to the creation of the European Research Area by facilitating practical initiatives to coordinate regional, national and European research programmes in specific fields, and to pool fragmented human and financial resources in order to improve both the efficiency and the effectiveness of Europe's research efforts. It provided support for creating close, long-term links between national research programmes with shared goals. The scheme's participants were programme funders and managers working in national and regional ministries and funding agencies rather than universities or enterprises. Ultimately ERA-NETs were expected to lead to collaboration of major significance, including the strategic planning and design of joint research programmes, the reciprocal opening of national research programmes to researchers from other member countries, and the launch of fully transnational programmes jointly funded by more than one country. Interest in the scheme was large, which is evidenced by the fact that by 2006, 229 proposals involving 2003 participants had been submitted. ERA-NETs were active in a wide range of fields.[120]

---

[117] European Commission, *Towards a European Research Area*, Communication from the Commission to the Council, the European Parliament, the Economic and Social Committee and the Committee of the Regions, COM(2000) 6 final, 18 January 2000.

[118] CREST stands for: Scientific and Technical Research Committee.

[119] This paragraph is based on the series of publications (four so far): European Commission, *ERA-NET – Networking of National Research Programmes in the European Research Area – Project Synopses.*

[120] For instance, Aeronautics; Agriculture; Astronomy; Biodiversity; Cancer clinical practice; Computational biology, Systems engineering, Informatics, mathematics; Coordination of national and regional activities; Energy; environmental science, environmental technology and Policy development; Ethical,

Another coordination mechanism supported through the 6th Framework Programme was Article 169 of the EC Treaty, which enables the participation of the Community in Member States' joint national R&D programmes. The use of this mechanism was approved by the Council and the European Parliament for the first time in 2003 and this for the European & Developing Countries Clinical Trials Partnership (EDCTP). The European Commission is supporting this long-term partnership between Europe and Developing Countries by providing €200 million for the development of new medicines and vaccines against HIV/AIDS, malaria and tuberculosis. The EDCTP brings together 14 EU Member States, Norway and Switzerland, Developing Countries, and other donors and industry. The main objective is to combat poverty-related diseases through more and better-structured R&D that meets the needs of the African populations in need.[121]

6. ASSESSING THE WIDER IMPACTS OF THE FRAMEWORK PROGRAMME

The above sections have dealt mainly with impacts that are produced within a relatively short period of time, and that are "manageable" from a data collection point of view because they are situated at the project level. Much more complex, however, is the assessment of the Framework Programme's wider impacts. It is well known that the effects of R&D can be spread out over very many years, and it is therefore important that research projects and programmes are not judged narrowly in terms of there immediate outputs. Moreover, the benefits of research are not confined to the R&D actors themselves, or to the beneficiaries of public funding (such as Framework Programme participants), but spill over across many actors, sectors of society, and the economy.

6.1. Wider economic impacts of community research policy

Measuring the economic impact of research is extremely difficult. Among the complicating factors are the long lead time between R&D activity and the eventual economic exploitation of the results, and the problem of unambiguously attributing effects on a firm's turnover to specific research projects. Even if one manages to

---

Legal and Social Aspects of Genomics; Humanities; Impact of and adaptation to climate change; Innovation policy; Integrated Water Resources Management; Marine Fisheries Research; Marine pollution; Marine water quality/food standards; Materials Science; Metrology; Nanoscience; Nanotechnology, micro-systems, microtechnology; Natural sciences, medical and life science, engineering, social sciences and humanities; Nursing; Occupational Safety and Health; Optics, photonics, optoelectronics and laser technology; Polar research; Population ageing; Renewable energy sources; Research, development and innovation; Road research; Shipping, Ship Building, Marine Engineering; SMEs: Collective Research; Social science; Transport.

[121] See www.edctp.org

estimate such effects, this will not provide the whole picture because there may be various indirect effects, as well as spillover effects on other actors.[122]

Nevertheless, efforts have been made to try and evaluate the wider economic impacts of the Framework Programme. One approach that has been usefully employed in the past is mathematical modelling. For example, a recent UK study estimated the impact of the Framework Programme on the United Kingdom's total factor productivity using a model developed at the OECD. It was found that application of the formula generated an estimated annual contribution to UK industrial output of over £ 3 billion, a manifold return on UK Framework activity in economic terms.[123]

The European Commission's Joint Research Centre at Ispra subsequently extended the analysis to all Member States for which data were available. The results seem to indicate significant effects on total factor productivity (Fig. 4.21). For example, for Finland, first estimates suggest that 0.9 per cent of the value added of industry per annum is attributable to funding from the Framework Programme, while many Member States record even higher contributions. It is estimated that on average €1 of Framework Programme funding leads to an (long-term) increase in industry added value of between €7 and €14, depending on the assumptions and

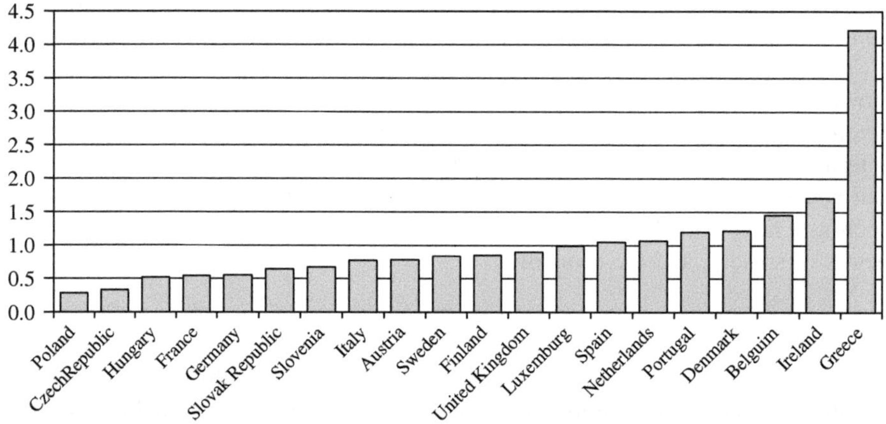

Fig. 4.21. FP5 returns to industry expressed as impact on total factor productivity (%)
Source: JRC, ISPRA

[122] For instance, calculating net economic effects at the macro-level on the basis of private, gross effects measured at the participant level would require taking full account of indirect effects (intended and unintended) and carry-over effects to arrive at private net effects, and taking full account of spillovers (effects on non-participants) and crowding-out effects to arrive at net economic effects – see Joanneum Research Forschungsges.m.b.H *et al.*, *Austrian Participation*, pp. 26–27.

[123] DTI - Office of Science and Technology, *Targeted Review*, p. 8. This study uses the model developed at the OECD by Guellec and van Pottelsberghe, which is presented in Dominique Guellec and Bruno van Pottelsberghe de la Potterie, *R&D and Productivity Growth: Panel Data Analysis of 16 OECD Countries*, STI Working Papers 2001/3, 14 June 2001.

parameters used. This increase will be spread over a number of years, because there is always a time lag before R&D spending has its economic effects. Of course, it is also important to stress that econometrics is not an exact science, and that such results must be treated with a large degree of caution. Other evidence on the positive macro-economic impact of the Framework Programme comes from the projections regarding the macro-economic impact of the 7th Framework Programme carried out via the Némésis model and discussed in a later chapter in this book (Chapter 6), as well as from BRITE-EURAM impact studies on a selection of projects. In the latter, it was found that nearly 1 600 new jobs were created and another 1 000 safeguarded.[124]

6.2. Wider social and environmental impacts of community research policy

It is not possible to present – like in the case of macro-economic impacts – in an aggregate way the many social and environmental societal challenges to the meeting of which FP-funded research projects have contributed directly. Only an extensive enumeration of case studies would be able to provide a proper insight into the vast range of areas in which valuable contributions have been made. It is important to note, however, that ever since the 5th Framework Programme the Framework Programme has been more explicitly positioned as having to contribute to the solution of such societal problems. In other words, ever since the 5th Framework Programme, research has become more mission oriented than curiosity driven.

Some examples of the contribution made by the Framework Programme to societal and environmental goals include the following:

- Developing the sophisticated analytical tools needed to predict and optimise the net environmental, economic, and social impacts of new policy measures. For example, 3E – a powerful computer model of interdependent economic, environmental, and energy variables – enabled EU negotiators at Kyoto to calculate the cost implication of different policy scenarios on the spot;
- Producing forecasting systems to give early warning of flooding, improving water treatment technologies, and planning tools to protect coastal zones against erosion and pollution;
- Creating many of the new technologies on which sustainable development will depend – for instance, new sensor systems to monitor levels of pollutants in the rivers that supply our drinking water;
- Strengthening the environmental research infrastructures which underpin European work in priority areas like climate change, encouraging coordinated transnational use of complex data sets, and state-of-the-art facilities.

The inclusion of socio-economic research in the Framework Programme has surely made it easier for European researchers to help meet the many social and environmental challenges Europe is facing. Through the Targeted Socio-economic

[124] European Commission, *Brite-Euram – Making a Lasting Impression on Europe*, Luxembourg, 2002.

Research (TSER) Programme under the 4th Framework Programme, the Key Action "Improving the Socio-economic Knowledge Base" under the 5th Framework Programme, and the Thematic Area "Citizens and Governance in a Knowledge-based society" under the 6th Framework Programme, efforts have been made to improve the understanding of the major structural changes taking place in European society, to identify ways of managing these changes, and to involve European citizens more actively in shaping their own future. Issues related to the emergence of a knowledge-based society and new forms of relationships between citizens on the one hand and between citizens and institutions on the other are being addressed. The activities carried out in these programmes mobilise the wealth and diversity of European research in the economic, political, and social sciences, and the humanities.

Wider social and environmental benefits have also been generated by the fact that policy-makers are increasingly well-informed and policies are increasingly evidence-based. This is the result of three different trends. First, there is the aforementioned inclusion of socio-economic research in the Framework Programme. Second, there is the inclusion in the Framework Programme of the so-called "scientific support" for policies. Third, there is the increasing willingness to base the development of new policies on the results of FP-funded research projects. Research serves more and more as the knowledge-base referred to in key policy documents.[125] The key internal provider of such scientific support for EC policies is the Joint Research Centre (JRC). Its role as a Community reference for EU policy in scientific and technological questions is epitomised by the increasing volume of legislation which is based on the work of (and which mentions) the JRC.[126] Thus European BSE (Bovine Spongiform Encephalopathy) research, to which the Community contributed €90 million in the period 1996–2003, provided the basis for close to 300 scientific opinions in support of almost 40 pieces of Community legislation in the fields of consumer protection, public health, and risk management. This research was proactive and explorative in nature, and thus enabled a rapid response to increased consumer food threats. In fact, the announcement in March 1996 of the links between BSE and the new variant Creutzfeld-Jacob Disease was the result of Community-funded collaborative research. Even more recently, with the outbreak of the SARS epidemics, Community action proved extremely adaptive and prompt. Some 70 projects under the Environment and Sustainable Development Programme of the 5th Framework Programme were explicitly referred to in various

[125] See notably European Commission, *The Overall Socio-Economic Dimension of Community Research in the European Framework Programme,* Luxembourg, 2003.

[126] In 2003, the JRC provided scientific and technical support to over 80 pieces of EU legislation. More widely, it should be noted that the JRC implements its mission through Direct Actions under the nuclear and non-nuclear Framework Programmes (FP). Thematically, the focal points are spelled out in the Specific Programmes (nuclear and non-nuclear parts) and are further defined in the Multi-Annual and Annual Work Programmes. To a lesser degree the JRC also uses Indirect Actions under the FPs, additional work for customer DGs, Enlargement Actions and Third Party Work to implement its mission. See also: http://www.jrc.ec.europa.eu

EU policy documents. This provides an indication of the actual exploitation of research results for policy support.[127] A case in point is also the European Climate Change Programme (ECCP), which explicitly recognised the role of research in the energy and environmental fields. The preparatory work that led to the adoption of the Directive establishing a scheme for greenhouse gas emission allowance trading was supported by the results of the PRIMES and POLES models developed under previous Framework Programmes.[128]

A number of societal issues (ethics, gender issues, environmental issues) have also been put on the agenda via the Framework Programme proposal evaluation process. Since the 5th Framework Programme, each project's environmental targets and objectives must be detailed in the project proposal. This plays an important role in the evaluation and selection of projects to be funded. Nonetheless, the detail provided on these targets and objectives is seldom sufficient to allow for subsequent systematic monitoring and evaluation at the programme level. Therefore, it is difficult to quantitatively assess the environmental effects of these projects (e.g. in the form of reduced emissions, or decreased health risks, etc.) at an aggregate level. Even so it is widely recognised that most RTD projects funded by the Framework Programme generate either directly or indirectly positive environmental impacts.

6.3. The Framework Programme and the EU's international standing

As discussed in the beginning of this chapter, many industrialised countries outside Europe participate in the Framework Programmes. Some – including the United States, Canada, Israel, and Australia – have already signed S&T agreements with the EU. For these countries cooperation with Europe is attractive. A study commissioned by the Australian Government Department of Education, Science and Training (DEST) on Australian science and technology cooperation with Europe, for instance, concluded that Australian researchers want to work with Europe because it is seen as a site of leading-edge collaboration, and that there is a large potential for expanded collaboration.

International S&T cooperation with developing countries via the Specific International Scientific Cooperation Activities (INCO) programme of the Framework Programme established in 1983 starts from the concept of sustainable development and the idea that poverty and social marginalisation can be overcome successfully by investing in human and institutional resources. The INCO programme is

---

[127] Among those, a few (ca. 15) relate to Written Questions from MEPs or/and to Annual Reports of JRC, while the majority point to actual policy documents (13 different policy documents often citing several research projects). For more details see Ricci Report, 2004.

[128] As mentioned in European Commission, *Green Paper on Greenhouse Gas Emissions Trading within the European Union*, Presented by the Commission, COM(2000) 87 final, 8 March 2000 and European Commission, *Proposal for a Directive of the European Parliament and of the Council Establishing a Scheme for Greenhouse Gas Emission Allowance Trading within the Community and Amending Council Directive 96/61/EC*, Presented by the Commission, COM(2001) 581 final, 23 October 2001.

based on dialogue and promotes the development of long-term durable research partnerships, with four main regions: Latin America, Asia, and Africa; the Mediterranean countries; the Western Balkans; and the Newly Independent States.[129] Though difficult to quantify, the impact of the Framework Programme in these regions is fourfold. First, research projects contribute directly to meeting local challenges. Positive impacts concern, for instance, scientific job creation, substitution of imported goods, improved food quality, and environmental protection from rapid industrialisation and urbanisation. Impacts of Asian and African INCO-funded projects in the fields of agriculture, food, and health concern improvements in farmer incomes, gender equality, public health, education, employment, protection against erosion, conservation of natural resources, and training and professional improvement. Second, Framework Programme projects also strengthen the local capabilities to meet local challenges.[130] Third, Framework Programme projects inform local policy development.[131] And fourth, Framework Programme projects promote regional collaboration.

CONCLUSION

Assessing the impacts of research programmes is complex, and even more difficult in the case of the Framework Programme. Nevertheless, a number of definite things can be said about past Framework Programmes. Across Framework Programmes, the collaborative research component has been characterised more by growth in the number of projects and participations than by growth in the average size of the project, raising the issue of the optimal size of projects and critical mass. The degree of industrial participation has become smaller, while that of universities and research institutes has increased. The by-country ranking in terms of participations and funding is more or less in line with that in terms of GDP or number of researchers, although smaller countries prove to be more "efficient" in obtaining funding than larger ones. The by-country ranking in terms of Framework Programme participations has not changed very much across Framework Programmes. The same finding is valid for the regional distribution of participations and funding: the regions with the highest income levels and/or the highest research activity are the front runners. However, a large number of peripheral regions are allocated a larger share of European research funds than could be expected on the basis of their R&D effort, economic importance, and so on. Regarding individual participants, it was shown that a relatively small number of organisations participate in a relatively large number of projects across calls and across Framework Programmes, collecting most

---

[129] See also the INCO portal: http://cordis.europa.eu/inco/home_en.html

[130] For example, under FP5, 344 research projects mobilised 461 teams from 19 Non-EU partners from the Mediterranean Basin together with their European peers. The projects addressed the key challenges in this extremely water-scarce region.

[131] See INCO impact study and high impact projects analysis carried out in support of the five-Year Assessment.

participations and most funding. Collaborative research projects establish an ever-increasing number of cross-sectoral and cross-border collaborative links. Networks remain remarkably stable after the end of projects and have a multiplier effect on a country's returns from the Framework Programme. Collaborative research has been characterised by an ever-increasing number of thematic priorities, avoiding major gaps in terms of coverage, though this does not seem to have caused fragmentation. Framework Programme collaborative research is generally considered to be of high quality, though not sufficiently risky and at the frontier. Community-funded collaborative research is characterised by a very high degree of additionality. In the absence of Framework Programme funding, projects would not have been carried out at all or would have been carried out in a very different manner with reduced objectives and a smaller number of partners. Framework Programme collaborative research has clear impacts on capabilities, behaviour, and competitivity, and on scientific and innovative performance. The Framework Programme also contributes to human resources development and has a structuring effect through its coordination of national research programmes.

We have seen in this chapter that learning from past experiences and identifying the strengths and weaknesses of previous Framework Programmes draws upon the evidence of ex-post evaluations. And such evaluations rely to a large extent on good and robust data. In the past few years, the Commission has undertaken important efforts to improve the completeness and consistency of cross-Framework-Programme-participation data. These ongoing efforts need to be continued and consolidated. Special attention should also be paid to developing innovative, including qualitative, approaches for assessing project outcomes.[132]

When conceiving its new Framework Programme, the Commission did not only look backwards to the experiences gained with previous Framework Programmes. It also started an open broad-ranging dialogue with the key actors who would be affected or involved. This will be the subject of the following chapter.

---

[132] In this context, it needs to be noted that the Commission is supported by the Member States. For instance, in early 2006, a group of 20 Member States – on the initiative of the German Ministry for Science and Research – submitted a joint paper to the Commission outlining ideas of how the reporting system on the Framework Programme implementation could be improved. See letter from the German Ministry for Science and Research to DG Research, together with a joint paper on "Future Supply of FP Data and Statistics to Member States and Associated Countries." 05.01.2006. Internal document.

CHAPTER 5

ENGAGING STAKEHOLDERS FOR BETTER POLICY-MAKING

An important element in the design of a good research programme is to listen to the stakeholders. These are the key players which are affected by the programme – be they researchers in the field, policy makers, users of research results, or interested individuals – , coming from a variety of sectors including public administrations, research institutes, universities, large companies, SMEs, international organisations, and so on. Canvassing their views not only increases the openness and transparency of policy-making, but also strengthens the quality of the policy itself.

This chapter describes how stakeholders were involved in the preparation of the 7th Framework Programme proposal. The Commission gathered their views through a combination of consultation tools. Alongside conferences, workshops, meetings with advisory committees, and so on, it used online stakeholder consultations. Although the consultations did not always provide clear-cut solutions, they did reveal different opinions and concerns which could be addressed further within formalised groups or expert meetings. Overall, the process helped to improve the dialogue between European science and its stakeholders and created ownership of the policy design process. This in turn made it easier to build consensus in the formal decision-making process (Chapter 7).

After a general introduction to consultation mechanisms employed at Community level (Section 1) and an overview of stakeholder involvement in the design of the Framework Programmes (Section 2), this chapter analysis the responses to three online stakeholder consultations.[1] Four specific areas are selected: the need to strengthen research at European level (Section 3), the call for improved accessibility and simplification (Section 4), the debate on the European Research Council and Technology Platforms (TPs)/Joint Technology Initiatives (JTIs) (Section 5), and finally, the challenge of improving the research performance throughout the Union (Section 6).

---

[1] Consultation on *Science and Technology, the Key to Europe's Future – Guidelines for Future European Union Policy to Support Research*, Consultation on *Research Themes in FP7*, Consultation on *FP7: Rules for Participation/Simplification Measures*. See: http://ec.europa.eu/research/ consultations/list_en.html (last accessed on 28 August 2006).

1. CONSULTATIVE MECHANISMS AT COMMUNITY LEVEL

1.1. Reinforcing the link between the Commission and its stakeholders

Consultation mechanisms form part of the legislative work of all European Institutions, from policy-shaping prior to a Commission proposal to the implementation of a legal act.[2] First and foremost, decision-making in the EU is formalised by the interaction of the legislative institutions which transform proposals in legal decisions.[3] Both the Council, representing Member States, and the European Parliament, representing EU citizens, possess democratic legitimacy.

However, this formalised institutional interaction does not exclude direct contact between the Commission and its stakeholders. In fact, it is one of the Commission's obligations according to the Amsterdam Treaty "[to] consult widely before proposing legislation and, wherever appropriate, publish consultation documents".[4] Over the last decade, the input from stakeholders has become even more and more important as a means to tackle a perceived "democratic deficit" of the EU.[5] As a result of this pressure, a number of measures have been taken to reinforce this interaction.[6]

Open dialogue between the European Commission and its stakeholders has many advantages. It helps ensure that rules and regulatory proposals keep pace with the reality of rapid change and allows the Commission to draw from a wide source of information, facts, and opinions to meet the challenge of policy-making under conditions of increasing complexity, policy interdependence, and time pressure.[7] This exchange of ideas not only improves the quality, effectiveness, and responsiveness of their policies, but also meets the growing stakeholders'

---

[2] European Commission, *General Principles and Minimum Standards for Consultation of Interested Parties by the Commission*, COM(2002) 704, 11 December 2002, p. 4.

[3] The EC Framework Programme is adopted under the co-decision procedure (Chapter 7).

[4] Protocol (No 7) on the application of the principles of subsidiarity and proportionality, annexed to the Amsterdam Treaty.

[5] The term "democratic deficit" is usually used to refer to "organizations which are democratic to some extent, but are not as democratic as they could be". Communication is still too much considered as a "Brussels affair", focussed too much on institutions instead of people and limited to telling what the EU does ("deficit model") instead of listening to people's views (two-way models and the notions of "upstream" public engagement). http://www.reference.com/browse/wiki/Democratic_deficit (last accessed on 28 August 2006).

[6] Among others, European Commission, *European Governance: A White Paper,* COM(2001) 428, 25 July 2001. European Commission, *Action Plan "Simplifying and Improving the Regulatory Environment",* COM(2002) 287, 10 June 2002. European Commission, *White Paper on a European Communication Policy*, COM(2006) 35 final, 1 February 2006. European Commission, *Action Plan to Improve Communicating Europe by the Commission* SEC(2005) 985 final, 20 July 2005.

[7] OECD, *Citizens as Partners.* OECD Handbook on Information, Consultation and Public Participation in Policy-Making, Paris 2001, pp. 18–20.

expectation that their views are taken into account.[8] The early involvement of stakeholders in the preparation of Community actions also increases acceptance and ownership of proposed strategies by the stakeholders, which in turn make it easier to find consensus during the formal institutional decision-making process.[9]

1.2. The European stakeholders' consultation scene

The link between policy makers and stakeholders is not a straightforward, linear process with stakeholders providing policy-makers a collection of opinions that have a direct impact on policy-makers' decisions. Rather, the process is dynamic, complex, and interactive. These interactions take many forms and are influenced by cultural differences, institutional pressures, networks, relationships, powers, trust, and so on.[10] The multi-level structure of European governance poses additional challenges for the design and implementation of participatory processes.[11] At national level alone, the Union brings together 25 Member States of different size, with diverse economic and environmental, social and cultural backgrounds, political cultures, participatory traditions, and so on. It is therefore challenging, to say the least, for policy-makers to build a consensus proposal within this complex setting and under severe budget and time constraints.

Moreover, there is generally no pre-defined set of actors who have to be involved in a participatory process. The Union considers "any person or organisation with an interest in or affected by EU legislation and policy making as a "stakeholder" in that process".[12] In other words, stakeholders are identified according to the problem at stake and the purpose of the policy action.[13] Some consultations therefore canvass the attitudes of European citizens in general, while others prefer to focus on a particular group of actors with professional knowledge of a particular topic. Accordingly, each Commission Directorate-General sets up its own mix of (complementary) consultation tools to sound out stakeholders' opinion (Box 5.1).

---

[8] OECD, *Integrating Multiple Interests into Policy,* Meeting of senior officials from centres of government on integrating multiple interests into policy (30 September–1 October 2006), PUMA/MPM(96)2, 2006, p. 3.

[9] OECD, *Integrating Multiple Interests into Policy*, p. 5.

[10] Pronéos in collaboration with Corvinus University of Budapest, Comenius University and Observatoire des Sciences et des Techniques, *Private Sector Interaction in the Decision Making Processes of Public Research (Draft)*, Study for the European Commission – DG Research, 14 March 2006, p. 27.

[11] Henning Banthie *et al., Governance of theEuropean Research Area: The Role of Civil Society.* Conference Proceedings, Brussels, 20 October 2003, pp. 47–48.

[12] http://europa.eu.int/abc/eurojargon/index_en.htm

[13] Banthien *et al., Governance of the European Research Area*, p. 8.

Box 5.1. The Commission's Consultation Toolbox

The oldest instrument giving voice to the concerns of public opinion is the *Eurobarometer opinion poll system.*[14] Since 1973, the Eurobarometer system monitors the evolution of public opinion in the Member States on major topics concerning European citizenship including enlargement, environment, the Euro, culture, and so on with the help of this system. The standard Eurobarometer report is published twice yearly and consists of approximately 1, 000 face-to-face interviews per Member State. If needed, it can be supplemented with more in-depth thematic studies (Special Eurobarometer), quick ad hoc thematic telephone interviews (Flash Eurobarometers), and qualitative studies investigating in a more in-depth way the motivations and feelings of a particular group on a certain topic.[15]
While the Eurobarometer system is organised in a centralised manner at Commission level and (indirectly) provides useful information from citizens for the policy-making process, each Commission Directorate-General also has direct contact with interested parties in their respective fields. This decentralised structure has the advantage that the specific nature and conditions of different policy fields can be taken into account. The classical approach to consult interested parties is *formal stakeholder consultation* which is based on the publication of a consultation document (notably Green paper or Commission staff working paper). This document sets out the Commission's policy orientation and/or possible courses of actions and invites the parties to comment on the ideas expressed in the paper. The Commission has also expanded the use of Internet-based tools as a means to collect and analyse reactions from a broad audience. Most formal stakeholder consultations now make use of the *Interactive Policy Making (IPM) initiative* (also referred to as *online stakeholder consultations*).[16] The IPM tools are available via a web portal "Your Voice in Europe", the Commission's single access point to all consultations as well as a wide variety of discussions and other tools.[17]
In addition, workshops, conferences, and seminars are regularly organised, while expert groups and advisory committees foster an exchange of information and opinions between stakeholder groups and the Commission on a particular issue.

[14] To read more about Eurobarometer surveys: http://europa.eu.int/comm/public_opinion/index_en.htm

[15] A Special Eurobarometer exists on "Europeans, Science and Technology" and is intended to gather information on Europeans' public understanding of and attitude towards S&T. The most recent Eurobarometer on Europeans, Science and Technology was published in June 2005: European Commission, *Europeans, Science and Technology*, Special Eurobarometer (224/Wave 63.1 – TNS Opinion & Social), June 2005.

[16] The IPM initiative involves the development of two Internet-based mechanisms: a feedback mechanism which collects spontaneous reactions, and a consultation mechanism which is designed to receive and store structured collection of reactions to new initiatives. The latter is implemented through questionnaires via which the stakeholder is confronted with several multiple-choice questions and is given the opportunity to provide a free-text response to certain more specific questions.

[17] http://europa.eu.int/yourvoice/ (last accessed on 28 August 2006).

Although the online stakeholder consultations are increasingly used and provide a rich pool of opinions, a major challenge of those consultations is the quality and quantity of the responses as they cannot ensure that the target population will respond.[18] Indeed, given the "digital divide", governments should not rely exclusively on ICT in reaching stakeholders, and this raises questions on how to provide equal access for everybody.[19] All together, these hurdles can cause a bias towards particular groups of respondents. This is why the Commission tends to set up a mix of complementary tools to communicate and interact with stakeholders.

1.3. Stakeholders' consultation and their impact on decision-making

Consultations often reveal different views on how issues should be addressed, with different actors seeking recognition of their specific interests and proposals. Not all stakeholders and their representatives are in an equal position either.[20] The opinion of advisory groups, government bodies, industry associations, or well-organised lobby groups may carry more weight than the opinion of unorganised minorities, while the views of a large multinational may have more influence than those of a small unknown company or of an individual citizen. The example of the REACH proposal starkly illustrates how industrial interest groups used the stakeholder consultation to express their concerns. Apart from the opposing interests and different power-constellations between stakeholder groups, there is also a large diversity within the stakeholder groups themselves. The opinion of "the" large companies, "the" SMEs, or "the" government bodies again do not exist as such. These labels only regroup actors sharing some basic characteristics, agreeing on some points but maybe diverging on many others. While this regrouping is an attempt to cover up this diversity, this can also be considered as a first step towards consensus-building where common interests are identified and points of divergence can be taken into account.

Overall, consultation is a democratic process allowing actors to express their opinions. However, because of the complexity outlined above, more involvement does not always mean better decisions.[21] These difficulties should not be seen as arguments against stakeholders' participation, but rather as a plea to confront this diversity.[22] Policy-makers must make a conscious decision to what extent they

---

[18] Pronéos *et al.*, *Private Sector Interaction in the Decision Making Processes*, p. 40.

[19] Most commonly, the digital divide describes "the gap between individuals, households, businesses and geographic areas at different socio-economic levels with regard both to their opportunities to access ICT and to their use for a variety of reasons" (OECD, 2001, Understanding the digital divide). OECD, *Promise and Problems of E-Democracy – Challenges of Online Citizen Engagement*, Paris, 2003, p. 60–64.

[20] Pronéos *et al.*, *Private Sector Interaction in the Decision Making Processes*, p. 21.

[21] Pronéos *et al.*, *Private Sector Interaction in the Decision Making Processes*, p. 2.

[22] Banthien *et al.*, *Governance of the European Research Area*, p. 72.

want to base their policy decision on expressed needs and/or its policy proposals. The wide spectrum of policy objectives and public budget constraints often imposes compromises in which not all comments or suggestions can be taken into account.[23]

But it would be wrong to evaluate consultation processes only in terms of the "direct" impact on decision-making. They can also have an "indirect" impact by stimulating communication or debate, setting the political agenda, removing a blockade within the policy process, contributing to an informed decision-making process or testing the acceptance of draft policy actions. So, even if no direct impact can be discerned or no real representative sample of opinions was gathered, dialogue and participation still contribute to a valuable culture of problem-solving, for example by changing attitudes within stakeholder groups.[24]

2. STAKEHOLDERS' INVOLVEMENT IN THE DEVELOPMENT OF THE 7TH FRAMEWORK PROGRAMME PROPOSAL

2.1. The 7th Framework Programme: A broad consultation

Until the 5th Framework Programme, there was only limited input of civil society and its organisations. The Framework Programme was mainly based on institutional consultation (Parliament and Council) and the consultation of experts and corporations.[25] However, the debate on the reform of governance in the EU as well as the desire of the research community and the public to become more involved in S&T policy design had significant consequences for research policy-making.

In 2000, a Commission working document was published on "Science, society and the citizen in Europe" focussing on the question, among others, "how to fully involve society in shaping the research agenda?"[26] Following this, the Science and Society Action Plan – contributing to the implementation of the White Paper on European Governance – was issued in 2002 by the Directorate-General for Research. It formulated 38 actions to make science more accessible to European citizens.[27]

---

[23] OECD, *Integrating Multiple Interests into Policy*, p. 11.

[24] Banthien *et al.*, *Governance of the European Research Area*, p. 68.

[25] Banthien *et al.*, *Governance of the European Research Area*, p. 83.

[26] European Commission, *Science, Society and the Citizens in Europe,* SEC(2000) 1973, 14 November 2000.

[27] Of particular importance are actions 22 and 23, which suggest stimulating a closer interaction between stakeholders and policy-makers. Action 22: "The Commission will organise, through workshops and networks, an exchange of information and best practice between Member States and the regions on the use of participatory procedures for national and regional policies". Action 23: "The Commission will organise regular events enabling civil society to participate (in the form of public hearings, consensus conferences or interactive online forums) in specific issues (biotechnology, environment information technologies, health, innovation, etc.), in cooperation with the Economic and Social Committee and the Committee of the Regions" (European Commission, *Science and Society Action Plan,* COM(2001) 714, 4 December 2001).

In general, citizens express their wish to see more public involvement in decisions about science and technology.[28] Communication with civil society not only has advantages for citizens, who will be better informed on ongoing research activities, but also for the researchers themselves, who will be better informed about societal needs and will benefit from a high level of confidence within society.[29] Involving the research community – that is, the (potential) participants in the Framework Programme – in designing European research policy helps the Directorate-General Research to improve its policy actions and to translate the Specific Programmes into effective Work Programmes.[30]

Against this background, the 6th Framework Programme marked the start of a new period aiming at harmonious and better-structured relations between science and society. An innovative element was the broad consultation of the research community through "Expressions of Interests".[31] To sound out the acceptance and popularity of the New Instruments, the Commission invited the research community to submit Expressions of Interests on research actions using the New Instruments prior to the adoption of the 6th Framework Programme proposal. It received not less than around 11, 700 Expressions of Interests which provided relevant information to draft a sound Work Programme. It also helped researchers to become familiar with the characteristics of the New Instruments and their applicability in the 6th Framework Programme.[32]

In the preparation of the 7th Framework Programme proposal, the views of the scientific community, industry and other interested parties have been gathered through a combination of consultation tools – alongside the more institutional fora (Fig. 5.1). While the advisory committees made up of government officials and (internal and external) experts provide a permanent contribution to the development of RTD policy at Community level (such as the Economic and Social Committee (EESC), Committee of the Regions (CoR), European Research Advisory Board (EURAB), Scientific and Technical Research Committee (CREST), etc.),[33] the Directorate-General Research also tried to engage the broad range of stakeholders who are involved in research activities, research policies, and the use of

[28] 58 per cent of citizens do not agree with the statement that "the public is sufficiently involved in decisions about S&T" (European Commission, *Europeans, Science and Technology*, Special Eurobarometer, June 2005).

[29] Banthien *et al.*, *Governance of the European Research Area*, p. 106.

[30] Work programmes are plans drawn up by the Commission for the implementation of the specific programmes of the Framework Programme. They comprise detailed descriptions of the activities planned and (instruments used, evaluation procedures and criteria, deadlines, roadmaps) provide all information necessary to launch calls for proposals.

[31] Banthien *et al.*, *Governance of the European Research Area*, p. 83.

[32] European Commission, DG Research, *Report on the Analysis of Expressions of Interest Received in Response to the Invitation published on 20 March 2002 with a deadline of 7 June 2002*, Preparing for FP6 (Identifier EOI.FP6.2002), 2002.

[33] The European Economic and Social Committee and the Committee of Regions are institutionalised advisory bodies established to assist the Commission, the Parliament, and the Council. They consist respectively of representatives of the economic and social components of the organised civil society and representatives of regional/local authorities.

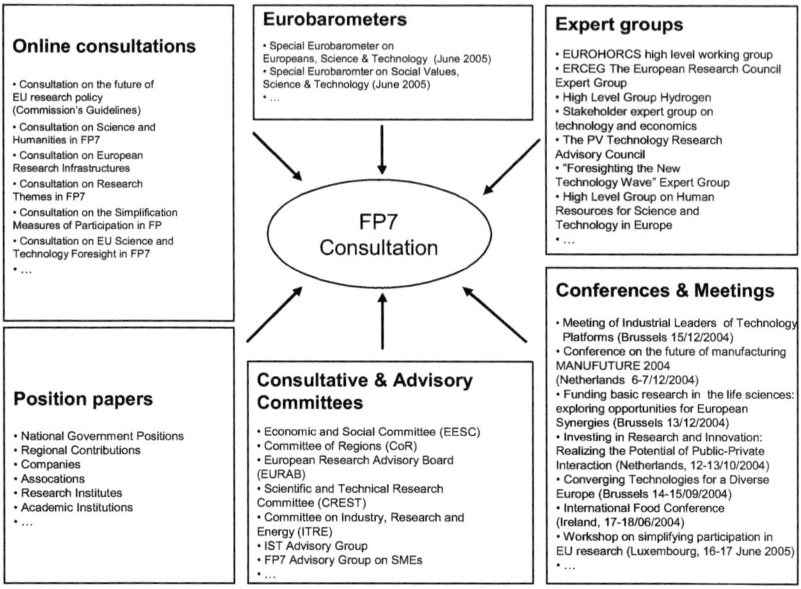

Fig. 5.1. The 7th Framework Programme – a combination of consultation tools
Source: DG Research

research results, including public administrations, research institutes, universities, large companies, SMEs, associations, international organisations and interested individuals. Several conferences and workshops were organised on a regular basis and for the first time the innovative Internet tools set up by the Interactive Policy Making (IPM) initiative were used to complement the formal stakeholder consultations. IPM complements the traditional formal consultation where stakeholders have the possibility to submit written responses (position papers) to the publication of a consultation document. With the help of a structured questionnaire, online consultations confront respondents with several multiple-choice questions and give them the possibility to write free-text responses on a particular issue.

2.2. *Roadmap to the 7th Framework Programme*

In June 2004, the Communication "Science and Technology, the Key to Europe's Future – Guidelines for Future European Policy to Support Research" launched the extensive debate on the future of European research (Fig. 5.2).[34] Within the Institutions, the Guidelines were debated in the Council (in particular the September and November Competitiveness Councils in 2004), the European Parliament (Committee on Industry, Research and Energy (ITRE)), the Committee of Regions, and the

[34] European Commission, *Science and Technology, the Key to Europe's Future – Guidelines for Future European Union Policy to Support Research,* COM(2004) 353 final, 16 June 2004.

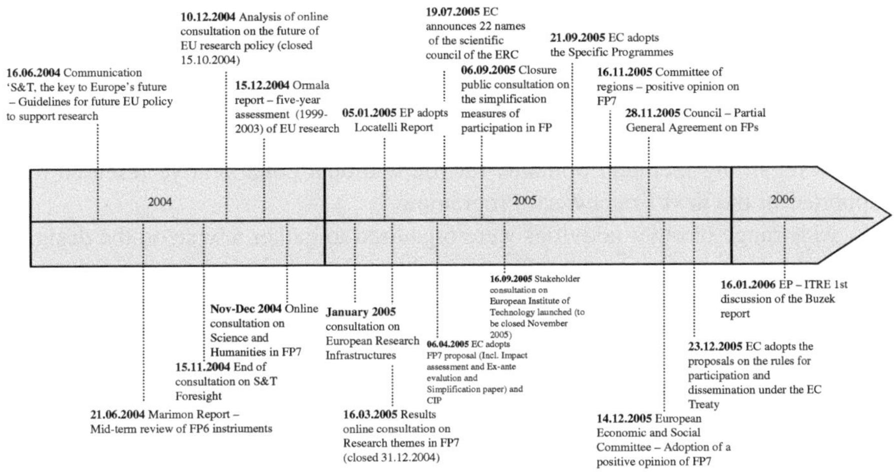

Fig. 5.2. Roadmap to the next Framework Programme
Source: DG Research

European Economic and Social Committee. To allow other individuals and organisations with an interest in European research to provide their views, the Commission launched an online stakeholder consultation on the Commission's Guidelines from July to October 2004. An online questionnaire asked respondents to rank the importance of the objectives set out in the Communication, to point out their level of agreement with the text of the Communication and to indicate their view on the potential impact of the proposals. For individuals, organisations, and associations who wished to provide more detailed views, it was facilitated to submit position papers and other written contributions. To further promote and inform the debate, a number of additional working documents were made available on the Internet during the autumn of 2004, for example on Technology Platforms and the principles of a European Research Council.[35]

Following the more general consultation on the Commission's Guidelines, the Commission organised a number of focussed consultations to identify those thematic areas of research to be funded under the 7th Framework Programme.[36]

---

[35] European Commission, *Technology Platforms - From Definition to Implementation of a Common Research Agenda,* Luxembourg, September 2004. European Commission, *Commission Interim Working Document on the Implementation of a Funding Mechanism for Basic Research*, 29 September 2004.

[36] To aid this process, a website was established providing links to all relevant consultations and setting out criteria (contribution to EU policy objectives; EU research potential; and EU added value) used by the Commission to identify which areas of research would be given priority in the 7th Framework Programme proposal. More information on the research themes for FP7 and the criteria used to identify thematic domains can be found at: http://europa.eu.int/comm/research/future/themes/index_en.html (last accessed on 28 August 2006).

Internet-based consultations were organised in the areas of information and communication technologies, nanotechnology, science and technology foresight, and the social sciences and humanities. In addition, another online thematic stakeholder consultation was undertaken by the Commission during November and December 2004 to allow interested parties (individuals and organisations) to submit contributions regarding thematic domains for transnational collaborative research to be supported in the next Framework Programme.

A wide range of other activities were organised to gather advice on the design of the new Framework Programme. For example, an expert group was appointed in the summer of 2004 to advise the Commission on its preparations for a new mechanism to fund basic research; the Commission organised conferences around diverse thematic areas (Manufacturing, Life Sciences, Food, etc.); a major conference was held under the Dutch presidency on "Investing in Research and Innovation" in October 2004, and a large number of meetings were held at national level on issues such as TPs.

The consultation process has not been brought to a standstill after the adoption of the 7th Framework Programme proposal by the Commission in April 2005. During the institutional negotiations on the new Framework Programme (Chapter 7), stakeholders were invited by the Commission to take an active role in the further development and implementation of the new Framework Programme. A working paper dealing with a first range of simplification measures – accompanying the 7th Framework Programme proposal – was the start to a more detailed debate on simplification.[37] In March 2005, a sounding board with representatives of SMEs and small research teams was set up to identify barriers to their participation and a workshop on simplifying participation in EU research was hosted jointly by the Luxembourg presidency and the European Commission in June. To gather more comments, an online stakeholder consultation on the proposed simplification measures was organised from June until September 2005. This consultation also welcomed input from all participants on other issues that would be addressed by the Rules for Participation such as Intellectual Property provisions, evaluation criteria, financial collective responsibility, and other participation principles.

More than 3000 institutions and individuals replied to the three online consultations – on the Commissions' Guidelines, Thematic Areas and Simplification – alone (see Annex 1 for a detailed breakdown by country and type of respondent). From this rich feedback many important issues emerged. The following sections will look at four of themes which were especially significant and resonant for the community of research stakeholders: the need to strengthen support for research at the European level; the calls for simplification of the Framework Programme;

---

[37] European Commission, *Simplification in the 7th Framework Programme,* SEC(2005) 431, 6 April 2005.

the importance of supporting both basic and applied research; and the debate on excellence versus cohesion addressing the challenge of improving the research performance throughout the Union.[38]

3. A STRONGER AND BIGGER PROGRAMME THAN EVER BEFORE

The widespread consensus among the institutional bodies (European Commission, Council and Parliament) that research, and the 7th Framework Programme in particular, was essential to achieve the Lisbon objectives was also reflected in the opinion of the research community, citizens, and Member States, who broadly agreed that the next Framework Programme should be stronger and bigger than ever before.[39]

3.1. Stakeholders agree that strengthening European research will have a major impact

In terms of general attitudes, a majority of European citizens who participated in the consultation agreed that their national government should spend more money on scientific research and less on other things.[40] They also felt that research collaboration at the EU level was very effective, and tended to agree that it brought a number of important advantages compared with research at national level (Fig. 5.3).

In the consultation on the Commission's Guidelines an overwhelming majority (97.6 per cent) agreed that the support for research at European level should be strengthened.[41] Concerning the impact of strengthened European level research support, over 90 per cent agreed that it would be important and would contribute significantly to Europe's competitiveness, social welfare and sustainability. All the

---

[38] Although the online consultations have the particular advantage that its results are easily quantifiable and accessible, other forms of consultation (conferences, workshops, etc.) are considered equally important. In what follows, the discussion of the results will focus mainly on the feedback received from these online consultations, but where possible it will be supplemented with useful results from Eurobarometers, expert groups and reports, discussions, workshops, conferences, and so on.

[39] Although the FP7 proposal only entered the co-decision process after its adoption by the Commission, the institutional debate started well in advance. See for example: European Parliament, *Report on Investing in Research: An Action Plan for Europe*, Linkohr R. (Rapporteur), FINAL A5-0389/2003, November 2003. European Commission, Community Research, *Five-Year Assessment of the European Union Research Framework Programmes 1999–2003*, December 2004. Presidency Conclusions of the European Council of 22 and 23 March 2005. Committee on Industry, Research and Energy, *Report on Science and technology - Guidelines for Future European Union Policy to Support Research*, Locatelli P.E. (Rapporteur), FINAL A6-0046/2005, February 2005. European Research Advisory Board (EURAB), *The Financial Perspective for Framework Programme 7 and Criteria for the Selection of Topics for the Work Programmes*, May 2005.

[40] In a Special Eurobarometer survey on Science and Technology in June 2005, 57 per cent agreed with this statement. The survey gathered information on Europeans' public understanding of and attitude towards Science and Technology.

[41] Only 1.4 per cent disagree or mostly disagree.

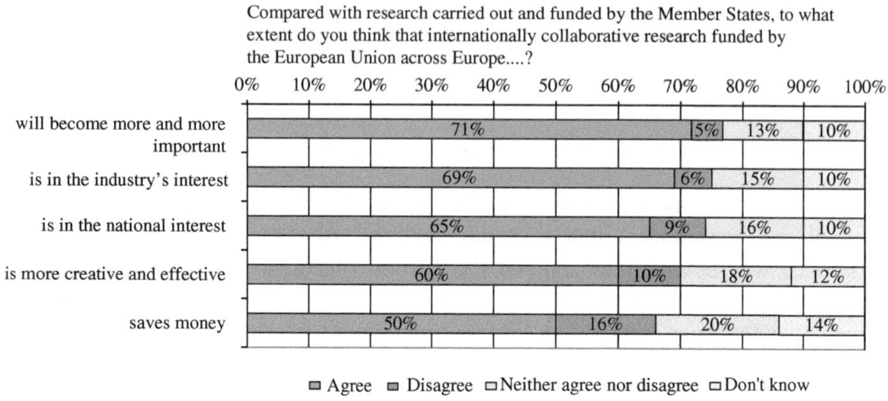

Fig. 5.3. Effectiveness of scientific research in the European Union
Source: European Commission, Europeans, Science and Technology, p. 107

Commission's orientations expressed in the Guidelines – in particular in the fields of human resources, collaborative research, involvement of SMEs, realising the potential of an enlarged EU, identifying topics of major European interest – received a high degree of backing.[42]

Strengthening EU research and human resources (making Europe more attractive to the best researchers) received strong support from all type of respondents (Fig. 5.4). European Technology Initiatives were particularly appreciated by large companies (and SMEs) and collaborative research by universities and government bodies. While basic research received strong support from universities and government bodies, it received below average support from large companies (and SMEs) (Section 5). On the whole, government bodies appeared very optimistic about all objectives and were particularly positive about the complementarity of the Framework Programme with Structural Funds, coordination of national programmes, and developing research infrastructures.

[42] For all the six major objectives proposed, more than 80 per cent of the respondents considered them as important or very important and agree or mostly agree with the text in the Communication (only for basic research 78.6 per cent of respondents agree or mostly agree with the text); more than 75 per cent of respondents considered the specified impact would be substantially greater or greater. For more detailed analyses, please consult: European Commission, DG Research, Analysis of the stakeholder consultation on the Commission Communication *Science and Technology* (COM(353)2004), December 2004 http://ec.europa.eu/research/fp7/home_en.html (last accessed on 28 August 2006). Since the online consultation and corresponding questionnaire were promoted on CORDIS, the majority of respondents had a background or a specific interest in research. Not surprisingly, these respondents appeared to be very positive when confronted with general questions on whether to strengthen government support for research.

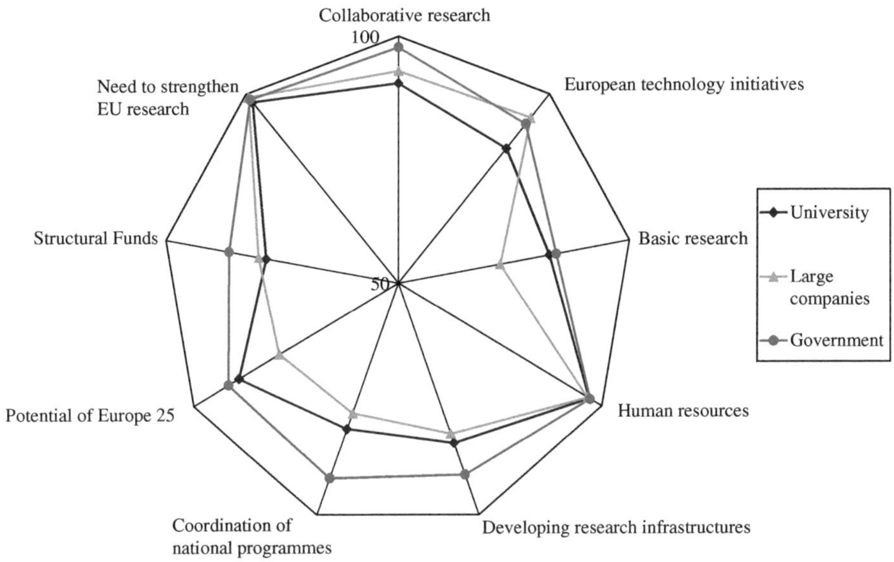

Fig. 5.4. Results of the stakeholders' consultation based on COM(2004) 353 by type of respondent – percentage considering this objective as "very important" or "important"[43]
Source: DG Research

Analysing the results across countries, there are relatively small differences in support for collaborative research and human resources, but large variations (over 14 percentage points) when it comes to European technology initiatives, basic research (European Research Council), research infrastructures, and coordinating national programmes, as well as complementarity with Structural Funds and realising the potential of EU-25.[44] Regrouping the respondents coming from old and new Member States, we see that actors coming from new Member States are consistently more positive about the initiatives put forward in the Commission Guidelines compared to those coming from old Member States (Fig. 5.5). They particularly welcome the complementarity of the Framework Programme with Structural Funds (Section 6).

Regrouping the Member States according to their R&D intensity, we see that all Member States – regardless of their R&D intensity – consider human resources, collaborative research, and the need to strengthen EU research as (very) important (Fig. 5.6). It is also apparent that the lower the R&D intensity the higher the support generally for the initiatives and actions expressed in the Commission Guidelines.

[43] European Commission, *Science and Technology.* The need to strengthen EU research refers to the percentage answering "mostly agree" or "agree".
[44] Results from selected (more than 40 responses) Member States (per cent answering "important" or "very important").

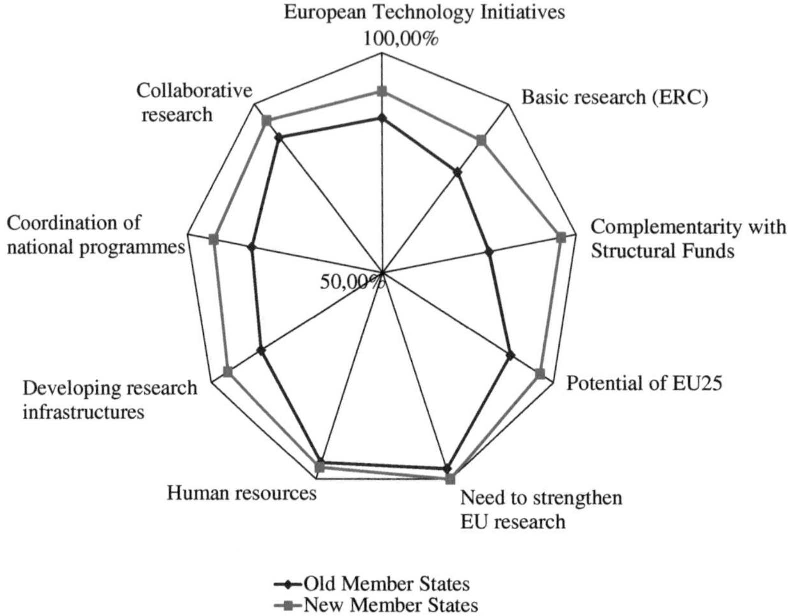

Fig. 5.5. Results of the stakeholders' consultation based on COM(2004) 353 by type of member state (old vs new) – percentage considering this objective as "very important" or "important"[45]
Source: DG Research

Not surprisingly, respondents from countries with a low R&D intensity particularly welcome the complementarity with Structural Funds and the development of research infrastructures.

Further to the online questionnaire, the Commission also received position papers from Member State governments in response to the Guidelines. Of the 21 Member State position papers received, all supported the Commission's proposal to strengthen the European effort in R&D, 6 explicitly supported the doubling of the 7th Framework Programme budget, and 11 agreed that its budget should be significantly increased and be given priority in the context of the Lisbon and Barcelona objectives.[46] A number of Member States indicated that the discussion on this budget had to be placed in the context of the financial perspectives' debate (the EU's budget plans setting out the revenue and expenditure ceilings for the period 2007–2013).[47] This 'One Per cent Group' (Germany, UK, France, Austria,

[45] European Commission, *Science and Technology*. The need to strengthen EU research refers to the percentage answering "mostly agree" or "agree".

[46] This number excludes duplicate responses. Position papers can be downloaded from: http://cordis.europa.eu.int/fp7/debate.htm (last accessed on 28 August 2006).

[47] The financial perspectives determine the Community's spending programme for R&D and the budget for the Framework Programme. Chapter 7 deals in detail with the debate on the financial perspectives 2007–2013 and its impact on the 7th Framework Programme negotiations.

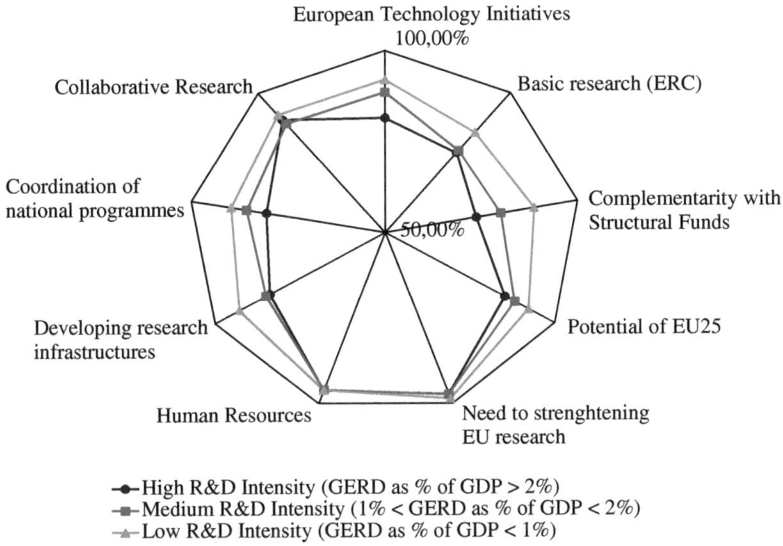

Fig. 5.6. Results of the stakeholders' consultation by country and R&D intensity (GERD as % of GDP) – percentage of respondents considering this objective as "very important" or "important"[48]
Source: DG Research

Netherlands, Sweden, and Lithuania) stressed that the funds for the 7th Framework Programme should be increased, but at the same time the budget must be in line with their adherence not to exceed the one per cent limit of EU GDP which represents the Member States' contribution to the EU budget. The Member States' positive attitude towards the importance of European research and the 7th Framework Programme budget appeared to be very promising for the upcoming financial perspectives debate. However, as will be seen in Chapter 7, the debate in Council proved to be very difficult.

3.2. An ambitious Framework Programme requires a careful design

While stakeholders generally supported increased spending on research, they raised two further considerations.

First, there was a widespread concern that this increase in funding should not replace national research funding. As the Research Councils UK point out in their position paper: "Given the requirements to increase national R&D spending, it would be clearly inconsistent, contradictory and counterproductive if increases in

[48] Based on European Commission, *Science and Technology*. MS with high R&D intensity (>2 per cent): AT, BE, DE, DK, FI, FR, SE; MS with medium R&D intensity: CZ, ES, IE, IT, LU, NL, SL, UK; MS with low R&D intensity: CY, EE, EL, HU, LT, LV, MT, PT, PL. The need to strengthen EU research refers to the per cent answering mostly agree or agree.

EU R&D funding would be offset by reductions in domestic research support within Member State."[49]

In this context, stakeholders pointed out that it was important that EU level spending does not overlap with national support schemes, but rather that it has its own clear justification in terms of European added value (Chapter 6). Traditional Framework Programme pillars, such as transnational collaborative research, human resources, and infrastructures have demonstrated that they complement rather than compete with national initiatives, and these actions received much support in position papers and the reactions to the online stakeholders' consultation. However, the proposed extension of the Framework Programme to include new elements and relating instruments needs to be introduced carefully "as a complement and not as a replacement for other actions".[50] The assessment of the European added value of new initiatives is not an easy task, as can be illustrated by the debate on the European Research Council where the traditional pillar of collaborative research in research consortia is replaced by Europe-wide competition (Section 6).

In line with the commitment taken in Barcelona, there are also reassuring signs that Member States are not planning to reduce their national R&D spending in the coming years. They have already reaffirmed their political support to the new Partnership for Growth and Jobs, and have submitted National Reform Programmes translating EU common challenges and policy guidelines into national challenges and priorities for action. Table 5.1 shows the distribution of the challenges identified according to a broad classification. One sees that R&D and innovation as the most frequently cited challenge for the 22 Member States covered in the analysis.[51] In addition, 18 Member States have established ambitious quantitative targets that would allow them to reach an average R&D intensity of 2.6 per cent by 2010.[52]

The second consideration raised by stakeholders was that the increase in funding, and the introduction of new actions, should go hand in hand with the greater efficiency and effectiveness of the Framework Programme. This implies the simplification of rules and procedures, along with adequate capacity to manage the increased budget. The detailed views of stakeholders on how to make the bureaucratic procedures leaner and simpler are presented in the next section.

---

[49] Research Councils UK, *RCUK response to OST FP7 consultation.*
See http://www.rcuk.ac.uk/international/documents/fp7_finalresponse.pdf (last accessed on 28 August 2006).

[50] See for example the position papers of Finland and the United Kingdom. Also several Member States recognize the difficulties in defining the EAV concept. See for example the position papers of the Czech Republic and the United Kingdom.

[51] The 22 Member States covered in the analysis are : BE, CZ, DK, EE, EL, ES, FR, IE, IT, CY, LV, LT, HU, MT, NL, AT, PT, SI, SK, FI, SE and UK.

[52] European Commission, DG Research, *Progress Towards the 3 Per cent Target on the Basis of the Information in the NRPs*, Directorate M – Investment in research and links with other policies, December 2005.

Table 5.1. Broad classification of the challenges identified in the national reform programmes

	Number of countries
Macro-economic challenges	
Sustainability of public finance	18
Budgetary stability (short-term challenges)	3
External account deficit	1
Others	7
Employment challenges	
Low employment and labour supply	19
Education and skills (including LLL)	18
Unemployment (structural and long term)	7
Regional employment disparities (including mobility)	4
Others	6
Micro challenges	
R&D-innovation	22
Business environment/ entrepreneurship	19
Sustainable development	17
Competition	15
Infrastructure	11
ICT	11
Efficiency of public administration/services	8
Others	4
Total number of challenges identified in the 22 NRPs	190

Source: DG Research

4. ENSURING ACCESSIBILITY FOR ALL

In Chapter 4 it was seen that over the last two decades, the Framework Programme has expanded significantly in terms of budget, scale, scope, and ambition. In line with this expansion, different types of support instruments have been developed to target an increasingly broad range of beneficiaries from a growing number of countries. Accordingly, complexity has arisen from the multiplicity of instruments, forms of contributions, requirements for submission and reporting, rules of implementation, and internal and external financial control procedures. This complexity also affects the participation in the Framework Programme.

4.1. Stakeholders ask for initiatives to simplify participation in EU research

While the Framework Programme benefits from an excellent reputation as far as the scientific contents are concerned, it suffers from a rather widespread perception that its administration is overly complex. Researchers feel strongly that there is a significant bureaucratic cost associated with applying for EU research funding, and participating in EU projects. However, when we come to the crucial question "do the benefits of Framework Programme participation outweigh the costs?" participants are broadly of the view that they do.[53]

But why is there such a negative public image of the Framework Programme management? Keeping a sense of perspective, it is important to note that this tension between researchers and the public administration is inherent in the system of publicly funded research. While both parties pursue the same objective – excellent research – they naturally have different views of administrative obligations. Researchers tend to consider the administrative paperwork as an inevitable but tedious burden which draws their time and attention from their actual activities. On the other hand, the public administration has the task to watch over that the taxpayers money is duly spent. This tension can be found at national as well as at EU levels. It might be more accentuated in the case of the Framework Programme due to the innate complexity of international programmes. However, there is currently no clear evidence base for comparing national and EU schemes. It is also not widely known that the Commission's room for manoeuvre is limited when it comes to simplifying procedures because the fundamental parameters of Community expenditure are set in legal acts adopted by the Parliament and the Council which do not leave complete discretion to the Commission.

Nevertheless, it is not without reason that the Commission has given a top priority to the simplification of the Framework Programme management. There are signals that the cost and workload for submitting a proposal to the Framework Programme, as well as managing the whole process, are considered too high and that some participants opt out because of the extensive bureaucracy. As demonstrated in Chapter 4, smaller actors such as SMEs, new Member States, small research centres and labs seem to be particularly affected. These actors feel that they encounter significant barriers to participation.[54] According to the United Kingdom's position paper, only 55 per cent of SMEs said that the benefits of participating in the Framework Programme exceed the costs. Likewise, the European Association for small and medium sized enterprises and craft (UEAPME), backed up by many other SMEs, urged a reduction in bureaucracy. In their position paper they recalled that the average cost of administrative burden is 6–30 times higher for SMEs than for larger businesses.[55] The new Member States felt they faced special obstacles when it came to accessing

---

[53] The academic community is usually more positive about this trade-off than the industrial one (Chapter 4). Reported benefits are: networking, enhanced skills of RTD personnel, enhanced reputation, development of new tools and techniques, and so on. Almost no work has been done to estimate the aggregate cost of FP calls for proposal and/or participation. See also: Arnold, Erik, *What the Evaluation Record Tells Us about Framework Programme Performance*, May 2005.

[54] See for example the position paper of The Institution of Electrical Engineers.

[55] In particular, they proposed to trim down the lengthy "time to contract" to a maximum of six months, to develop a central database with key data on participants, and to make use of lump sum and timely payments.

Framework Programme funding, and advocated rebalancing the funds towards traditional instruments, reducing the contract negotiation period, and providing more extensive feedback to researchers whose proposal has been turned down in order to assist them to improve the quality of their proposal in future calls.

However, the smaller actors are not alone in complaining about the Framework Programme's red tape. Large companies and research institutes also tend to get entangled in the web of rules and procedures. For instance, Philips, which has participated in nearly 1000 research projects under successive Framework Programmes, calls for "a fundamental overhaul [. . .] of the entire procedural, financial and administrative implementation system of the Framework Programme".[56] Fraunhofer-Gesellschaft, another experienced partner, maintained that the complex administrative and legal rules require more specialised staff to ensure a professional handling of the project.[57] They pointed out that smaller actors cannot afford to employ *"administrative EU-specialists"* and are therefore obliged to pay considerable amounts to external consultants and auditors. They also highlighted that the average processing time for a proposal from the submission to the end of contract negotiations took 368 days during the 5th Framework Programme.[58] In the 6th Framework Programme, the duration has in some cases increased up to 420 days. This means that in total it takes around two years or more from the beginning of proposal preparation until project start. Considering that time (and money) is relevant for the transfer of new findings to the market, and is especially important for smaller actors, this might jeopardise the success of the project.

Large companies also complained about intellectual property rights. It was suggested that the current restriction on the transfer of intellectual property to affiliates and third parties in the 6th Framework Programme acts as an obstacle to participation by global industry players in EC-funded research activities in Europe.[59] According to the Union of Industrial and Employers' Confederations of Europe (UNICE), these issues impede industries' participation in the Framework Programme because they create uncertainty among the industry participant in those cases where the research shows commercial potential, but cannot be properly exploited. More generally, it has been observed that intellectual property rights tend to become more complicated with larger and heterogeneous consortia.

---

[56] Philips, Philips' comments on Commission's proposal for FP7, p. 6.

[57] Fraunhofer-Gesellschaft, Comments by the Fraunhofer-Gesellschaft on the Commission Communication "Science and Technology", October 2004, p. 4–5.

[58] Court of Auditors, *Special Report No 1/2004 on the Management of Indirect RTD Actions under the Fifth Framework Programme (FP5) for Research and Technological Development (1998 to 2002)*, Together with the Commission's Replies (Pursuant to Article 248(4) Second Subparagraph EC) (2004/C 99/01), 23 April 2004.

[59] The general principles laid down in the model contract, Annex II, states that "The Commission may object to the granting of access rights to third parties, in particular to those not established in a Member State or an Associated State, if such grant is not in accordance with the interest of developing the competitiveness of the dynamic knowledge-based European economy, or is inconsistent with ethical principles." See for example the position papers of Nokia and UNICE.

The Marimon report of July 2004, prepared by a panel of high-level experts set up to assess the effectiveness of the New Instruments of the 6th Framework Programme, also pointed out that over-subscription has increased from the 5th to the 6th Framework Programme.[60] Although on average not unacceptably high, certain thematic priorities – such as "nanotechnology and nanosciences" and "citizens and governance in a knowledge based society" – and Marie Curie Fellowships were particularly affected by this problem. To deal with "the current mismatch between high preparation cost and low success rates" some have suggested increasing the general Framework Programme budget (Section 3), reallocating resources to high-demand programmes and traditional instruments (Section 4.3) and increasing the focus of the calls for proposals.[61] Another way to deal with high preparation costs is the use of the two-stage submission procedure in all calls. This procedure includes a first stage where a relatively short outline proposal is submitted and evaluated, followed by a second stage of submission and evaluation of a full proposal only for the outline proposals evaluated positively. Although most of the stakeholders had no problem with the principle itself, several respondents drew attention to the fact that its success would depend on how it was implemented. Stakeholders pointed out that the first stage needs to be administratively light and address a limited set of evaluation criteria. To cite the UK Research Councils, "If this approach is to avoid deterring good applicants, there must not be an excessive burden in preparing first stage submissions and any two-stage procedures must allow enough time for those invited to submit full proposals to do so." Moreover, not all respondents were in favour of the "general" use of two-stage submission. Some felt that its application should only be on a "fit for purpose basis" as they were concerned that its general use would take longer than one-step procedures and cause additional work for proposals admitted to the second step.[62] For the same reasons, some did not deem this procedure appropriate for fast-developing areas.[63]

4.2. What did stakeholders think about the simplification measures proposed?

The need for simplification and rationalisation was already raised in the Marimon report.[64] Although a series of corrective measures have been taken as a result of this assessment, the stakeholder consultation (and particularly that devoted to the rules of participation and simplification), as well as the position papers and other contributions conveyed a clear message: more can and should be done to

---

[60] European Commission, *Report - Evaluation of the Effectiveness of the NewInstruments of Framework Programme VI*, Report of a High-level Expert Panel Chaired by Professor Ramon Marimon, 21 June 2004, p. 32.

[61] See for example the position paper of Germany and Philips.

[62] See for example the position papers of France, Belgium, and Philips.

[63] See for example the position paper of Hungary.

[64] European Commission, *Evaluation of the Effectiveness of the New Instruments of Framework Programme VI.*

make the Framework Programme easier for non-administrators and smaller actors in particular. Against this background, the preparation of the 7th Framework Programme represented both an opportunity and a challenge to further rationalise and simplify the participation in EU research. Accordingly, the Commission established an inter-services working group to bring forward proposals for simplification and a Sounding Board of smaller actors with the aim of eliminating or reducing the barriers faced by small participants.

Taking account of the expressed concerns, the Commission presented in April 2005 its proposal for the new Framework Programme, together with a working document on simplification setting out 10 concrete measures to simplify the implementation modalities.[65] Not only should administrative rules and financial procedures be simplified, but also the Framework Programme and its projects should be managed so as to make participation more attractive for all type of actors. Table 5.2 is the result of the stakeholders' feedback received on the simplification measures proposed by the simplification working document. Overall, very positive feedback was received on the proposed measures, especially concerning points 2 (consistent, high-quality communication) and 3 (rationalisation of the requests for information addressed to the participants). It should be noted that a lot of comments could not be classified in simply "for" or "against" as they contained comments which did not relate to the proposed measures. The vast majority of respondents suggested their own simplification measures, which were not mentioned in the working document (e.g. intellectual property right issues) and were not always in the interest of all the participating actors (e.g. 100 per cent funding for non-profit and public organisations). A number of these additional ideas or considerations are added in the last column of Table 5.2. Large companies strongly supported the need for adequate protection of intellectual property, while the university/higher education sector was especially in favour of the full cost model (100 per cent funding), and SMEs were mainly concerned about receiving funding on time and restoring the balance towards traditional instruments which are more SME-friendly. There was also a very strong gender lobby, especially in universities/higher education and government bodies, which stressed that gender issues should be taken more into account when selecting projects, for example by including an expert in this field in the evaluation panel.

4.3. New instruments: An opportunity to develop the European Research Area or an obstacle to traditional Framework Programme research?

Some research actors expressed the view that it is not only administrative complexity that can narrow the accessibility of the Framework Programme, but also the type of instruments it employs. In particular, the so-called "New Instruments" introduced in the 6th Framework Programme – Networks of Excellence (NoE) and Integrated Projects (IP) – were the subject of much attention.

[65] European Commission, *Simplification in the 7th Framework Programme,* SEC(2005) 43, 6 April 2005.

Table 5.2. Results of the consultation on the 10 measures proposed in the working document on simplification

Proposed measure	For/ Against	Comments
1. Simple set of funding schemes allowing continuity with the instruments of FP6 and providing a broad flexibility of use	For: 36 Against: 2	* Some suggested the abandonment of work programmes and free choice of research topic instead * The freed funds of cancelled projects should be available for other RTD activities and not go back to overall EC funding * Funding sets should not disrupt continuity
2. Consistent, high-quality communication	For: 44 Against: 5	* All relevant materials have to made available well in advance of the FP and should be consistent (correct translation in all EU official languages) * Contracts and annexes should be interpreted in a consistent way by different DGs and Services * Information on Calls for proposal, Call for expression of interest and so on should be broadcasted sufficiently so all actors aware of possibility of participation * A central appeal authority (like an ombudsman) should be set up to comment on practice by participants and Commission staff
3. Rationalisation of the requests for information addressed to the participants	For: 54 Against: 6	* Broad support for registration system for participants * Support for wider application of two-stage submission with stage 1 being sufficiently light
4. Guaranteeing the protection of the Community's financial interest	For: 25 Against: 7	* Positive feedback especially from SMEs * Use of external evaluators is questioned on the grounds of experts' objectivity, criteria for their selection, use of experts makes the task time-consuming, more experts from industry should be involved * Liability should be restricted to own share of the project. Collective financial responsibility is a major disincentive especially for smaller SMEs * Necessity to provide bank guarantees is a serious obstacle, especially for smaller SMEs. This form of protection of the Community's interest should be used only in clearly defined circumstances, i.e. only when EC contribution is higher than a particular amount and only for technical aspects
5. Full operational autonomy entrusted to consortia	For: 32 Against: 4	* Frequent abuses from dominant parties make a revision of the existent practice for consortium agreements necessary * Clear assignment of tasks, costs and EC contribution both inside the consortia agreements and in the contracts, as well as more flexibility in the planning process which should be left to the consortium

		* Establishment of a helpdesk dealing with project management issues * Possibility of external management of projects has been praised and very welcomed, especially in the context of increased SME participation. However, some participants expressed doubts as to who will pay these costs
6. Streamlining the selection process	For: 32 Against: 10	* Abandoning expression of interest has been signalled in several messages * Proposal to reduce time-to-contract was very welcomed. The contract negotiation should start immediately after evaluation of the projects. It was recognised that large consortia slow the time-to-contract * A number of participants do not approve to decrease the involvement of Programme Committees
7. Most effective possible use of the budget dedicated to the research policy	For: 22 Against: 1	* A broader flexibility of the research budget, similar to that of the Structural Funds, may be advantageous * In case financial support is impossible because of a lack of funds, the most interesting and promising projects should be included in a reserve list to be co-financed by the European Investment Bank * The Commission could deal with the supply of guarantees to small players in the same way that covers the risk guaranties scheme run by EIB * The rules should foresee a possibility of several stages of a project, with different sets of partners for each stage * Increased possibilities for funding of participants from third countries (India, China, etc.)
8. A more extended use of flat-rate financing	For: 26 Against: 11	* Different opinions concerning increased application of lump-sums and flat rates. In general, increased usage of flat rates and scale of unit cost has been deemed appropriate for Marie Actions and research project should be financed by means of reimbursement of actual costs * There should be a choice between the reimbursement of eligible costs and lump sum/flat rates * Overheads should be based on personnel costs only. Flat-rate should depend on the size of organisations. * According to some comments, lump-sum financing for fraud and does not encourage participants to perform in their most efficient way
9. Removing the need for complex cost reporting models and clarifying definition of eligible costs	For: 27 Against: 6	* Proposal to eliminate cost models has been praised by a number of respondents, however others are concerned that changing them would work against continuity and actually make it more complex * 100% funding for non-profit organisations is supported by some (but not by others)

(Continued)

Table 5.2. (Continued)

Proposed measure	For/ Against	Comments
		* Financial support and possibilities for participation for individual researchers are far insufficient * Several requirements in the cost model should be simplified
10. Simplified support rates per type of activity	For: 24	* The maximum of 7% of the EC financial contribution for management costs in FP6 is absolutely insufficient taking into account audit costs which need to be paid within this category
	Against: 2	* EC contribution to public and/or non-profit bodies should be increased to 100% funding. Maximum support ceilings laid down in State Aid framework should apply solely to profit making bodies * Flat rate for indirect costs is currently too low and should be revised * Overhead rate should be dependant on the project type and on activity * No need to have many different levels of support rates per type of activity when accepting cost reporting models based IFRS and GAAP.

Source: DG Research[66]

The Marimon report detected a bias in the New Instruments towards large consortia which seemed more adapted to large organisations than to smaller research teams.[67] This would appear to be supported by participant data which show that the average number of participants per project doubled from the 5th to the 6th Framework Programme (from 6.5 participants to over 13) and that the average funding per project surged (from €1.4 million to €4.3 million) (Chapter 4). Furthermore it seemed as if a lack of clarity in the definition of critical mass led to artificial pressure to set up big projects.[68]

The European Association for small and medium sized enterprises and craft (UEAPME) confirmed that the New Instruments were often too big for SMEs. In fact under the 6th Framework Programme, just 22 per cent of SME proposals of "very high standard" received funding, compared with 50 per cent of all projects with the same rating. UEAPME pointed out that this imbalance was not due to a general lack of funds for SMEs, but because 80 per cent of the SME budget

---

[66] European Commission, *FP7 Simplification – Feedback from actors and stakeholders on Commission Staff Working Document*, 2005. Internal Commission document – DG Research (A3).

[67] European Commission, *Evaluation of the Effectiveness of the New Instruments of Framework Programme VI.*

[68] European Commission, *Evaluation of the Effectiveness of the New Instruments*, p. 22.

was allocated towards low-demand New Instruments and only 20 per cent to high-demand SME programmes. Consequently, higher SME participation should be sought by providing instruments and objectives that meet the need of this group.[69] This might include open-topic application procedures, smaller projects with few partners and a flexible use of attractive instruments.[70] Some suggested that the co-operative research projects (CRAFT), Collective Research and Economic and Technical intelligence programmes would fit SME wishes well.[71] Surprisingly, some large companies also considered Networks of Excellence as not very appealing to industry and asked for small focussed projects.[72] According to the Marimon report, the main reasons for the lack of enthusiasm of industry seem to be a misunderstanding and reluctance in relation to the concept of durable integration, and the problems of handling of Intellectual Property within such networks.[73]

The new Member States were also convinced that a more balanced approach should be taken in financing bigger and smaller projects and that consequently more funding should be oriented to traditional instruments, especially Specific Targeted Research Projects (STREP) and cooperative research projects. The Czech Republic expressed its concern that Integrated Projects and Networks of Excellence were instruments that are not that well suited to countries with lower R&D capacity. This view was shared by Slovenia, Lithuania, and Latvia, as well as some old Member States, such as Germany, Ireland, France, Portugal, and the Netherlands. For Portugal, the New Instruments had become an obstacle to the traditional research carried out in the previous Programmes. Germany recommended that it must be the applicants primarily who select the instrument for a project, Belgium opined that maintaining excellence became harder as size increased, while according to Italy partnerships of 35–40 actors in Networks of Excellence very seldom allowed participant groups to achieve a real integration and led to efficiency problems.

Notwithstanding these comments, stakeholders were in broad agreement that the New Instruments should be maintained in order to ensure continuity from one Programme to another and to continue their significant role in developing the European Research Area.

5. FUNDAMENTAL AND APPLIED RESEARCH: HOW SHOULD THEY COEXIST?

An important debate during the preparation of the 7th Framework Programme proposal was about the way how basic and applied research should coexist. Two core actions were at the heart of this discussion: the proposed European Research Council (ERC) and the Joint Technology Initiatives/Technology Platforms (JTI/TP).

[69] See for example the position paper of Italy.
[70] See for example the position papers of Germany, Austria, and UEAPME.
[71] See for example the position papers of Austria and UEAPME.
[72] See for example the position paper of NOKIA.
[73] European Commission, *Evaluation of the Effectiveness of the New Instruments of Framework Programme VI*, p. 13.

While the majority of stakeholders were in favour of including both basic and applied actions in the new Programme, it would be fair to say that industry placed more emphasis on JTIs, while universities were strongly supportive of the European Research Council. Accordingly, stakeholders expressed different ideas about how to strike an adequate balance between these two elements in terms of volume of funding. Importantly, they urged to avoid creating segregated actions earmarked for universities on the one hand and industry on the other, arguing for the need to encourage synergies between "frontier" and "applied" research so as to stimulate the transformation of research into commercially viable products and services.[74]

5.1. European Research Council: Finding the right balance between competition and collaboration

In the space of two years, the European Research Council – "A new European-level funding mechanism to support the very best research carried out at the frontiers of knowledge" – evolved from a promising idea to a serious political proposal.[75] The European Research Council is expected to support investigator-driven research projects carried out across all fields by individual teams in competition at the European level. Projects will be funded on the basis of proposals presented by researchers on subjects of their choice and evaluated on the sole criterion of excellence as judged by peer review.[76] With this principle in mind, this Research Council is seen as an important building block both for the further development of the European Research Area and for an emerging European Innovation System, which Europe urgently needs to reinforce to strengthen its position as a knowledge-based economy.

As will be seen in Chapter 7, there was broad political support for the principle of a European Research Council. But how did this come about? The initial initiative for its creation came from the scientific community. An important early step occurred in November 2002, when an ERC expert group, set up under the Danish EU presidency, reported on options for the possible creation of a European Research Council using input from a number of sources (e.g. European Research Advisory Board (EURAB), European Science Foundation (ESF)). Following this impetus, the attention turned to the Commission as the body which was in a position to get things moving. Indeed, in 2004, it published a Communication on Europe and basic research which highlighted the "need to introduce a European level support mechanism for individual teams' research projects".[77] The Spring 2004 European Council welcomed this suggestion. Since then, reports and articles by prominent groups and individuals have argued for the need to create such a body.[78] Recently, a

---

[74] See for example the position paper of UNICE.

[75] European Commission, *Frontier Research: The European Challenge – High-Level Expert Group Report*, Luxembourg, February 2005, p. 11.

[76] European Commission, *Science and Technology*, p. 6.

[77] European Commission, *Europe and Basic Research,* COM(2004) 9, 14 January 2004, p. 12.

[78] Among many others, the European Science Foundation's position paper was very supportive of the idea and saw this as one of the structures to deliver the European Research Area.

high-level group (with experts from Europe and the US), set up by the Commission to examine the likely impact of a European Research Council, concluded that such a body would help to address Europe's poor performance in cutting-edge areas of research.[79]

In the consultation on the Guidelines, the initiative "to stimulate creativity of basic research through competition between teams at European level" received support from 81 per cent of respondents. Analysing the results across types of actor, we notice that after governments (84 per cent) universities (83 per cent) backed this initiative the most, followed by SMEs (78 per cent) and associations (77 per cent). Perhaps not surprisingly, larger companies were the least enthusiastic about this idea (72 per cent). Member State governments' position papers welcomed the initiative. They considered the creation of a European Research Council as an unavoidable and very promising consequence of the political process, and particularly welcomed the bottom–up approach. Of the 22 Member State position papers received, Italy was perceived "to be most at odds with recent policy statements on the ERC by the Commission and other Member States".[80] The Italians felt that the initiative raised 'strong perplexity' for a number of reasons. However, the Italian concerns (subsidiarity, collaborative nature, the creation of a new body, the excellence criterion, involvement of Commission and Member States, simplification) were shared by others as well. Apart from the general need to clarify the concept of the Research Council, actors appeared to have different expectations on what the Council would deliver, and there were questions on how it would be administered and funded.[81]

At the same time, other stakeholders (larger companies and SMEs) felt that efforts should be focussed less on basic research and be more applied and business-oriented. According to many industrial respondents, projects should not only be selected on the basis of their scientific excellence, as assessed by peer review but also on the basis of their business relevance and likely impact. In their view, the main problem was not the quality of basic research, but rather its transformation into commercial success. They therefore maintained that science and industry should collaborate closely to ensure the effective liaison between research areas and needs, and some industrial stakeholders proposed the creation of an advisory board where the main players – researchers, the Commission, the Member States, and industrial players – could work together.

Another challenge was to find the right balance between competition and collaboration. The statements in the Guidelines that "open competition between individual research teams and support for them would boost dynamism, creativity and excellence of European research" and that "projects would then be selected, without any obligation of transnational collaboration" provoked a lot of comments as not all

---

[79] Cordis Press Release, *Expert Group Backs ERC Vision*, 22 March 2005.

[80] Cordis News, *Italy's contribution to FP7 debate raises doubt over the need for an ERC*, 6 October 2004.

[81] The Scientist, *European Research Council Doubts*, 23 January 2004.

Table 5.3. Competition versus collaboration: Some quotes from stakeholders

"Competition leads to extinction while cooperation leads to success!"
"The competition is good for sport activities but not for science, in this case collaboration is better"
"I think it is better to stimulate competition with countries outside of Europe and favour the European collaboration"
"Wild competition could be disappointing"
"Intellectual forces and resources in danger of being preoccupied by competition instead of being used for research"
"There is already an international competition, why should a European competition increase creativity?"
"The concept of competition needs to be clarified. Given a primarily economic performance related meaning and rationale, it might threaten collaborative efforts that address overarching social issues"
"Brilliant ideas can not be forced by competition, but competition may distract people with brilliant ideas from their work"

Source: DG Research

stakeholders were convinced that competition without the collaboration requirement would be a good thing (Table 5.3). Several stakeholders pointed out that this could conflict with another major objective mentioned in the Guidelines: "Creating European centres of excellence through collaboration between laboratories".[82] By stimulating competition without requiring transnational collaboration, the Research Council might challenge one of the important guiding principles of successive Framework Programmes. Some stakeholders were afraid that this mechanism might weaken the focus of the European Research Area of which coordination and collaboration are key pillars.[83] There was also the fear that competition could have negative effects for smaller actors. Some respondents forecasted that this mechanism would simply funnel more funds towards the big names/laboratories at the expense of small centres of excellence in less high-profile institutions. They felt this was a real possibility because competition depended on the budget available, and smaller players (including poorly resourced countries) would have difficulties in keeping up the pace.

Certain respondents were concerned that the European Research Council might infringe the subsidiarity principle. They pointed out that the main beneficiaries of national funding were also likely to be the excellent teams selected for support by the Research Council, and that consequently there was a danger of Council funds

[82] The Guidelines state that "Programmes to support transnational collaboration between research centres, universities and companies have an observable impact on the quality of research in Europe and the dissemination of knowledge and results within the Union, and the ability of researchers to become more involved in high-level projects." European Commission, *Science and Technology*, p. 5.

[83] While there is agreement that collaborative research across frontiers has clear European added value, the initiative of the ERC challenges the traditional idea that European added value only lies in the collaborative nature of research.

partially replacing national support. This risk would be significant if the Research Council's actions were insufficiently distinct from those projects supported by national schemes. One way to ensure this complementarity would be to allocate funding to projects with a clear European added value, for instance, towards projects which are too large, too complex, or too risky to be financed by one country alone or to maintain the transnational requirement.

While the majority agreed that projects should be selected on the basis of excellence, some raised questions about how this excellence would be measured and how research proposals would be compared with each other.[84] Given that the proposals would not be subject to thematic constraints, it might be very difficult to compare objectively the merits of very different projects. How would the eligibility criteria be set up for completely different projects from different research areas competing for the same funding? Some were of the view that it would be hard for organisations to judge their likelihood of success as they would not know with whom they would be competing. Care was advised in designing an evaluation system that would ensure that relevant fields of research were granted a similar chance.

There was also the general concern that the European Research Council would become entangled in Commission bureaucracy and politics. While the majority agreed that the success of the Research Council "will depend on a clear definition of its strategic mission and on a firm political commitment to ensure its autonomy and adequate resources to attain its goals", it was a matter for discussion how far this autonomy should go.[85] While some favoured management of the ERC by the scientific community alone, others believed that the Commission and Member States should be included in the steering bodies, since choosing scientific priorities would imply political decisions that could not originate from people personally engaged in the research work in a particular area. Nor was everybody convinced that the creation of a new structure was the most appropriate solution for increasing the support to basic research. According to some, such funding could also be provided by enlarging the scope of existing areas such as New and Emerging Science and Technology (NEST) or Future and Emerging Technologies (FET). Some respondents (especially industrial respondents) stressed that the ERC should come in addition to existing Framework Programme activities and not at their expense.

5.2. Technology platforms and joint technology initiatives: A way to stimulate industrial participation or to develop industrial dominance?

The concept of Technology Platforms was introduced under the 6th Framework Programme in order to better align EU research priorities to industry's needs. Under industrial leadership, the platforms bring together companies, research institutions, the financial world, and regulatory bodies at European level to define a common research agenda on strategic issues. Through public and private sector partnerships,

[84] Stakeholders also pointed out that the FP should also be aware of the pitfalls of today's very poor methods to measure scientific performance.

[85] European Commission, *Frontier Research: The European Challenge*, p. 15.

they mobilise a critical mass of national and European resources. Once an agenda is defined, it can be implemented by means of Integrated Projects, or in a limited number of cases – where a pan-European approach appears more appropriate – by large-scale "Joint Technology Initiatives (JTI)".[86] Technology Platforms also bring together stakeholders in fields such as regulation, finance, and markets – which often create barriers in the innovation process – so as to facilitate innovation and promote the delivery of the benefits from European R&D.

In the consultation on the Guidelines, it is not surprising that industry was one of the strongest supporters of Technology Platforms and Joint Technology Initiatives (TP/JTI). Indeed, after human resources (97 per cent), TP/JTI was the second most popular measure (94 per cent of large companies and 92 per cent of SMEs were supportive).[87] Member State government position papers were very positive. Many felt that the initiatives would strengthen the link between research and industry and bring greater stakeholder involvement in future research agenda-setting, thus helping to align fragmented RTD efforts at several levels (Community, intergovernmental, national, and regional). However, respondents raised a number of issues to be addressed if TPs and JTIs were to be successful.

First, many thought that the concepts needed to be clarified. They expressed their confusion as to what actually were a Technology Platforms, a Joint Technology Initiative and an Integrated Project and what were the differences between them. They also regretted the lack of information on the modalities for implementing the Platforms and Technology Initiatives. A variety of Platforms had already been established, but it was felt by some that the procedures had not always been transparent and well communicated. As a new instrument, several respondents preferred that the TPs should first be tested through pilot projects in a few selected areas.[88] Stakeholders were also cautious regarding the potential application of article 171 EC Treaty.[89] Because it was considered to be a heavy procedure, it was proposed that it should be used only in limited, well-justified cases, requiring public–private partnerships, when the usual instruments were not

[86] European Commission, *Science and Technology*, p. 5. Annex II (Funding schemes for FP7) of the 7th Framework Programme proposal, speaks about "Collaborative Projects".

[87] Percentage of respondents answering "very important" or"important"

[88] It was suggested that these projects could make use of the extensive experience gained in national projects and in EUREKA cluster projects, thereby taking into account the lessons learned and bringing together existing initiatives.

[89] Article 171 (TEC): "The Community may set up joint undertakings or any other structure necessary for the efficient execution of Community research, technological development and demonstration programmes." The Commission pointed out, "Article 171 TEC offers a wide range of implementation structures for community research and development programmes, of which the most prominent is a Joint Undertaking. The main advantage of a Joint Undertaking is that it creates a strong and efficient coordination mechanism, able to structure and handle contributions coming from different fields and sectors." European Commission, *Report on European Technology Platforms and Joint Technology Initiatives, Fostering Public-Private R&D Partnerships to Boost Europe's Industrial Competitiveness*, Commission Staff Working Document, SEC(2005) 800, 10 June 2005, p. 19.

suitable. To further inform the debate, the Commission disseminated a working document via the Internet during autumn 2004.[90]

Regarding the subject of research, stakeholders favoured a bottom–up process for the establishment of Technology Platforms, with objectives clearly defined from the beginning. Some believed that they should be created in all areas of S&T that were vital to the future EU competitiveness (the majority refers of course to their own particular technological field). Others thought they should be limited to areas where political and societal/legal aspects are paramount (renewable energy, electric vehicles, water and waste treatment, etc.) and exclude areas where technological development is satisfactory (e.g. electronics). This latter group suggested demonstrating – a priori – the need for such an activity and its European added value.

Many respondents were convinced that industry should play a leading role in the creation and operation of Technology Platforms. However, some feared that the Platforms and Technology Initiatives might be dominated by large companies and multinationals, and that access could be difficult for smaller actors (especially SMEs, smaller research teams, and regions). They warned of possible misuse of the Platforms by dominant lobbies to advance strategies that favour the strongest industries or to allow one organisation to control an entire sector. One of the respondents to the online consultation expressed his fear that "the control from large companies can lead to holding back truly leading technological contributions [...] to commonly exploited platform to reserve the "latest generation" for their own commercial exploitation". This group also advocated corrective measures, such as simplification of existing procedures or specific measures for SMEs, to avoid a bias towards subsidizing large European industrial groups instead of SMEs and smaller actors.

Technology Platforms as well as Integrated Projects were conceived as very large entities with many partners. A lot of respondents feared that efficient cooperation would be impossible with so many partners involved. In their view, managing these Platforms would be very complicated and would increase the level of bureaucracy (and cost), thus reducing transparency and limiting participation. In this context, the adequate protection of intellectual property was a concern for all, and needed to be further elaborated if industries with innovative ideas were to participate in these schemes.

A last issue concerned the funding available for the projects. It was envisaged that TPs should be financed from many sources (Commission, industry, European Investment Bank, etc.), and that the budget granted should allow them to achieve a critical mass and strategic impacts. In this regard, Platforms could align Community support with other important funding programmes (transnational, national, regional, and other). However, some respondents (especially universities) expressed concern that the current Framework Programme was focussed too much on applied research, and are afraid that money will be missing for basic research. Indeed, a respondent

---

[90] European Commission, *Technology Platform – From Definition to Implementation of a Common Research Agenda.*

to the online consultation on the Guidelines even viewed Platforms and Technology Initiatives "as a way for industry to reap goodies and have its engineering paid for by public money".

5.3. European Research Council and Technology Platforms/Joint Technology Initiatives: Mixed views – what next?

Many questions were raised about both initiatives – European Research Council and Technology Platforms/Joint Technology Initiatives – regarding their organisation and funding. Different actors had different views on the importance of the initiatives, what they should deliver, and how they should be administered and funded. Since these actions were new, this debate was both normal and healthy. The consultation did not reveal general consensus – on the contrary – but it did provide the Commission with a rich pool of views which were fed into a number of more focussed meetings, events, and publications.[91] While not resolving all issues, it elucidated the many areas where there was agreement, and moved the conceptual design closer to a workable solution.

The proposal that emerged for the 7th Framework Programme took into account this debate. The next step was to try and achieve an agreement in Council and the European Parliament, and this will be presented in detail in Chapter 7.

6. EXCELLENCE VERSUS COHESION: IS EFFICIENCY COMPATIBLE WITH SOLIDARITY?

One important issue raised in the Communication on the Guidelines was the need to realize the full research potential of the Europe of 25 which is not yet the case. Europe needs to make better use of its "untapped" high research potential, in particular in less advanced regions remote from the European core of research and industrial development. Indeed, at the time of preparing the 7th Framework Programme, this theme was more relevant than ever with the arrival of the ten new Member States. The Communication proposed initiatives in relation to human resources, exchanges, and networking. It also mentioned the need to provide opportunities for young teams and less powerful institutions to participate in small-sized projects with excellent laboratories and countries, which would allow them to benefit from transnational collaboration and acquire the necessary experience to become involved in more complex projects in the future.[92] The Guidelines also set out for consultation the objective of supporting the development of research capability by taking full advantage of complementarity with the Structural Funds.[93] It was proposed that this complementarity should be strengthened and the combined use

[91] For information on the TP debate, see: http://cordis.europa.eu.int/technology-platforms/home_en.html (last accessed on 28 August 2006). For more information on the development of the Basic Research debate, see: http://europa.eu.int/comm/research/future/basic_research/index_en.html (last accessed on 28 August 2006).

[92] European Commission, *Science and Technology*, p. 8.

[93] European Commission, *Science and Technology*, p. 8.

of Structural Funds and the Framework Programme should be made possible for research projects in "Convergence objective regions".

Most respondents (88 per cent) agreed with the objective of realising the potential of a Europe of 25. They also accepted the importance of integrating new Member States in the ERA and using and developing their human, scientific, and technical potential. The highest support came from associations (93 per cent) and governments (91 per cent), closely followed by universities (89 per cent) and SMEs (83 per cent), and large companies (79 per cent) provided the lowest support. Not surprisingly, respondents from new Member States (98 per cent) attached even more importance to this objective than those from old Member States (92 per cent).[94]

In their position papers, however, Member State governments made few direct comments on this objective.[95] Several Member States indirectly referred to the importance of the excellence criterion for project selection in the Framework Programme and only some explicitly support special measures to stimulate the integration of the new Member States.[96]

Nevertheless, many stakeholders, while recognising the importance of realising the full potential of the EU, felt that the budget of the 7th Framework Programme should be spent primarily on "excellent research" and not on initiatives towards developing and structuring regions. There was a minority who proposed that centres of excellence should be established in countries currently lacking them, and that some positive discrimination may be needed. However, the majority clearly advocated a Framework Programme that did not deviate from the notion of excellence. It was their view that the less advanced regions did not need special treatment, and that if one offered the right opportunities over a prolonged period then in the end good research institutes would arise by themselves. Moreover, some pointed out that there were wider European initiatives for tackling these problems and that the basic principle of the Framework Programme should not be deflected by this larger issue.

Many thought that diluting the principle of excellence could weaken research efficiency. In their view, trying to raise some of the poorer institutions to an average level instead of making good institutes excellent could weaken the Union's global competitive position because resources and attention would be diverted from those institutions and regions that can best deliver research.

---

94 The difference was even larger when respondents were asked for their level of agreement with the text: 96 per cent of the new Member States agreed or mostly agreed, compared with only 86 per cent of the old Member States.

95 Since "realising the potential of Europe of 25 or more" was not one of the six major objectives of the Communication on the Guidelines but one of the cross-cutting issues, many Member States did not devote a separate section or paragraph to it.

96 See for example the position papers of Poland, Latvia, and Lithuania. Some ideas on supporting the involvement of new Member State participants were given, including mobility actions, specific calls for small projects for new research players, incentives for established players to involve emerging institutions, and attention to priorities of new Member States.

The objective to support projects of limited size was warmly welcomed (see also Section 5.2). This is potentially a means of improving access for young researchers or smaller institutes, and of them gaining experience of complex projects. While some respondents were in favour of introducing quotas for new Member States or Associate Countries in Network of Excellence, Integrated Projects, and Specific Targeted Research Projects, others felt this would not help to promote research excellence. Some also stressed the need to reduce administrative barriers for small projects in order to encourage excellent young researchers with innovative ideas to apply, and felt that, in general, simplified rules for proposal submission and project management were essential if less powerful groups and new Member States were to participate fully in the future Framework Programme.

It was not surprising then, given the strong views on the 7th Framework Programme supporting excellence, that when it came to question about taking advantage of complementarity with the Structural Funds to develop research capability, the majority agreed (79 per cent). The highest support came from governments (86 per cent), followed by associations (81 per cent), large companies (80 per cent), and SMEs (79 per cent). Respondents from new Member States (98 per cent) supported this objective more than those from old Member States (78 per cent). However, among the old Member States, there were significant differences: over 90 per cent of the respondents from Italy, Portugal, and Greece supported this objective, compared with more modest support from German (70 per cent), United Kingdom (64.9 per cent), and Swedish (65 per cent) respondents. Member State government position papers generally welcomed the complementarity between the two sources of funding.

Many respondents from the new Member States were unfamiliar with how to tap into the funds and opportunities in the Structural Funds. Some were also afraid that the combined use of Structural Funds and Framework Programme' funds could complicate the administration and the execution of projects; to avoid this funding procedures should be simple, explicit, and direct. Transparency was another issue of concern. It was felt that the overlapping of resources from different sources was to be avoided, and the inclusion for political reasons of specific partners from convergence regions with no or little added value should be prevented. More than once it was stressed that Structural Funds and the Framework Programme are, and should remain, two separate instruments, and that the funding of the 7th Framework Programme should not be redirected into regional development.

CONCLUSION

Without the input from the broad community of stakeholders, the proposal for the 7th Framework Programme would not have taken the shape it did. The extensive consultations launched by the Commission were a valuable tool for capturing opinions from a wide range of actors. The resulting stock of information helped to design the proposal around key areas of consensus and to make it responsive to the practical concerns of the actors on the ground.

For the Commission, the consultations proved how challenging it is to balance the different, often opposing views and to formulate a proposal acceptable to the vast majority of stakeholders. As seen in this chapter, some topics found broad support (such as "Strengthening the Budget" and the "Need for Simplification"), while others were more subject to debate (such as the European Research Council, Technology Platforms/Joint Technology Initiatives, and "Excellence versus Cohesion").

One of the insights gained during the consultations concerns the need for methodological development of future consultations despite the considerable efforts made by DG Research to set up a dialogue with all stakeholders. It would be useful to learn more about how to do such consultations in practice, their strengths and weaknesses as well as possible blind spots. For example, what are the underlying reasons for the low response rate of particular groups or countries (especially the new Member States, see Annex)? Are the consulted stakeholders satisfied with the online consultation system? And, though difficult and rarely achieved, how can the Commission provide stakeholders with a feedback of how their inputs had an impact on policy-making? A lack of transparency may cause cynicism and fatigue among stakeholders.[97] Finally, it seems necessary to develop a coordinated approach towards the involvement of stakeholders in the process of designing future Framework Programmes. At the same time, it has to be noted that the consultation process is ongoing, permanent, and that it is not – and would lose much from being – circumscribed to a tightly structured and streamlined process. In effect, one can present this rich and heterogeneous dialogue as based on two strands. A decentralised strand consisting of various advisory groups and of the very open consultations run by the thematic directorates with the actors of the research and innovation system and a centralised strand structured along Commission-wide guidelines to ensure maximum openness and comparability.

To conclude, the consultations certainly improved the communication and dialogue between science and civil society, and helped to create ownership of proposed strategies. This can be confirmed by the fact that the 7th Framework Programme proposal should find broad support by the decision-making institutions (Chapter 7).

ANNEX

Analysis of respondents to the online consultations

Europe's research community welcomed the opportunity to take part in an extensive debate on the role of research and its future in Europe. A total of 1727 responses were received to the online consultation on the Guidelines, of which just over 30 per cent were from individuals, 26 per cent from universities, 8.3 per cent from SMEs, and a further 6.7 per cent from larger companies (Table 5.4). In terms of country balance (Fig. 5.7), there was generally a good response from old Member

[97] Parliamentary Office of Science and Technology, *Debating Science*, Postnote No 260, March 2006.

Table 5.4. Result of the stakeholders' consultation on the guidelines by type of respondent

Type of respondent	Number	%
Individual person	540	31.2
Government body	141	8.2
University/higher education	456	26.4
Commercial organisation (including consultancy) more than 250 employees	115	6.7
Commercial organisation (including consultancy) less than 250 employees	144	8.3
Association (e.g. trade association, trade union, employers association, chamber of commerce, NGO)	113	6.5
Other	218	12.6
Total	**1727**	**100**

Source: DG Research

States (e.g. over 100 responses from Germany, the United Kingdom, Italy, Belgium, France, and Spain). However, new Member States only represented 5.9 per cent of the overall response.

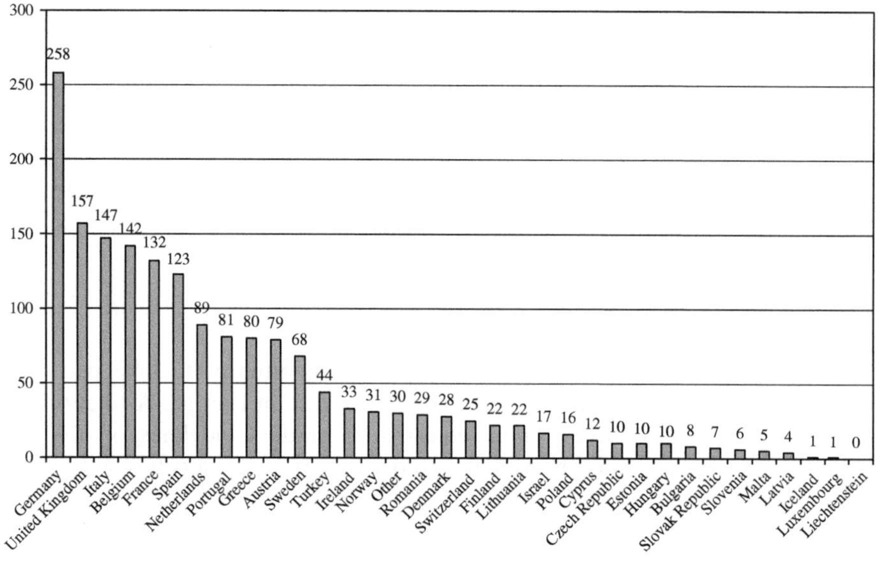

Fig. 5.7. Result of the stakeholders' consultation on the guidelines by country
Source: DG Research

Table 5.5. Result of the "thematic" stakeholders' consultation by type of respondent

Type of respondent	Number	%
Individual person	902	49.5
Government body	98	5.4
University/higher education	234	12.8
Commercial organisation (including consultancy) more than 250 employees	51	2.8
Commercial organisation (including consultancy) less than 250 employees	43	2.4
Research institute	182	10.0
Association (e.g. trade association, trade union, employers association, chamber of commerce, NGO)	246	13.5
Other	68	3.7
Total	**1824**	**100**

Source: DG Research

The Commission also received over 150 position papers and other written documents in response to the Guidelines. These included contributions from many of the leading research organisations, national and regional (S&T) government bodies, and research users (including industry). The contributions generally reflected the views put forward in the online consultation and confirmed the broad support for the guidelines set out by the Commission and the need to strengthen European support for research.

For the online thematic consultation alone, a total of 1824 contributions were received of which approximately half came from individuals and half from organisations (Table 5.5).[98] The highest number of organisational responses came from associations (13.5 per cent), closely followed by universities/higher education (12.8 per cent), and then research institutes (10.0 per cent). The highest number of individual responses came from universities/higher education (42.6 per cent). Contributions came from a wide spread of countries (Fig. 5.8), including all EU Member States (except Luxembourg).

A high response came from the United Kingdom, Germany, the Netherlands, France, and Spain, which collectively processed over 1000 responses. All together, only 6.4 per cent of the total came from the 10 new Member States. In addition, many hundreds of responses were sent to the consultations in specific areas of research. The research agendas developed by Technology Platforms also provided a valuable contribution to identifying priority areas for future FP support. These

[98] Excluding duplicate responses.

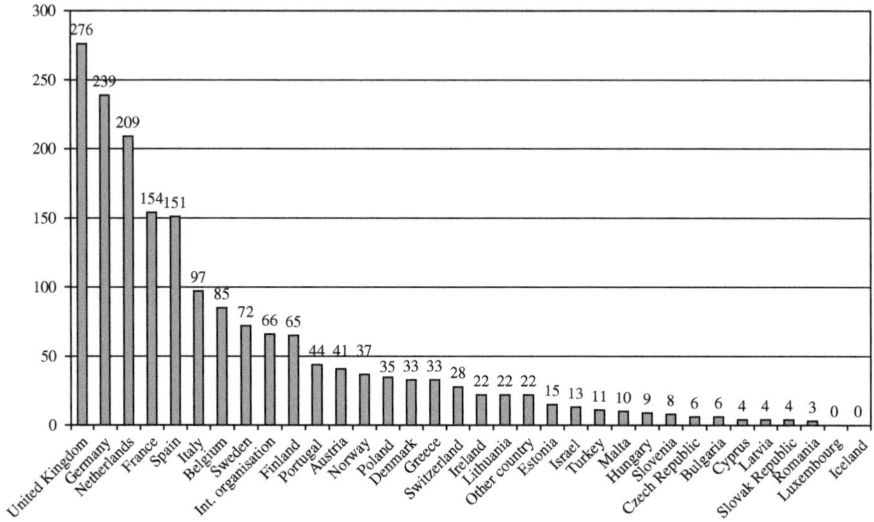

Fig. 5.8. Result of the "thematic" stakeholders' consultation by country
Source: DG Research

inputs were used in formulating the content of the high-level themes set out in the 7th Framework Programme proposal and its Specific Programmes.

For the online consultation on the rules for participation and simplification, around 180 respondents sent their comments.[99] Analysing the response by type of respondents (Table 5.6), we notice that approximately 30 per cent came from individuals and 70 per cent from organisations. The highest number of individual responses came from universities/higher education (44.2 per cent), followed by research institutes (23.1 per cent). The highest number of organisational responses also came from universities/higher education (17.8 per cent), followed by associations (14.5 per cent) and government bodies (12.2 per cent). Compared to the two other consultations, it is interesting to point out the much higher involvement of large companies (11.1 per cent) which appeared to be greatly concerned about simplifying the FP.

While online stakeholder consultations provide a rich pool of opinions, a major challenge is to ensure that the responses representative of the target population.[100] For the 7th Framework Programme consultation, the response reflected the wide range of actors involved in research activities. Individuals, SMEs, large companies, research institutes, public administration and so on all expressed their opinion and concerns for the future of research in Europe and the design and implementation

[99] Comments sent after 9 September 2005 are also taken into account. It is not possible to give a detailed analysis by country because this information is not available.
[100] Pronéos *et al.*, *Private Sector Interaction in the Decision Making Processes.*

Table 5.6. Result of the stakeholders' consultation on simplification by type of respondent

Type of respondent	Number	%
Individual person	52	28.9
Government body	22	12.2
University/higher education	32	17.8
Commercial organisation (including consultancy) more than 250 employees	20	11.1
Commercial organisation (including consultancy) less than 250 employees	9	5.0
Research institute	11	6.1
Association (e.g. trade association, trade union, employers association, chamber of commerce, NGO)	26	14.5
Other	8	4.4
Total	**180**	**100**

Source: DG Research

of the 7th Framework Programme. Overall then the consultation achieved a good degree of representativeness. However, some sources of bias in the sample of respondents should be mentioned.

Some stakeholders would inevitably have been under-represented because of the "digital divide".[101] This factor is especially important in the new Member States.[102] Countries also differ in the extent to which they have a "consultation culture", and therefore in the extent to which people are used to replying to such questionnaires. Other factors which may have introduced some bias in the results include the (technical) language barrier.

[101] For a definition of "digital divide" and the problems associated in the context of consultations, see section 2.

[102] Although around half of the households have Internet access at home (53 per cent in EU15 and 48 per cent in EU25) and their access has increased significantly over the last years, there are still large differences across the Member States. Especially new Member States and southern old Member States have a lower degree of Internet access. Whereas more than 60 per cent of households in countries such as the Netherlands (78 per cent), Sweden (73 per cent), Denmark (75 per cent), Germany (62 per cent) and the United Kingdom (60 per cent) have access to the Internet at home, less than 30 per cent of the households in countries like Slovakia (23 per cent), Hungary (22 per cent), the Czech Republic (19 per cent), and Lithuania (16 per cent) have Internet access. The situation in the Candidate countries is worse: 10 per cent or less of the households have access to the Internet (Bulgaria: 10 per cent, Turkey: 7 per cent, and Romania: 6 per cent). Analysing the share of individuals using the Internet for interacting with public authorities, we notice that only around 20 per cent of individuals (22.9 per cent in EU15 and 20.7 per cent in EU25) use the Internet to obtain information from public authorities' web sites, only 10.5 per cent download official forms, and only about 5.5 per cent send filled forms to public authorities via the Internet (Source: Eurostat)

Inevitably, if one asks more complex questions the response rate may fall. In the online consultation, responses to one or two issues of a more technical nature may have been limited to those familiar with these topics. One for example required some basic knowledge of the EU Structural Funds.

Since the consultation was promoted on CORDIS – "an innovation space devoted to European R&D and innovation activities" – the majority of respondents also had a background or a specific interest in research.[103] These respondents may therefore be more positive when confronted with general questions on whether to strengthen government support for research (see for example Section 5.2).

One should be aware of these problems in interpreting the results. Because of these limitations, Internet consultations should not be used as the only tool to consult stakeholders but should be complementary to other consultation methods.

---

[103] CORDIS visitors usually have a specific interest in research, contrary to Eurobarometers which consult the general public (see Section 1). The online consultations are accessible on http://europa.eu.int/yourvoice/ (last accessed on 28 August 2006).

CHAPTER 6

EUROPEAN ADDED VALUE, OPTION ANALYSIS, AND THE COMMISSION'S PROPOSAL FOR THE 7TH FRAMEWORK PROGRAMME

As discussed in earlier chapters, a broad range of inputs helped shape the Commission's new Framework Programme proposal. Account was taken of the European societal challenges to be addressed, the role that S&T could play in addressing those challenges, Europe's S&T weaknesses, the scope for public intervention, the European (research) policy context since Lisbon, the lessons learned from the implementation of past Framework Programmes, and the views of stakeholders.

Two additional inputs underlying the formulation of the Commission's 7th Framework Programme proposal were the consideration of European added value, and the analysis of options and scenarios. The former consisted of making sure that the objectives of the actions proposed for inclusion in the Framework Programme were indeed best pursued at European rather than at national/regional level. The latter consisted of making sure that the implementation options selected were the best among the available options.

The chapter is organised as follows. Section 1 considers the issue of European added value. Section 2 considers option analysis with regard to three issues: maximising the macro-economic benefits of the Framework Programme, raising the quality and riskiness of European basic research, and raising the competitiveness of Europe's industries through research. Section 3 recapitulates all inputs discussed so far that helped shape the Commission's 7th Framework Programme proposal and provides a brief overview of the 7th Framework Programme proposal as adopted by the Commission. And Section 4 provides a brief overview of the expected impacts of the proposal.

1. CONSIDERING EUROPEAN ADDED VALUE

During the formulation of the Commission's 7th Framework Programme proposal, the issue of European added value had to be considered. In accordance with the Commission's impact assessment guidelines, evidence had to be brought forward that the objectives of the actions proposed for inclusion in the Framework Programme were best pursued at European rather than at national/regional level. The issue of European added value relates to that of subsidiarity and to the "vertical" division of labour between the Union and the Member States and regions. It is not an easy concept. In the field of research, however, it has been possible to define relatively precisely the nature of actions characterised by European added value.

1.1. The concept of European added value

The concept of European added value underlies practically all policy areas and activities in which the EU is involved, and relates closely to the principle of "subsidiarity" which is neatly expressed in the Treaty: "In areas which do not fall within its exclusive competence, the Community shall take action, in accordance with the principle of subsidiarity, only if and in so far as the objectives of the proposed action cannot be sufficiently achieved by the Member States and can therefore, by reason of the scale or effects of the proposed action, be better achieved by the Community".[1] A protocol on subsidiarity and proportionality annexed to the Treaty of Amsterdam provides more precise guidelines for examining whether the subsidiarity condition is fulfilled. This can occur when "the issue under consideration has transnational aspects which cannot be satisfactorily regulated by action by Member States; actions by Member States alone or lack of Community action would conflict with the requirements of the Treaty (such as the need to correct distortion of competition or avoid disguised restrictions on trade or strengthen economic and social cohesion) or would otherwise significantly damage Member States' interests; action at Community level would produce clear benefits by reason of its scale or effects compared with action at the level of the Member States".

The European added value of Union action concerning the internal market, for instance, is easy to understand: it concerns a matter with transnational aspects; if left to the Member States alone there might be damaging effects; and if done at EU level there will be clear benefits for all. In the same vein there is European added value in the EU responsibility for trade negotiations since the Union representing 460 million consumers has more clout in trade discussions than the government of any single Member State.

1.2. European added value in the field of research

Actions implemented under Articles 163–173 of the Treaty, which describe the objectives of EU RTD policy and define the Framework Programme as the basic implementation mechanism, must be characterised by European added value.[2] As a result, European added value has been the guiding principle across Framework Programmes, and research collaboration through transnational consortia a key vector. The idea is that financing at EU-level research teams composed of researchers coming from different Member States will generate Europe-wide benefits of scale and scope, leverage additional funding, pool scarce resources, enable the undertaking of "big science", strengthen research capabilities, foster excellence, enable the tackling of pan-European challenges, encourage multi-level policy coordination, and make knowledge available to a much wider constituency. The kinds of objectives giving rise to European added value can be grouped into

---

[1] Treaty Establishing the European Community, *Official Journal C 325* of 24.12.2002.

[2] For those interested in the Treaty articles, the full text can be viewed at: http://eur-lex.europa.eu/en/treaties/index.htm.

three main categories: (1) pooling and leveraging resources; (2) fostering human capacity and excellence in S&T; and (3) better integration of European R&D.

1.3. Pooling and leveraging resources

EU funding impacts upon the scale and scope of research efforts. It is able to bring together what might otherwise be fragmented R&D resources to achieve a "critical mass" not attainable at national level, and is effective in leveraging new resources which would not have been mobilised without EU intervention.

EU intervention can achieve critical mass, enhancing economies of scale and scope
Some research activities are of such a scale and complexity that no single Member State can provide the necessary financial or personnel resources. They need to be carried out at EU level in order to achieve the required "critical mass". This occurs where a large research capacity is needed and resources must be pooled to be effective, or where there is a strong requirement for complementary knowledge and skills (e.g. in highly inter-disciplinary fields). Given the rising costs of carrying out R&D, economies of scale and scope are increasingly important.[3]

Framework Programme projects tackle this problem by establishing international consortia that bring together resources and expertise from many Member States and research actors. The average FP6 shared-cost project has a budget of € 4.6 million and involves more than 14 participants coming from at least 6 Member States, often combining universities, public research centres, SMEs and large enterprises. The 6th Framework Programme projects are substantially larger than those under the 5th Framework Programme, and one can now speak of an even greater critical mass effect than before. Such multinational and multidisciplinary research projects would be difficult to manage at national level, whereas intervention at EU level seems more adequate.

ITER is another example where the construction and operating costs are so high, that it is inefficient for countries to duplicate these investments, if not impossible.[4] ITER is "big science", a research and development project that aims to demonstrate the scientific and technical feasibility of fusion power. Fusion power can be used as an energy source to produce electricity in a safe and environmentally benign way. It is practically inconceivable for any single country to carry out such research on its own as the technical requirements to achieve this goal are enormous, and the

[3] For example, in pharmaceuticals, studies put the cost of researching and developing a new chemical entity (NCE) at € 895 million. Research is also highly risky: on average, only one out of 5000 to 10,000 promising substances will survive extensive testing in the R&D phase to become approved as a quality, safe and efficient marketable product. (European Foundation of Pharmaceutical Industries and Associations (http://www.efpia.org – accessed 8 September 2006)).

[4] Direct construction costs for ITER are estimated at $2755 million. These costs do not include staff costs for constructing ITER. Also to be added are the annual costs of operating ITER in the order of $188 million and totalling $3760 million over 20 years. These estimated costs include personnel costs (~32%), energy and tritium fuel costs (~20%), and capital improvements, spare parts and materials, and waste management operations (48%). For more info see: http://www.iter.org (accessed 8 September 2006).

budget needed runs to many billions of euros. This is why ITER currently involves a consortium of six partners: the European Union (represented by EURATOM), Japan, the People's Republic of China, India, the Republic of Korea, the Russian Federation, and the United States. After the International Space Station, it is the largest global research and development collaboration.

The scale of research is also important when investigating rare diseases. Since they affect less than one person in 2 000, there are often only a few cases per country. EU collaborative research provides a substantial added value by increasing the number of patients available for each research project and bringing together the scattered specialists with complementary expertise, for developing new diagnostics and treatments. For example, the European cluster on the Genetic resolution of Myopathies (the 5th Framework Programme project involving 13 partners from 6 different countries) merged the individual cohorts of congenital muscular dystrophy (CMD), thus achieving the largest collection of CMD families in the world.

EU intervention has a leverage effect on private investment There is considerable evidence that public funding of R&D carried out by enterprises leads to what is called a "crowding-in" effect on investment: in other words, it stimulates firms to invest more of their own money in R&D than they would otherwise have done. A recent study estimated that a € 1 increase in public R&D investment induced € 0.93 of additional private sector investment. In the case of the Framework Programme, there is evidence that many projects would not have been carried out at all without EU funding. Table 6.1 summarises results from a number of recent studies. The consistent picture is that in approximately 60–70 per cent of cases the Framework Programme enables research activities to take place that would not otherwise have occurred.

EU support for R&D encourages a particular type of research project, in which private companies can collaborate with foreign partners at a scale not possible at national level, in projects tested for excellence, and gain valuable access to complementary skills and knowledge. It is therefore reasonable to conclude that the attractiveness of EU schemes induces firms to invest more of their own funds than they would under national funding programmes.

Table 6.1. Additionality of the Framework Programmes: Participants that would have abandoned the project without funding (%)

ATLANTIS 5 year assessment (2004)	**58**
NIFU *et al.* (2004) FP5 Norway	**95**
AFSK (2000) – FP4 Denmark	**90**
Technopolis (2001) FP4 Austria	**70**
GOPA – Growth programme (2003)	**65**
Technopolis (2004) – FP5 UK	**70**
Technopolis (2001) FP4 Ireland	**82**
Uotila *et al.* (2004) FP5 Finland	**75**

Source: DG Research

Large-scale European projects enable participants to access a much wider pool of firms in their own industry than would be possible at purely national level. This mechanism offers clear advantages to enterprises compared with national level schemes. It broadens the scope of the research, and allows for a division of work according to each participant's field of specialisation.

It also considerably reduces the commercial risk, because involving key EU industry players helps ensure that research results and solutions are applicable across Europe and beyond, enables the development of EU- and world-wide standards and interoperable solutions, and offers the potential for exploitation in a market of 450 million people.

1.4. Fostering human capacity and excellence in S&T

A second group of factors underpinning European added value relates to the ways in which EU intervention can help to improve the quality of research and of researchers. This notably includes EU actions to develop and better exploit the human resources in S&T through Europe-wide training and mobility schemes. It also covers the promotion of quality and excellence in S&T which can result from cross-border knowledge transfers and from EU level competition for funding.

EU intervention offers unknown training possibilities to researchers while enhancing Europe's research capacity EU actions through the Marie Curie programme provide a coherent framework to address research training and mobility at all stages of researchers' careers in both academia and industry, so helping to make Europe as a whole more attractive for researchers. National schemes have not played a strong role in promoting such transnational actions, partly because they are difficult to organise at Member State level. Framework Programme activities with coordination at EU level have therefore been the principal driver. From a cost-effectiveness point of view, EU-level measures avoid the higher costs of bilateral arrangements between Member States, while providing a common framework which promotes reciprocity between countries.

Carrying out such actions at EU level provides a more harmonised approach than would be possible through purely national schemes. It offers a considerable structuring effect throughout the European Union on the organisation, performance, and quality of research training, researchers' career development, the participation of women in research and knowledge sharing in all domains of research addressed under the Treaty, explicitly including sustainable pathways between academia and industry (including SMEs) and between disciplines.

The actions, based on transnational as well as on intersectoral mobility, underpin the development of a genuine European labour market for researchers with good career perspectives, in support of a beneficial 'brain circulation'. Participants of EU-funded training projects indicated that Commission funding has been valuable in two ways: providing training in new areas of research, and bringing specialised groups together that were not collaborating before.

EU action aiming at the transfer of skills and knowledge through people across frontiers offers Europe-wide synergies which transcend the capabilities of individual Member States.

EU intervention as a means to promote competition in research while harvesting excellence The level of competition in research varies according to the Member State or region, in both basic and applied research. In basic research, many fields of science are highly specialised, and there may be only a handful of top-level experts in a country, especially in small Member States. Thus, there can be limited competition between research teams at national level. Similarly, many industries are highly concentrated, and one needs to search at a global level for specialist firms who can compete in a given market or area.

Public R&D funding at national level can only provide effective competition in such cases when it is opened up to researchers from abroad. However, experience to date would suggest that the opening up of national programmes has been very limited and not easy to implement.

Intervention at the EU level, however, has proven to be an effective way of promoting more intense competition in research, leading to higher quality and excellence. The calls for proposals launched under the Framework Programme have stimulated competition between universities, companies, and research centres across a wide geographical area: 30 countries participated in the 2nd Framework Programme, rising to 140 countries in the 5th Framework Programme. Preliminary figures for the 6th Framework Programme show that 68 countries participated in the first call alone.

The high levels of over-subscription for the Framework Programme funds are further evidence of the huge interest of European researchers, resulting in an intense competition induced by EU-wide calls for proposals. First data for the 6th Framework Programme indicate that around one in five proposals are retained for funding for the "Integrating and strengthening" area.

1.5. Better integration of European R&D

A third group of objectives giving rise to European added value relates to the better integration of research and development in Europe.

EU intervention as a way to deal with pan-European policy challenges Public policy challenges have increasingly taken a global dimension (e.g. environment, health, food safety, climate change) and can be faced only on the basis of a common scientific endeavour. Given the shared interest and the scale on which these issues arise, such research activities are more effectively carried out at EU rather than at national level. The need has also been identified to support – by excellence in science, technology, and innovation – community policies such as the all-embracing maritime policy.

For example, the EU Common Fisheries Policy requires a strong scientific base for fisheries management and for technical advice. Similarly, the EU needs an efficient and sustainable transport system if it is to become a more competitive global region. Inputs from Community RTD projects have played an important role in shaping such EU policies.

Encouraging the coordination of national policies As analysed earlier, there is still considerable overlapping and compartmentalisation of national research efforts. Better coordination of policies can help to target public investments more efficiently and reduce fragmentation. EU funding exercises a "catalytic" effect on national initiatives and improves the coordination of the activities of the Member States in areas of common interest (natural hazards, climate change).

Carrying out research at EU level to enable comparative research Certain fields of research, by their nature, are best explored in a comparative international context. For example, research in the social sciences often seeks to explore issues against the different historical, institutional, and cultural backgrounds prevailing in different countries (e.g. European social diversity, migration, comparative economic studies). EU research mechanisms provide the opportunity to assemble international research teams, each with its own national expertise and insights, and to set up pan-European research tools such as cross-country surveys.

Dissemination of research results Dissemination is one of the most important arguments for research at EU level. It is a critical complement to research itself. If results are not well disseminated, the value of carrying out research is seriously diminished. Indeed, one of the key justifications of public expenditure on R&D is to maximise the social return through ensuring the widest possible dissemination of research results. Moreover, in the context of the Lisbon objectives, research can only contribute to economic growth, to competitiveness, and to job creation if the results are disseminated to the European business sector so that they can be transformed into new products, processes, or services.

Carrying out this dissemination at an EU level – to users, industries, firms (SMEs in particular), citizens, and so on – is more efficient and leads to a better exploitation of research, with a larger impact than would be possible only at Member State level. Given the classical obstacles of language, proximity, and so on, when research is carried out at national level, it can be difficult for researchers abroad to access this new knowledge if no special incentives are provided. In addition, a country may generate important results in a particular area of science, but if it has no industrial activity corresponding to this discipline, then commercial exploitation may be severely hampered. EU-level research teams provide a powerful mechanism for disseminating results internationally. Unique and easy access to results through CORDIS enables knowledge to be shared with firms across all Member States and associated states, and to be exploited to commercial advantage. In 2004 alone, the CORDIS website had more than 2.5 million different users.

2. ANALYSING POLICY OPTIONS

The analysis of policy options and scenarios consists of making sure that the options selected are the best among the available ones. The best policies are made through an iterative process, which involves analysing alternative options and understanding the pros and cons of each before finally selecting the preferred option. The emerging practice of impact assessment in the European Commission formalised this process by requiring an assessment of the expected positive and negative impacts and risks associated with each one of a limited set of credible policy options.[5] In this section we look at options in relation to three issues: maximising the macro-economic benefits of the Framework Programme, raising the quality and riskiness of European basic research, and raising the competitiveness of Europe's industries through research.

2.1. Options for maximising the macro-economic impact of the 7th Framework Programme

The Framework Programme constitutes the Union's main policy tool in the field of research, which together with education and innovation is situated at the heart of the Lisbon strategy. The 7th Framework Programme is therefore expected to contribute as much as possible to the achievement of the Lisbon objectives. That is why it was important to maximise the macro-economic impact of the 7th Framework Programme.

Formulating options The determinants of the macro-economic impact of the 7th Framework Programme are many: the size of the programme, the crowding-in rate, the economic multiplier, the behaviour of the Member States, and so on. That means that in reality thousands of policy options and sub-options could be considered for a programme of the size and range of the Framework Programme.

There was a need to focus, and the Commission opted for programme size as the key variable around which to build different scenarios. Views on what constitutes the most optimal level of EU policy intervention in the area of research differ widely. That is because ultimately it touches upon the question of what research governance model to envisage for Europe. The division of public R&D resources and responsibilities between EU, national and regional levels will be one of the most important issues for EU policy in the coming decades because it will dictate how effectively Europe uses its scarce resources for R&D.

---

[5] "The growing use of scenarios as anticipatory planning tools results from the robustness and generality of the approach. Articulating an explicit story about how the future may unfold forces strongly held, but vaguely defined viewpoints into written specificity" (A. Branscomb, Scenario Development, In: David Hulse *et al.* (Eds.), *Willamette River Basin Planning Atlas – Trajectories of environmental and Ecological Change*, Corvallis, 2002, p. 84). In the FP7 impact assessment work three alternative futures of European research policy were analysed whereby these three policy options were considered "plausible alternative policy sets" (Idem, p. 84).

It is with these considerations in mind that the following three main policy options were chosen for carrying out the 7th Framework Programme impact assessment:[6]

Option 1 – "Business-as-usual Framework Programme": This scenario consisted of launching the 7th Framework Programme as a continuation of the 6th Framework Programme, with perhaps a marginally larger budget, but with essentially the same objectives, the same institutional actors, the same research priorities, the same instruments, and so on. The inclusion of such a business-as-usual or no policy change baseline scenario against which the impacts of other policy options could be compared and assessed was fully in line with Commission Impact Assessment guidelines.[7] The premise underlying this option was that the 7th Framework Programme could adequately address the major challenges facing Europe in the next few years without introducing any major changes to its size, structure, and organisation. This option responded most clearly to the important concerns about the continuity and stability of EU research actions. And it represented the view that Europe has already attained, or even gone beyond the optimal level of funding for EU research intervention, and that the Framework Programme should either be frozen at its current size or scaled back. For example, some work on alternative models for the governance of technology and innovation policies in the EU outlines several future scenarios, one of which involves a greater decentralisation of policies, particularly towards the regions.[8] It is argued that this trend might be driven by the increasingly heterogeneous membership of the enlarged EU, in terms of their economies, politics, and culture, and a functional breakdown stemming from too many contradictory regional, national, and sectoral interests.[9]

Option 2 – "A much larger Framework Programme": This was the option of radical change (corresponding to the 7th Framework Programme proposal finally submitted). It consisted of launching a restructured Framework Programme, twice as large as the 6th Framework Programme, and designed so as to better respond to the targets set at Lisbon via old as well as new actions (European Research Council, JTIs). It started from the observation that circumstances have changed very significantly since the launching of the 6th Framework Programme, and proposed an action that built upon the accomplishments of the 6th Framework Programme, but was characterised by a new scale, scope, and ambition. It was based upon

---

[6] European Commission IA guidelines recommend that a maximum of four or five promising policy options be short-listed (European Commission, *Impact Assessment Guidelines*, SEC(2005) 791, 15 June 2005). In practice, however, most Commission impact assessments analyse just three basic policy options. Outside the European Commission too, working on the basis of just three basic policy options seems to be accepted practice. See, for instance, the UK Regulatory Impact Assessments accessible at http://www.dti.gov.uk/consultations/ria/index.html (accessed 8 September 2006).

[7] European Commission, *Impact Assessment Guidelines*, pp. 18–19.

[8] Karen Siune, *Convergence in European Research Policy: Strength or Weakness*?, Note from The Danish Centre for Studies in Research and Research Policy 2005/2, Aarhus, 2005.

[9] Stefan Kuhlmann and Jakob Edler, Scenarios of Technology and Innovation Policies in Europe: Investigating Future Governance, In: *Technological Forecasting & Social Change*, 70, 2000, pp. 619–637.

the premise that there is scope for funding more research that has a higher added value by virtue of being carried out at an EU level. By doing so, more impacts can be generated and the taxpayer will get better value for money than would be the case if research spending was performed at 25 separate national levels. It was in accordance with the new policy context and the demands of stakeholders.

Option 3 – "No 7th Framework Programme": This scenario consisted of having no intervention at EU level in the field of science and technology, with compensating policy measures being provided at national level – in other words this was the extreme scenario of a (partial or complete) re-nationalisation of EU research policy. Commission Impact Assessment guidelines state that the option of "no EU action" (i.e. no new EU intervention, but also the possible abandoning or downgrading of existing EU policies) should always be considered, but not where there is an obligation to act laid down by the Treaties (which is the case for the Framework Programme).[10] There were important reasons for nevertheless including a "no 7th Framework Programme" option. First of all, setting aside the Treaty obligation, a scenario entailing the re-nationalisation of public support for research is not completely unrealistic when, as was seen above, there is substantial discussion on the proper division of labour between the EU and the national levels and significant concern about issues of subsidiarity. Perhaps more importantly, the Treaty obligation cannot by itself guarantee the long-term survival of the Framework Programme. The case for the Framework Programme has to be made again and again and in a convincing manner. Within this context it was expected that the difference between the results of the assessment of the "no policy" option on the one hand and those of the assessments of the "business-as-usual" and "new policy" options on the other hand would serve to demonstrate the full added value of the Framework Programme. The inclusion of this option was done not only to give the Euro-sceptics a voice and assess their argumentation for "less Europe" seriously, but to force the policy-making community to think out of the box, to make implicit arguments explicit, and by doing so offer greater transparency in the policy formulation process and producing, in the end, better quality legislation.[11]

Choosing a model There are different possible approaches to estimating the aggregate macro-economic impacts of programmes. A simple approach consists of obtaining from the ex-post evaluation literature the average extra turnover generated, the average number of jobs saved or created, and so on per project, estimating the future number of projects, and carrying out the multiplication. Another approach consists of estimating the macro-economic impact of the proposed programme on the basis of what the literature tells us about the crowding-in effect of public funding for research and the macro-economic impact of R&D investment.

A third approach consists of making use of a mathematical model based on known, inferred, and assumed parameters. Over the past few years, the use of

[10] European Commission, *Impact Assessment Guidelines*, p. 24.
[11] European Commission, *Impact Assessment Guidelines*, p. 23.

mathematical models for the ex-ante estimation of policy effects has increased significantly within the Commission, and also at national level.[12] Though such use is sometimes viewed with concern by policy-makers, who fear that models cannot capture the complexity of the real world, it is a heuristic tool that can play an important supporting role in the design of well-targeted research policies. With all mathematical modelling a degree of uncertainty is associated, yet scenario development for mathematical modelling makes explicit the assumptions underlying and the risks associated with policy proposals. While not exact in the narrow sense, ballpark impact figures are obtained for each scenario allowing for policy optimisation.

Mathematical models have already been used for assessing the macro-economic impact of research projects and programmes. One example was presented in chapter 4 of this book and concerns the use by the United Kingdom's government of a model developed at the OECD to estimate the macro-economic impact of the Framework Programme on the UK economy, an approach subsequently extended by Directorate-General Research to estimate the Framework Programme's impact on the whole of the European economy.[13] The United States' "Advanced Technology Programme" (ATP) has also experimented with the use of mathematical models to estimate the impact of research programmes and projects. A recent study, for instance, used a regional econometric model to examine the economic impact of the ATP-funded CONSAD project to improve control of dimensional variation of stamped metal parts in the production of automobile bodies and other discrete-parts manufacturing. The study projected increases in GDP of at least US$ 3 billion and employment gains of several hundred thousand jobs. These estimates took into account inter-industry effects of a predicted small increase in market share for US automobile producers attributable to a quality improvement resulting from the ATP project.[14]

The model approach is the one used in the 7th Framework Programme impact assessment. The model chosen was Némésis, a "general equilibrium model" (Box 6.1). This was the first time that the macro-economic impact of a European Commission expenditure programme in the field of research had been estimated in

[12] For instance, to analyse the impact of Lisbon-type structural reforms (European Commission, *The Economic Costs of Non-Lisbon – A Survey of the Literature on the Economic Impact of Lisbon-type Reforms*, Commission Staff Working Document, SEC(2005) 385, 15 March 2005); to assess the macro-economic effects of the Community Support Frameworks (European Commission, *Unity, Solidarity, Diversity for Europe, Its People and Its Territory. Second Report on Economic and Social Cohesion*, Luxembourg, 2001); to estimate the macro-economic consequences of reaching the 3 per cent objective in 2010 (Dorothée Brécard *et al.*, *A 3 per cent R&D Effort in Europe In 2010: An Analysis of the Consequences, Using the Némésis Model*, January 2004). A recent Dutch scenario building and modelling exercise is: Ruud de Mooij and Paul Tang, *Four Futures of Europe*, Den Haag, 2006.

[13] DTI – Office Of Science And Technology, *Targeted Review Of Added Value Provided By International R&D Programmes*, London, May 2004, p. 8. This study uses the model developed at the OECD by Guellec and van Pottelsberghe, which is presented in the following paper: Dominique Guellec and Bruno van Pottelsberghe de la Potterie, *R&D and Productivity Growth: Panel Data Analysis of 16 OECD Countries,* STI Working Papers 2001/3, Paris, 14 June 2001.

[14] http://www.atp. nist.gov/eao/gcr-709.htm (accessed 8 September 2006).

an ex-ante or prospective manner. Although there is always room for improvement, the pioneering status of this venture is a fact, and it represents a qualitative jump in the ex-ante impact assessment of research programmes unknown before.[15]

Box 6.1. The Némésis model

In the 7th Framework Programme impact assessment, use was made of the Némésis-model, initially an econometric model and now de facto a GEM at the macro- and sectoral levels built by a Commission-funded consortium of European research institutes under 5th Framework Programme.[16] The original Némésis-model was built to estimate the impact on the European economy of reaching the Barcelona 2002 objectives (R&D intensity of 3 per cent of GDP, two-thirds financed by the private sector). The Némésis-model can be applied in a wide range of fields, however, including science; R&D; competition policy; industrial policy and internal market; employment; energy; transport; agriculture and fisheries; enlargement; employment and social policy; taxation; external relations; environment and health protection; and so on. The model has a good track record. It has been used for numerous policy analyses for French institutions (Ministry of Environment, ADEME, SENAT, Chambre de Commerce et d'Industrie de Paris), international institutions (OECD), and the European Union (e.g. to make an assessment of the 3 per cent RTD objective). The Némésis-model is large scale. It comprises roughly 70,000 equations, and all behavioural equations are econometrically estimated. The model's geographical and sectoral/product coverage is wide. The model is of a multi-country nature, covering the EU-15 countries plus Norway. For the time being, other countries are treated as exogenous and grouped into one of ten different world regions.

[15] Better data collection is needed. The model can be further extended and refined. A bridge is also needed between policy and academics so that relevant questions are asked and results are correctly interpreted.

[16] Némésis was undertaken within the Environment and Sustainable Development Programme (project EVG1-CT1999-0014) A number of models are available for estimating the impact of increased R&D investment on the European economy. These fall into two categories: computable general equilibrium type of models and more traditional models with econometrically estimated parameters. The former category includes, for instance, GEM-E3 (General equilibrium model for energy, economy and environment), NiGEM (Macro-economic model of the National Institute for Economic and Social Research London), HERMIN (General equilibrium model for cohesion countries), and Némésis (New econometric model for environment and strategies implementation for sustainable development). The latter includes, for instance, QUEST (Macro-economic model of the European Commission). While each kind of model has its advantages and disadvantages it would seem that general equilibrium type of models are better suited for assessing the economic impact of increased R&D investment, though with some adaptation. Econometric models cannot be very complex, covering many sectors of activity and many disaggregated variables. Limited time series availability also hampers these models. General equilibrium models, on the other hand, are sometimes static, complex, and not easy to manage, and suffer from theoretical rigidity but since they are better able to take account of inter-sectoral linkages in the economy they constitute a better starting point.

But efforts are now being made to include into the model the new EU Member States, the United States, and Japan. An effort is also made to make the model applicable to NUTS2 and NUTS3 regional levels for key variables such as production, value-added, investment, R&D, and employment. The model also covers 30 production sectors and 27 consumption good categories.
The model is highly innovative. Its supply-side block incorporates some properties of new theories of growth, for example endogenous R&D decisions, process/product innovations, and technological/knowledge spillovers between sectors and countries. Five types of conversion matrices – for technological transfers, final consumption, investment goods, intermediate consumption, and energy-environment – are used for describing interdependencies between activities. The Némésis-model also includes an energy-environment module, which transforms activity indicators from the macro-model at a sectoral level into energy-relevant indexes with price effects and pollutant emissions: CO_2, SO_2, NO_x, HFC, PFC, and CF_6. Each individual country is linked to others by external trade.[17]

Comparing impacts across scenarios Since the aim was to forecast impacts up to the year 2030, different assumptions were tested about future rates of growth of research spending at EU and at national level. At the time this first phase modelling was carried out, the 7th Framework Programme was assumed to cover the period 2007–2010.

(1) Business-as-usual (the "baseline scenario"): This scenario was based on a continuation of 6th Framework Programme funding levels, with a constant real growth rate of 3.87 per cent per year until 2030 (this was the average growth between the 4th Framework Programme and the 5th Framework Programme, and between the 5th Framework Programme and the 6th Framework Programme).

(2a) The 7th Framework Programme doubling with modest growth in R&D spending after 2010: After a doubling of funding under the 7th Framework Programme for the period 2007–2010, this scenario assumed that subsequent Framework Programmes would grow more modestly (3.87 per cent per year) from 2011 to 2030.

(2b) The 7th Framework Programme doubling with strong growth in R&D spending after 2010: Following the doubling of budget under the 7th Framework Programme, this scenario assumed a continued strong growth from 2011 to 2030 (7.74 per cent yearly).

[17] More information on Némésis can be found at: http://www.nemesis-model.net (accessed 8 September 2006). See also A. Fougeyrollas *et al.*, *The Némésis Model: New Econometric Model for Environment and Sustainable Development Implementation Strategies*, Brussels, 2002; OECD, *Environment and Employment: An Assessment*, Paris, 2004.

(3) No 7th Framework Programme with re-nationalisation of EU funding: Under this option, there was no Framework Programme from 2007 onwards and all proposed 7th Framework Programme funding was now spent at Member State level on purely national projects.[18] From 2007 to 2030, this ex-Framework Programme funding would grow at 3.87 per cent per year (as under option 1) while national funding would grow at 1.8 per cent.

Results were obtained for impacts on "growth", "competitiveness", "employment", and "R&D intensity". The impact of each scenario was estimated for each year in the period 2010–2030. Results were presented as deviations in a positive or a negative direction from the "business-as-usual" (the 6th Framework Programme) scenario.

The impact of doubling the Framework Programme is large. Under the "7th Framework Programme doubling with modest growth after 2010" scenario, GDP, extra-EU exports, (research) employment and R&D intensity grow to a significant extent, while the reverse is true for extra-EU imports (Figs. 6.1 and 6.2). The Framework Programme's macro-economic effects are substantially larger if one assumes a continued strong growth of Framework Programme funding in the years after the 7th Framework Programme. The impacts of the Framework Programme are largest in the medium- to long-term. It takes several years between the beginning and

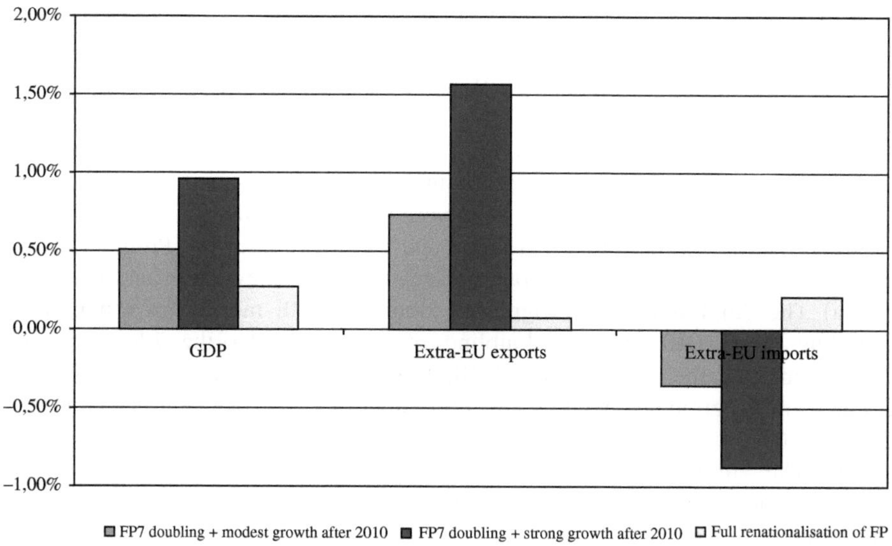

Fig. 6.1. Impacts by 2030 on Gross Domestic Product, exports and imports of Three Policy Options (% difference compared with the "business-as-usual" baseline scenario)
Source: DG Research

[18] Results are presented in this section for "full re-nationalisation" where all ex-FP money is spent nationally. Other "partial re-nationalisation" options were also examined, but are not presented here.

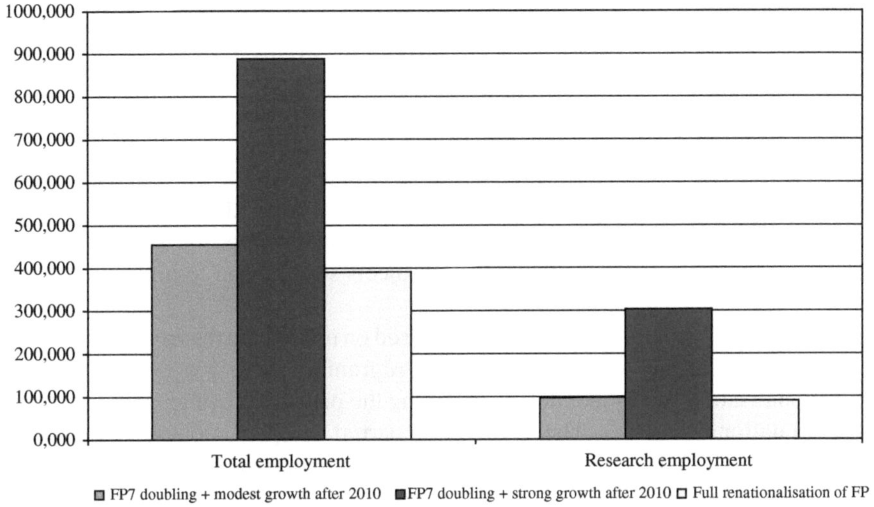

Fig. 6.2. Impacts on total employment and research employment in 2030 of Three Policy Options (change in number of jobs compared with the "business-as-usual" baseline scenario)
Source: DG Research

the end of the innovation cycle. Carrying out research, generating useful research results, transforming them into product and process innovations, and valorising them through higher turnover and so on takes time. That means that the maximum incremental effect of doubling the Framework Programme will take time to show. Clear positive effects will start to show from the years 2010–2015 onwards, however.

Figures 6.1 and 6.2 also show the very negative macro-economic effects that would be generated by fully re-nationalising the Framework Programme (with the same funding being spent instead on national schemes). In the beginning of this chapter, it has been argued that actions implemented at European level are characterised by European added value, which essentially means that they are more efficiently implemented at European than at national level. One can bring this into the model by assuming in the case of full renationalisation of the Framework Programme lower crowding-in factors and economic multipliers. Carrying out European added value actions at national level will make them less attractive to potential participants and draw in less additional research money directly and indirectly. At the same time, international collaboration will not be easily achieved, resulting in more limited dissemination of knowledge and more limited conversion of knowledge into new technology, resulting in smaller economic effects. On the basis of these assumptions, it is clear that there is a loss in terms of economic efficiency and macro-economic impacts obtained.

The above results are merely a subset of numerous variants of the three policy options which were simulated using the model. In some cases the assumptions of the model were modified in order to test the robustness of the findings. For example, one of the model parameters related to the distribution of the Framework Programme funding between countries. The results in Figs. 6.1 and 6.2 use an allocation based on each Member State receiving the same share of the Framework Programme funds as it obtained for the 5th Framework Programme. Another variant tested was to assume that Member States receive a share of the Framework Programme funding identical to its share of EU R&D. A third variant was based on distributing the Framework Programme amounts according to countries' S&T performance (share of scientific publications and patents). It was noteworthy that the 'performance based' allocation of funds showed significantly higher impacts, suggesting that targeting funds on excellent, output-driven projects should lead to higher economic returns.

Another set of simulations was performed based on modifying the re-nationalisation assumptions. Under the "no 7th Framework Programme" scenario, varying degrees of re-nationalisation were modelled by changing the per centage of ex-FP funds spent instead on national schemes. The results for a partial compensation at national level (< 100 per cent of ex-FP funds spent nationally, and the remaining funds being lost) generated still lower impacts than a full re-nationalisation scenario.

Other parameters included the distribution of the Framework Programme funding between basic and applied research, the distribution of the Framework Programme funding for applied research among different sectors in each Member State, the crowding-in effect of the Framework Programme funding, the economic multiplier of the Framework Programme funding, and the research investment of the Member States. Table 6.2 shows the values of the various parameters used in the different scenarios.

2.2. *Options for raising the quality of European basic research at the frontier*

Basic research is key to innovation. European basic research could be of higher quality and riskiness, however. An important reason for these weaknesses is the fragmentation of European basic research. The establishment of a European Research Council emerged as the best policy option to overcome the fragmentation of European basic research at EU level.

The importance of basic research for innovation High-quality basic research constitutes a critical input into the innovation process. It is, as it has been called, "the fuel for future technology development".[19] When combined with efficient and effective university–industry linkages, high-quality basic research ensures that companies have at their fingertips the input they need to be able to remain at the frontier of new product and process development. The previous, "linear" conceptual model of innovation, with a one-way and straightforward relation from basic research to technology to innovation, can no longer be accepted. Much more attention is paid these days to the interaction effects between basic research and technology. Basic research remains,

[19] Office of Science and Technology Policy – Executive Office of the President, *Promoting Innovation*, Final 1.31.03.

Table 6.2. Scenarios used for Némésis modeling of the 7th Framework Programme policy options

	Phase I scenarios			
	"Business-as-usual" Baseline scenario	FP7 doubling + modest growth after 2010	FP7 doubling + strong growth after 2010	Full renation-alisation of FP7
FP funding real growth rate 2007–2010	3.87%	Doubling	Doubling	No FP
FP funding real growth rate 2011–2030	3.87%	3.87%	7.74%	No FP
National funding growth rate 2007–2010	1.8%	1.8%	1.8%	1.8% (except for additional ex-FP funds which grow at 3.87%)
National funding growth rate 2011–2030	1.8%	1.8%	1.8%	1.8% (except for additional ex-FP funds which grow at 3.87%)
Allocation of FP funding to EU MS	FP5	FP5	FP5	FP5
Allocation of FP funding to basic and applied research	Same as FP 5	Same as FP 5	Same as FP 5	Same as FP 5
Allocation of FP applied research funding to sectors within MS	Share of each sector in MS' R&D expenditure	Share of each sector in MS' R&D expenditure	Share of each sector in MS' R&D expenditure	Share of each sector in MS' R&D expenditure
FP funding crowding-in factor for the private sector	1.1	1.1	1.1	1.1
FP funding crowding-in factor for the public sector	0	0	0	0
National funding crowding-in factor for the private sector	1	1	1	1

(Continued)

Table 6.2. (Continued)

	Phase I scenarios				
	"Business-as-usual" Baseline scenario	FP7 doubling + modest growth after 2010	FP7 doubling + strong growth after 2010	Full renation-alisation of FP7	
National funding crowding-in factor for the public sector	0	0	0	0	
Multiplier (Bèta) for R&D resulting from EC funding	Bèta higher for FP funding than for national research funding	Bèta higher for FP funding than for national research funding	Bèta higher for FP funding than for national research funding	Bèta higher for FP funding than for national research funding	
Multiplier (Bèta) for R&D resulting from national funding	Country-specific	Country-specific	Country-specific	Country-specific	
	Phase II scenarios				
	"Business-as-usual" Baseline scenario	Luxemburg proposal + business-as-usual MS R&D investment growth	Luxemburg proposal + substantial MS R&D investment growth	EC proposal + business-as-usual MS R&D investment growth	EC proposal + substantial MS R&D investment growth
FP funding real growth rate 2007–2013	3.5%	7.5%	7.5%	15%	15%
FP funding real growth rate 2014–2030	3.5%	5.5%	5.5%	9.25%	9.25%
National funding growth rate 2007–2010	3.2%	3.2%	6.4%	3.2%	6.4%
National funding growth rate 2014–2030	3.2%	3.2%	4.8%	3.2%	4.8%

Allocation of FP funding to EU MS	FP5	FP5	FP5	FP5	FP5
Allocation of FP funding to basic and applied research	Same as FP 5	Same as FP 5	Same as FP 5	Same as FP 5	Same as FP 5
Allocation of FP applied research funding to sectors within MS	Share of each sector in MS' R&D expenditure	Share of each sector in MS' R&D expenditure	Share of each sector in MS' R&D expenditure	Share of each sector in MS' R&D expenditure	Share of each sector in MS' R&D expenditure
FP funding crowding-in factor for the private sector	1.1	1.1	1.1	1.1	1.1
FP funding crowding-in factor for the public sector	0	0	0	0	0
National funding crowding-in factor for the private sector	1	1	1	1	1
National funding crowding-in factor for the public sector	0	0	0	0	0
Multiplier (Bèta) for R&D resulting from EC funding	Bèta 25% higher for FP funding than for national research funding	Bèta 25% higher for FP funding than for national research funding	Bèta 25% higher for FP funding than for national research funding	Bèta 25% higher for FP funding than for national research funding	Bèta 25% higher for FP funding than for national research funding
Multiplier (Bèta) for R&D resulting from national funding	Country-specific	Country-specific	Country-specific	Country-specific	Country-specific

however, of fundamental importance, especially as far as radical (as opposed to, for instance, engineering-based incremental) innovation is concerned, and increasingly so. As Tassey recently noted, "it is hard to imagine apoptosis, antisense, monoclonal antibodies, or other generic biotechnologies being developed through experimentation rather than derived from previous advances in microbiology".[20] To have an impact, such basic research has to be situated at the frontier.

Europe's weak basic research performance The picture regarding Europe's basic research performance is mixed. Ever since the mid-1990s – and to great alarm in the United States – Europe is substantially outperforming the United States in terms of the overall share of world scientific output.[21] In 2003, for instance, the EU-25 accounted for 38.3 per cent of the world's scientific publications, as compared to 31.1 per cent for the United States and 9.6 per cent for Japan.[22] On the other hand, the publication "intensity" is much lower in Europe than in the United States. In 2003, for instance, the number of scientific publications per million inhabitants amounted to 639 in the EU-25, as compared to 809 in the United States and 569 in Japan.[23]

The EU is catching up with the United States in terms of the overall share of world citations.[24] And as far as highly cited papers as a per centage of the total number of scientific publications are concerned, a substantial number of EU countries score above the world average. However, Europe achieves a far lower citation impact than the United States.[25] Furthermore, Europe performs badly in those new and emerging fields where it matters. In terms of disciplines, the EU is on a par with the United States in the physical sciences, engineering, and mathematics, but lags in life sciences.[26] Based

[20] Gregory Tassey, Policy Issues For R&D Investment In A Knowledge-Based Economy, In: *Journal of Technology Transfer*, 29, 2004, p. 173.

[21] See, for example, National Science Board, National Science Foundation, Division of Science Resources Statistics, *Science and Engineering Indicators 2004*, Arlington, VA, 2004. American Electronics Association, *Losing the Competitive Advantage? The Challenge for Science and Technology in the United States*, Washington, DC, 2005. *Rising Above The Gathering Storm: Energizing and Employing America for a Brighter Economic Future*, Statement of Norman R. Augustine, Retired Chairman and Chief Executive Officer Lockheed Martin Corporation and Chair, Committee on Prospering in the Global Economy of the 21st Century – Committee on Science, Engineering, and Public Policy – Division on Policy and Global Affairs, The National Academies, before the Committee on Science, US House of Representatives, 20 October 2005.

[22] European Commission, *Key Figures 2005: Towards a European Research Area: Science, Technology and Innovation*, Luxembourg, 2005, p. 59. See also David A. King, The Scientific Impact of Nations. What Different Countries Get for Their Research Spending, In: *Nature*, Vol. 430, 15 July 2004, p. 311, or US Slide in World Share Continues as European Union, Asia Pacific Advance, In: *Science Watch*, Vol. 16, No. 4, July/August 2005, pp. 1–2.

[23] European Commission, *Key Figures 2005*, p. 59.

[24] King, The Scientific Impact of Nations, p. 311; US Slide in World Share Continues pp. 1–2; European Commission, *Key Figures 2005*, p. 59.

[25] Andreas von Bubnoff, Asia Squeezes Europe's Lead in Science, In: *Nature*, Vol. 436, 21 July 2005, p. 314.

[26] King, The Scientific Impact of Nations.

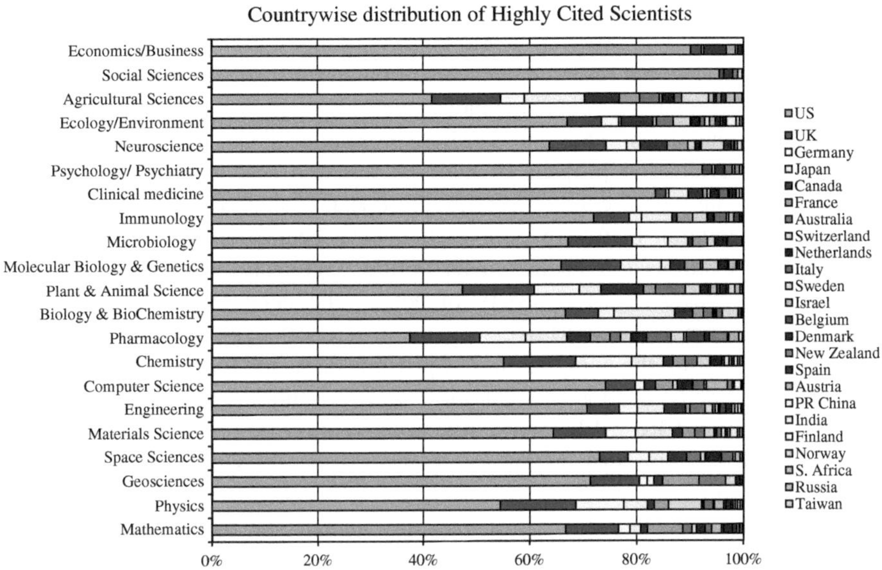

Fig. 6.3. Highly-cited scientists by country per field
Source and data: Aparna Basu, Institute of Genomics and Integrative Biology, New Delhi, India[27]

on Fig. 6.3, one can also question whether European basic research is situated at the frontier.

The basic challenge – the fragmentation of European basic research Europe compares rather well to the United States in terms of basic research investment. The bulk of the overall R&D investment gap between Europe and the United States is situated at the level of the private sector, while the investment gap in basic research is limited (Chapter 2). Thus the problem is situated elsewhere, namely in the extreme fragmentation of European basic research funding (Chapter 2). For instance, a majority of individual EU-15 Member States, as well as all new Member States combined, spend less on basic research than the United States's Johns Hopkins University alone. Nevertheless, most Member states, whatever their size, have their own special funding for all scientific disciplines. This results in very modest amounts of money being disbursed to individual fields in individual Member States (particularly in the smaller countries), and in the duplication of efforts across Member States. At the same time, especially in small Member States, the number of researchers active in a particular field may be overly small. This means that no meaningful competition can be organised at national level. A scaling up of the competition for funding is thus required.

[27] See also her paper Aparna Basu, Using ISI's "Highly Cited Researchers" to obtain a country level indicator of citation excellence, In: *Scientometrics*, Vol. 68, No. 3 (2006).

Options for overcoming fragmentation[28] One option for reducing the fragmentation of European basic research would be to build upon the collaboration already existing between national research agencies through, for instance, the further development of the ERA-NET scheme. Under this option, policies would aim at the reciprocal opening-up of national research programmes and the encouragement of convincing evaluation projects of national research activities by international panels made up largely of experts from the European countries. This would amount to the implementation in full of policies that have been proposed in the past by the European Commission itself.[29] This option would suffer from some major weaknesses, however: "The diversity of actors, the need to combine different statutory arrangements and objectives, and the differences in governance and instruments, not to mention the complication of possible demands for juste retour, would make this extended collaboration far more time-consuming and cumbersome to implement than an ERC". ERA-NET like collaboration is important, and there is no substitute for it, but since progress is piecemeal results will only be generated in the longer term.

Another option would consist of radically increasing the amount of funds allocated to the current "New and Emerging Science and Technology" (NEST) activity within the Framework Programme. While NEST goes a long way towards constituting "a responsive and excellence-driven funding mechanism open to multidisciplinary research", however, and valuable lessons can be learned from its implementation, it requires collaborative proposals, and has no independent governance structure.

The above considerations underpinned the conclusion that a new organisation needed to be created for funding European basic research, the projected impacts of which would be large. It would, for instance, be able to fund the research that some national governments cannot or do not wish to support (because too risky, too intrinsically international in nature, not seen as a priority at national level). It would be able to channel funds more quickly into fast-moving research areas. It would stimulate change among other research-funding agencies (though the risk of crowding-out national funding exists). And it would help create a visible community of leading European scientists, which would enable Europe to play its role in international research, and contribute to sustainable development and the solution of global problems.

---

[28] This section based on European Commission, *Frontier Research: The European Challenge – High-Level Expert Group Report*, Luxembourg, February 2005, pp. 22–24 unless indicated differently.

[29] European Commission, *Towards a European Research Area*, Communication from the Commission to the Council, the European Parliament, the Economic and Social Committee and the Committee of the Regions, COM(2000) 6 final, 18 January 2000, p. 12.

2.3. Options for raising the competitiveness of Europe's industry through research and innovation

As discussed in Chapter 1, Europe's feeble competitive position is clear from its weak trade performance, especially that at the high-tech end. Europe's weak high-tech trade performance reflects in turn its weak technological performance. Europe fails to create value through commercially valuable and successful innovations from science-based findings and new knowledge resulting from research and innovation.[30] With regard to raising the competitiveness of Europe's industry through research and innovation, the establishment of JTIs emerged as the best policy option.

Europe's weak technological performance After a debate going back a number of years, there now exists general consensus on the existence of a "European Paradox", under which Europe has a relatively strong research base, but fails to exploit this downstream when it comes to improving its competitiveness, developing new products and processes, and boosting productivity.[31] In relation to its population, Europe generates fewer patents with high economic value than the United States or Japan. In 2000, the EU-25 had 31 so-called "Triadic patents" per million population, versus 53 for the United States, and 93 for Japan.[32] The EU's share of Triadic patents is also less than that of the United States (31.5 per cent versus 34.3 per cent for the United States) (Fig. 6.4).[33] The United States also has a higher share of patent applications at the European Patent Office (27.5 per cent) than the EU has at the US Patent Office (16.3 per cent).[34] On a yearly basis, Taiwan makes more patent applications to the US Patent Office than either France or the UK. In 2003, only one European company featured in the top 10 recipients of patents granted by the US Patent Office. In key areas such as biotechnology and information and communications technologies, Europe also lags behind in its share of patents. In biotechnology, the EU-25 accounts for only 25 per cent of high-value "triadic"

---

[30] For an industry view on this see, for example, UNICE Message to the European Institutions 2004-2009, where they state that "successful research and innovation could quite possibly make the largest contribution towards improving Europe's competitiveness".

[31] See, for example, the discussion in 'Overcoming the European Paradox', 28 February 2002, from UK Prime Minister Tony Blair and Dutch Prime Minister Wim Kok to Prime Minister Aznar of Spain, ahead of the Barcelona Summit.

[32] European Commission, *Key Figures 2005*, p. 61. Triadic patents relate to inventions patented at all three major patent offices: the European Patent Office, the US Patents and Trademarks Office, and the Japanese Patent Office. It is thought that such patents are likely to be associated with a higher expected commercial return, since it is costly to patent in three different patent systems.

[33] European Commission, *Key Figures 2005*, pp. 60–61.

[34] European Commission, *Towards a European Research Area – Science, Technology and Innovation – Key Figures 2003–2004,* Luxembourg, 2003, p. 67. It is of course to be expected that, due to "a home advantage", European inventors tend to apply for more patents in their own system (EPO) than does the United States, and that US inventors tend to apply more than the Europeans at the US patent office (USPTO).

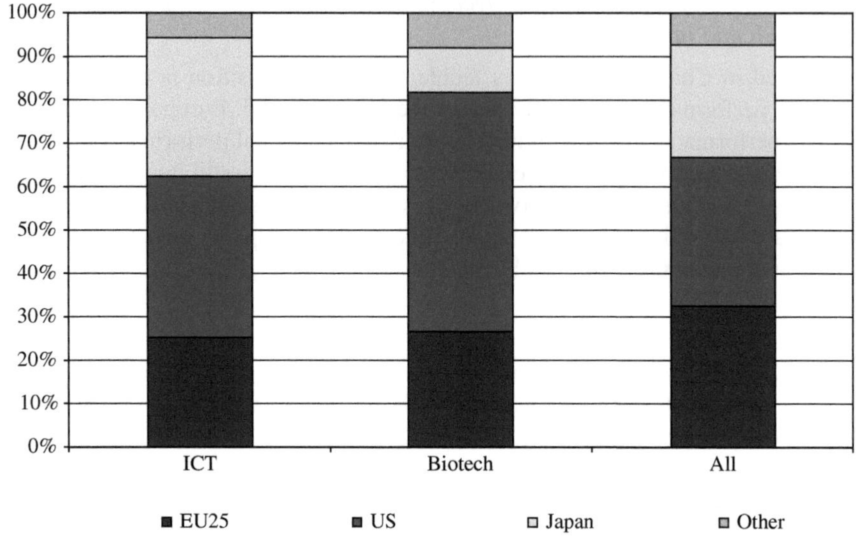

Fig. 6.4. Shares of triadic patents (%)
Source: DG Research
Data: OECD, Patent Database
Notes: (1) Triadic patents relate to those inventions for which a patent application is made in each of the three patent offices (EPO, USPTO, and JPO). (2) Data are for the priority year 1998, by country of the inventor.

patents, compared with 53.5 per cent for the United States, while in ICT the shares are 26 per cent for the Union versus 36 per cent for the United States.[35]

In terms of its broader innovation performance, the EU would also appear to be lagging behind the United States and Japan. Fig. 6.5 presents a composite index based on 26 separate innovation indicators, which is produced each year for the European Innovation Scoreboard. A gap can be seen between the United States and EU innovation performances, as well as between the EU and the Japan. The gap in innovation performance, thus measured, has been relatively stable between the EU and the United States, but has been increasing between the EU and Japan.

Europe's weak technological performance stems from its weakness in critical high-tech industries. The OECD has so far identified five high-technology industries: aerospace, pharmaceuticals, computers and office machinery, communication equipment, and scientific (medical, precision, and optical) instruments. High-tech industries are science-based and R&D intensive. R&D gives rise to innovation, and innovative firms gain market share, create new product markets, and use resources more productively. High-tech industries develop high value-added products, export

[35] European Commission, *Key Figures 2003–2004*, p. 71.

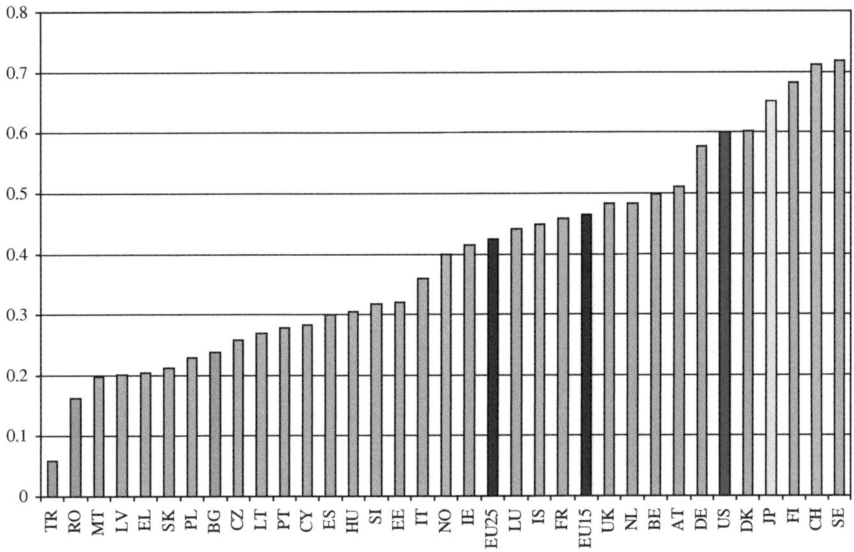

Fig. 6.5. Summary Innovation Index 2005
Source: *Innovation Scoreboard 2005*
Notes: (1) Details of the SII can be found at http://trendchart.cordis.lu/scoreboards/scoreboard2005/ (2) The SII is calculated using re-scaled values of the 26 indicators, where the highest value within the group of EU-25 countries, Iceland , Norway and Switzerland is set to 1 and the lowest value within the group of EU-25 countries to 0. The SII is then calculated as the average value of all re-scaled values and is by definition between 0 and 1 for the EU-25 countries. The SII is a relative not an absolute ranking. Having an SII twice that of another country does not mean that the absolute innovation performance is also twice as good. (3) The SII values for TR , US and JP are estimated because US data are only available for 16 indicators, Japanese data for 15 indicators and Turkish data for 13 indicators.

more, and, on average, pay higher salaries than other manufacturing industries. In addition, industrial R&D performed by high-technology industries benefits other commercial sectors as new products, machinery, and processes are developed that increase productivity and expand business.

High-technology industries drive economic growth around the world. Over the past 25 years, high-technology production grew at a real average annual rate of nearly 6.4 per cent, compared with 2.4 per cent for other manufactured goods. Output by the five high-technology industries represented 8.1 per cent of global production of all manufactured goods in 1980, while by 2003, it had doubled to 17.7 per cent (Fig. 6.6).

The share of high-technology value added in total manufacturing value added is lower in Europe than in the United States or Japan, and also stabilising, while it is increasing in the United States (and declining in Japan). In 2003, high-technology value added accounted for 13.4 per cent of total manufacturing value added in the EU (compared to 9.5 per cent in 1980, 11 per cent in 1990, and

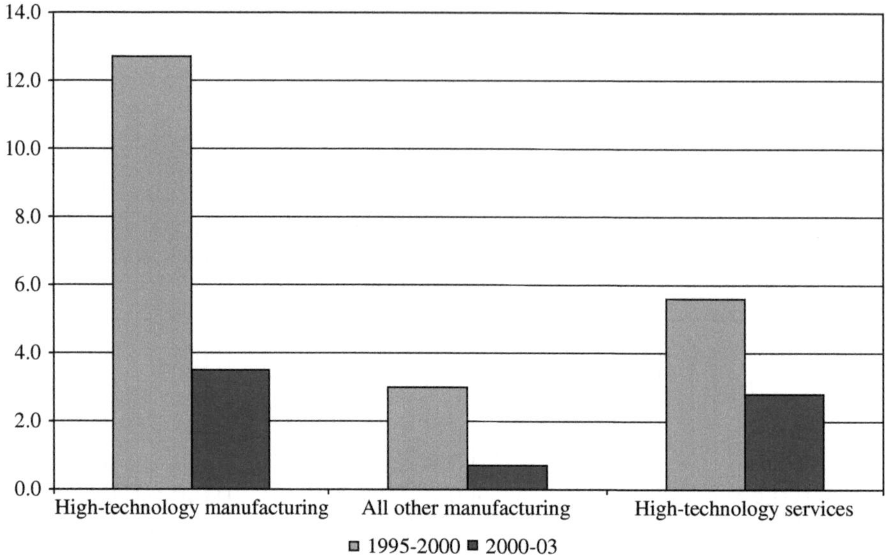

Fig. 6.6. Global industry sales, average annual growth rate, by sector: 1995–2003 (%)
Source: DG Research
Data: National Science Foundation, *Science and Engineering Indicators 2006*
Note: Growth Rates Calculated From Inflation Adjusted 1997 Dollars

13.2 per cent in 2000) vs 34.2 per cent in the United States, and 15.7 per cent in Japan. One European economy firmly on the high-technology track is the Irish one, where high-technology value added accounted for 12.4 per cent of total manufacturing value added in 1980, 26.4 per cent in 1990, and over 50 per cent since 1999. Such dynamism is also visible in East Asia. In 1980, high-technology value added accounted for 9.6 per cent of South Korea's total manufacturing value added, increasing to 14.8 per cent in 1990 and 21.5 per cent in 2003. The makeover of Taiwanese manufacturing is even more remarkable. High-technology manufacturing in Taiwan accounted for 9.7 per cent of total domestic output in 1980 and 15.9 per cent in 1990, and jumped to an estimated 28.5 per cent in 2003 (Fig. 6.7).

Europe accounts for a smaller share of global high-technology added value than the Unites States, and its share is rapidly declining. From 1980 through 1995, the EU contained the world's largest high-technology manufacturing sector. Beginning in 1996, however, and for each year thereafter, it was exceeded by the States. The Union accounted for the largest share of global high-technology value added in 1980 (34.2 per cent). Since then this share fell to 28.0 per cent in 1990, 20.1 per cent in 2000, and an estimated 18.4 per cent in 2003. The United States accounted for about one-quarter of global high-technology value added from 1980 to 1995. This share moved up sharply in the late 1990s, peaking at 43.1 per cent in 2002, and dropping slightly to 42.5 per cent in 2003. Japan accounted for its

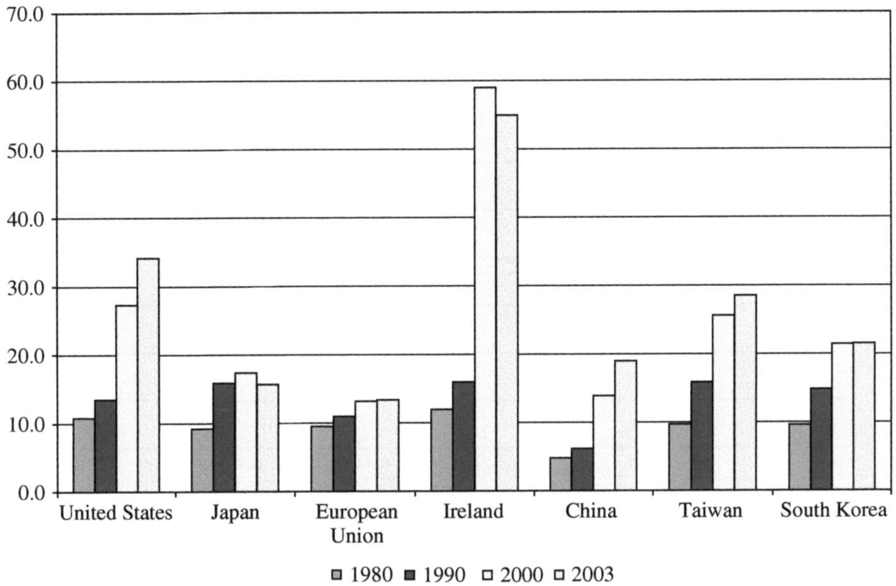

Fig. 6.7. High-technology value added as a share of total manufacturing, value added in selected countries/regions: 1980–2003 (%)
Source: DG Research
Data: National Science Foundation, *Science and Engineering Indicators 2006*

largest share ever (25.2 per cent) in 1991. By 2002, its share had fallen to 12.0 per cent. While other Asian countries like South Korea and Taiwan also account for sizeable shares, these have been surpassed by China, which may soon exceed Japan (Fig. 6.8).

The need for new instruments at European level[36]

In response to Europe's weak technological performance, many European industries have established Technology Platforms through which they are now formulating strategic research agendas. It is expected that existing EU instruments for research support through collaborative research (possibly with some adaptations) will suffice in most cases to implement these strategic research agendas. That means that in most cases Community support for implementation will be through open calls for proposals for collaborative research (e.g. Integrated Projects or other collaborative research instruments), research infrastructures, and so on. The participation of the Community in national research programmes, as provided for by Article 16910 of the Treaty, could also be envisaged.

[36] This section based on European Commission, *Technology Platforms – From Definition to Implementation of a Common Research Agenda*, Luxembourg, 2004.

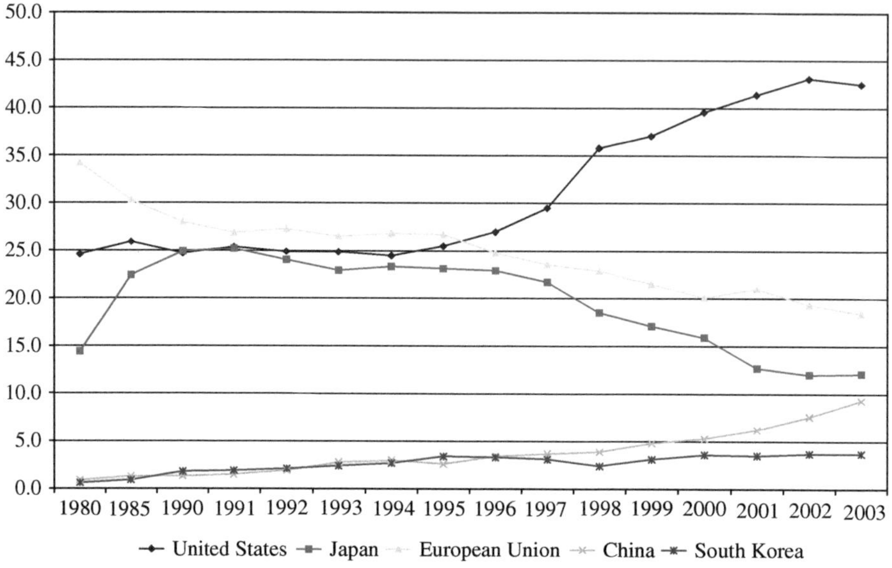

Fig. 6.8. Share of global high-technology value added, by country/region: 1980–2003 (%)
Source: DG Research
Data: National Science Foundation, Science and Engineering Indicators 2006

Nevertheless, a limited number of research agendas can be expected to be of such an ambitious scale that they will require the mobilisation of very high public and private investments, as well as a large critical mass of researchers throughout Europe and even beyond. In view of establishing and coordinating the necessary public–private partnerships to implement such research agendas, it can be anticipated that a mechanism would be needed which could enable coherent, large-scale structures to be set up specifically for this purpose.

The structures which the Community may set up under the terms of Treaty Article 171 may provide such a mechanism. Article 171 states that "The Community may set up joint undertakings or any other structure necessary for the efficient execution of Community research, technological development and demonstration programmes". It is these joint undertaking that are termed 'Joint Technology Initiatives'.

3. A SHORT OUTLINE OF THE COMMISSION'S 7TH FRAMEWORK PROGRAMME PROPOSAL

The preparation of the Commission's proposal for the 7th Framework Programme was a complex task which took place over a period of many months. Designing such a large and ambitious programme in consultation with interested parties across 25 Member States is not a simple linear process. Nevertheless, it is possible to highlight some of the key factors which helped to shape the proposal.

3.1. Recapping some of the factors shaping the proposal

The 7th Framework Programme proposal was conceived as a response to the challenges facing the EU in the coming years (Chapter 1). The proposed budget was aimed at injecting more resources into research in order to stimulate the development of the knowledge-based economy and to improve progress towards the Lisbon goals. The research areas chosen responded to numerous challenges from developing industrial technologies for growth, to tackling climate change and improving the health of EU citizens.

The proposal aimed to address some of Europe's key S&T weaknesses highlighted in Chapter 2, in particular the fragmentation and dispersion of European S&T. It took account of the new policy context, which allows for a more ambitious Framework Programme as knowledge is the designated heart of the Lisbon strategy and the role the Union can play is recognised (Chapter 3).

Lessons from the past were analysed and integrated in the proposal (Chapter 4). Collaborative research actions were continued because of their proven success in creating and disseminating knowledge, and their wide-ranging scientific, technological, and economic impacts. These and other actions such as training and mobility measures were also expanded to satisfy the unmet demand observed in past Framework Programmes. But past weaknesses also had to be addressed, notably the often complicated procedures for participating, which encouraged the design of a more user-friendly Framework Programme.

The views of stakeholders were a guiding element. As was seen in Chapter 5, there were many interactions with research players across the EU – be it policy-makers, entrepreneurs, academics, or the ordinary citizen – in order to gather their views and promote a useful debate on future options. The opinions expressed helped to move the proposal in important directions, for example the introduction of new actions relating to the European Research Council (ERC) and the Joint Technology Initiatives (JTIs) was an explicit response to stakeholder demands. Similarly, the continuity of cooperative research activities, the increase in the total Framework Programme budget, and efforts towards simplification were all responses to concerns raised by the research community. European added value was considered, and with respect to a number of issues several policy options were considered (this chapter).

In addition, the adoption of the 7th Framework Programme proposal by the Commission was preceded by a formal Commission Interservice Consultation. Interservice consultation is an obligation under Commission procedures, is part of the inter-departmental coordination process, and is initiated by the lead department. Through interservice consultation, the lead department seeks the formal opinion of all the Directorates-General and services with a legitimate interest in the contents of the proposal.[37] This consultation process provides a good means to ensure quality and consistency of the Commission's and its departments' work as well as reinforcing the sense of collective responsibility among the departments.

---

[37] European Commission, *The Commission in Close-up – How Decisions Are Made*, Brussels, 2005.

Directorate-General Research as the main sponsor of the Framework Programme proposals requested all other Commission services for their feedback on the proposal. In total, 34 departments replied. Main issues of concern included the lack of a draft budget (as the discussion on the Financial Perspectives was still going on at that time), and the proposed management modalities, degree of externalisation and organisation of the ERC, and degree of simplification. Inevitably, each Commission service argued in favour of its own particular sectoral or other interests and lively debates occurred during the preparation of the 7th Framework Programme proposal encouraging the argumentation of choices and preferences. This is exactly what the Better Regulation package intends to promote, namely a culture of dialogue and participation in order for everyone concerned to be heard during the lawmaking process; this includes views external to the European Commission as well as inter-institutional ones.[38]

3.2. The content of the Commission's 7th Framework Programme proposal

Before moving on to describing the inter-institutional decision-making process where the 7th Framework Programme proposal is amended by the European Parliament and the Member States through the Council of Ministers, this section summarises the main features of the 7th Framework Programme proposal as wished-for by the European Commission.

The 7th Framework Programme is organised in four specific programmes corresponding to four major objectives of European research policy: cooperation, ideas, people, and capacities (Table 6.3).

Cooperation The first specific programme will concern 'cooperation'. This programme will use the largest share, about 60 per cent, of the proposed 7th Framework Programme budget.[39] It will support transnational cooperation, mainly through collaborative research projects and networks among researchers, through JTIs bringing together enterprises to implement a joint research agenda, and the coordination of national and regional research programmes. It will support research collaboration across the EU as well as beyond as international cooperation between the EU and third countries is an integral part of this action.[40]

The thematic areas of research for which European funding will be available are: (1) Health; (2) Food, Agriculture and Biotechnology; (3) Information and Communication Technologies; (4) Nanosciences, Nanotechnologies, Materials and new Production Technologies; (5) Energy; (6) Environment (including Climate Change); (7) Transport (including Aeronautics); (8) Socio-economic Sciences and the Humanities; and (9) Security and Space.[41] These thematic areas have been identified on the basis of their contribution to EU objectives, including the transition

---

[38] European Commission, *The Commission in Close-up.*

[39] European Commission, *Building the Europe of Knowledge*, p. 10.

[40] ibid., p. 3.

[41] ibid., p. 4.

Table 6.3. The structure of the 7th Framework Programme

Cooperation	✓ Collaborative research ✓ Joint technology initiatives ✓ Coordination of non-Community research programmes ✓ International cooperation	✓ Health ✓ Food, agriculture and biotechnology ✓ Information and communication technologies ✓ Nanosciences, nanotechnologies, materials and new production technologies ✓ Energy ✓ Environment (including climate change) ✓ Transport (including aeronautics) ✓ Socio-economic sciences and the humanities ✓ Security and space
Ideas		European Research Council
People		Marie Curie actions
Capacities		Research infrastructures Research for the benefit of SMEs Regions of knowledge Research potential Science in society Activities of international co-operation

Source: DG Research

to a knowledge society, the relevant European research potential, and the added value of EU level intervention for these subjects.

On collaborative research among Europe's researchers Collaborative research continues to constitute the bulk and the core of EU research funding. The objective is to establish, in the major fields of advancement of knowledge, excellent research projects and networks able to attract researchers and investments from Europe and the entire world. This will be achieved by supporting collaborative research through a range of funding schemes: Collaborative projects, Networks of Excellence, and Co-ordination/Support actions.

The substantial increase in funding for collaborative research proposed under the 7th Framework Programme will be crucial in distributing more widely the significant effects on restructuring research in the EU, and on pooling and leveraging resources, and will move Europe closer to a real "single market" for research. The proposed simplification of rules and procedures of the 7th Framework Programme, notably in relation to proposals for research consortia, will also have a significant impact in making the Framework Programme easier for applicants and participants, and thus more attractive and useful to the research community.

Joint Technology Initiatives (JTIs) among Europe's enterprises Topics of industrial relevance have been identified relying, among other sources, on the work of different "European Technology Platforms" set up in fields where Europe's competitiveness, economic growth, and welfare depend on important research and technological progress in the medium to long term. European Technology Platforms bring together stakeholders, under industrial leadership, to define and implement a Strategic Research Agenda.

JTIs are closely linked to European Technology Platforms' Strategic Research Agendas which are generally supported via the Framework Programme's collaborative research. In a limited number of cases based on the scope of a research objective and the scale of the resources involved, the setting up of long-term public–private partnerships in the form of JTIs might be justified. These will focus on one or a small number of selected aspects of research in the TPs field, and will combine private sector investment and national and European public funding, including grant funding from the Framework Programme and loan finance from the European Investment Bank. JTIs may be decided on the basis of Article 171 of the EC Treaty or on the basis of the Specific Programme Decisions in accordance with Article 166 of the EC Treaty.

Potential JTIs will be identified on the basis of a series of criteria including: added value of European-level intervention; the degree and clarity of definition of the objective to be pursued; strength of the financial and resource commitment from industry; scale of the impact on industrial competitiveness and growth; importance of the contribution to broader policy objectives; capacity to attract additional national support and leverage current or future industry funding; inability of existing instruments to achieve the objective. Particular attention will be paid to the overall coherence and coordination between JTIs and national programmes and projects in the same fields.

The JTIs are an innovative element of the 7th Framework Proposal as the collaborative research method includes the bottom-up development of a joint research agenda among competitors in the marketplace.

On coordination of research programmes The ERA-NET scheme will develop and strengthen the coordination of national and regional research activities by providing a framework for actors implementing public research programmes to step up the coordination of their activities. This will include support for new ERA-NETs as well as for the broadening and deepening of the scope of existing ERA-NETs, for example by extending their partnership, as well as opening mutually their programmes. The ERA-NET scheme will also provide additional EU financial support to those participants that create a common fund for the purpose of joint calls for proposals between their respective national and regional programmes ("ERA-NET PLUS").

The participation of the Community in national research programmes jointly implemented on the basis of Article 169 is especially relevant to European cooperation on a large scale in "variable geometry" between Member States sharing common needs and/or interests. Such Article 169 initiatives will be launched in areas to be identified in close association with the Member States, including the possible cooperation with intergovernmental programmes, on the basis of a series of criteria: relevance to EU objectives; the clear definition of the objective to be pursued and its relevance to the objectives of this Framework Programme; presence of a pre-existing basis (national research programmes existing or envisaged); European added value; critical mass, with regard to the size and the number of programmes involved, the similarity of activities they cover; efficiency of Article 169 as the most appropriate means for achieving the objectives.

For the coordination of non-Community research programmes use will be made of two main tools: the ERA-NET scheme and the participation of the Community in jointly implemented national research programmes (Treaty Article 169). The action will also be used to enhance the complementary and synergy between the Framework Programme and activities carried out in the framework of intergovernmental structures such as EUREKA and COST.[42]

Actions in the field of international cooperation will concern the opening of all activities carried out in the thematic areas to researchers and research institutions from all third countries.[43] Actions also include those reinforcing third countries' research capacities, such as in candidate countries as well as neighbourhood countries, and addressing particular needs in fields such as health, agriculture, fisheries, and environment in developing and emerging countries.

Ideas The objective of this programme – which accounts for about 16 per cent of the overall 7th Framework Programme proposed budget – is to enhance the dynamism, creativity, and excellence of European research at the frontier of knowledge.[44] Support will be provided to investigator-driven (i.e. based on the initiative of the research community) frontier research projects. Research will be supported in all scientific and technological fields, including engineering, socio-economic sciences, and the humanities.[45] Individual teams will compete for funding at the European level, and projects will be funded on the basis of proposals presented by researchers on subjects of their choice and evaluated on the sole criterion of excellence as judged by peer review.

Investigator-driven frontier research is a key driver of wealth and social progress, as it opens new opportunities for scientific and technological advance, and is instrumental in producing new knowledge leading to future applications and markets. A Europe-wide competitive funding mechanism for frontier research executed by individual teams is expected to help reinforce the dynamism and attractiveness of Europe for the best researchers from both European and third countries, and for industrial investment. It will be implemented independently of the thematic orientations of the other parts of the Framework Programme through a European Research Council (ERC).

---

[42] This will include financial support for the administration and coordination activities of COST.

[43] Candidate countries; countries neighbouring the EU; Mediterranean partner countries; Western Balkans and the Newly Independent States; developing countries; emerging economies.

[44] European Commission, *Building the Europe of Knowledge*, p. 10.

[45] ibid., p. 3.

The ERC will consist of a scientific Council, supported by a dedicated implementation structure. The scientific Council will consist of representatives of the European scientific community at the highest level, acting in their personal capacity, independently of political or other interests. Its members will be appointed by the Commission following an independent procedure for their identification. The scientific Council will, inter alia, oversee decisions on the type of research to be funded and act as guarantor of the quality of the activity from the scientific perspective.

The dedicated implementation structure will be responsible for all aspects of implementation and programme execution. It will, in particular, implement the peer review and selection process according to the principles established by the scientific Council and will ensure the financial and scientific management of the grants.

People The objective of this programme in support of the individual researcher – which accounts for about 10 per cent of the overall 7th Framework Programme proposed budget[46] – is the strengthening, both quantitatively and qualitatively, of the human potential in research and technology in Europe. This is accomplished by stimulating people to enter into the researcher's profession, encouraging European researchers to stay in Europe, and attracting to Europe researchers from the entire world, making Europe more attractive to the best researchers. This will be done by putting into place a coherent set of "Marie Curie" actions, addressing researchers at all stages of their careers, from initial research training to life-long learning and career development, with a better focus on the key aspects of skills and career development and strengthened links with national systems.[47] The rationale is that abundant and highly trained qualified researchers are a necessary condition to advance science and to underpin innovation, but also an important factor to attract and sustain investments in research by public and private entities.

Activities will concern the initial training of researchers to improve their career perspectives, in both public and private sectors, including through the broadening of their scientific and generic skills, and attracting more young researchers to scientific careers. This will be implemented through Marie Curie Networks with the main objective to overcome fragmentation of and to strengthen at European level the initial training and career development of researchers.

Activities will also concern life-long training and career development to support the career development of experienced researchers with a view to complementing or acquiring new skills and competencies or to enhance inter/multidisciplinarity and/or inter-sectoral mobility. This action line will be implemented through both individual fellowships awarded directly at Community level and through the co-financing of regional, national, or international programmes.

Activities will concern support to longer-term cooperation programmes between organisations from academia and industry, in particular SMEs, aims at increasing knowledge-sharing through joint research partnerships, supported by the recruitment

[46] ibid., p. 10.
[47] ibid., p. 4.

of experienced researchers to the partnership, by staff secondments between both sectors, and by the organisation of events.

Activities will concern the international dimension, to increase the quality of European research by attracting research talent from outside Europe and fostering mutually beneficial research collaboration with researchers from outside Europe. This will be addressed through international outgoing fellowships (with an in-built mandatory return phase); international incoming fellowships; partnerships to support the exchange of researchers. Common initiatives between European organisations and organisations from countries neighbouring the EU and countries with which the EU has a Science and Technology agreement will also be supported.

Capacities This part of the Framework Programme – accounting for about 10 per cent of the overall 7th Framework Programme proposed budget[48] – will enhance research and innovation capacities throughout Europe and ensure their optimal use. This aim will be achieved through optimising the use and development of research infrastructures; strengthening innovative capacities of SMEs and their ability to benefit from research; supporting the development of regional research-driven clusters; unlocking the research potential in the EU's convergence and outermost regions; bringing science and society closer together for the harmonious integration of science and technology in European society; and horizontal actions and measures in support of international cooperation.

The activities undertaken in this part of the Framework Programme will also support the coherent development of policies, complementing the coordination activities under the Cooperation programme, and contributing to Community policies and initiatives that aim to improve the coherence and impact of Member States policies. This will include strengthening and improving the European science system, such as questions of scientific advice and expertise and contributing to 'better regulation'; monitoring and analysis of research-related public policies and industrial strategies; coordination of research policies, including transnational cooperation initiatives undertaken at national or regional level on issues of common interest.

On research infrastructures The objective of the research infrastructures component is optimising the use and development of the best research infrastructures existing in Europe, and helping to create in all fields of science and technology new research infrastructures of pan-European interest needed by the European scientific community to remain at the forefront of the advancement of research, and able to help industry to strengthen its base of knowledge and its technological know-how.

The rationale is that research infrastructures play an increasing role in the advancement of knowledge and its exploitation. For example, radiation sources, data banks in genomics and data banks in social science, observatories for environmental

[48] ibid., p. 10.

and space sciences, systems of imaging or clean rooms for the study and development of new materials or nano-electronics are at the core of research. They are expensive, need a broad range of expertise to be developed, and should be used and exploited by a large community of scientist and customer industries on a European scale.

The EU can and should play a catalysing and leveraging role by helping to ensure wider and more efficient access to, and use of, the infrastructures existing in the different Member States, by stimulating the development of these infrastructures in a coordinated way and by fostering the emergence of new research infrastructures of pan-European interest in the medium to long term.

On the participation of SMEs The objective of the research for the benefit of SMEs component is strengthening the innovation capacity of European SMEs and their contribution to the development of new technology-based products and markets by helping them outsource research, as well as increase their research efforts, extend their networks, better exploit research results, and acquire technological know-how.

The rationale is that SMEs are at the core of European industry. They should be a key component of the innovation system and in the chain of transformation of knowledge into new products, processes and services. Faced with an increasing competition in the internal market and globally, European SMEs need to increase their knowledge and research intensity, expand their business activities on larger markets, and internationalise their knowledge networks. Most Member States' actions relevant to SMEs do not encourage and support transnational research cooperation and technology transfer. Actions at EU level are necessary to complement and enhance the impact of actions undertaken at national and regional level. In addition to the actions listed below, the participation of SMEs will be encouraged and facilitated, and their needs taken into account, across the Framework Programme. SMEs will also receive support through the Competitiveness and Innovation Programme.

On the role of regions in Europe's research landscape The objective of the regions of knowledge component is strengthening the research potential of European regions, in particular by encouraging and supporting the development, across Europe, of regional "research-driven clusters" associating universities, research centres, enterprises, and regional authorities.

The rationale is that regions are increasingly recognised as important players in the EU's research and development landscape. Research policy and activities at regional level often rely on the development of "clusters" associating public and private actors. Based on the Pilot Action on "Regions of Knowledge", the new Regions of Knowledge initiative will involve and bring together regional actors involved in research: universities, research centres, industry, public authorities (regional Councils or regional development agencies).

Projects will cover joint analysis of research agendas of regional clusters (in coordination with other activities on the broader issue of regional innovation clusters) and the elaboration of a set of instruments to address them in specific research activities, including through "mentoring" of regions with less-developed research profiles by regions with highly developed ones. This will comprise measures aiming at improving research networking and access to sources of research funding as well as better integration of research actors and institutions in regional economies. These activities will be implemented in close relationship with EU regional policy and the Competitiveness and Innovation Programme and the Education and Training Programmes. In the context of the specific activity of "Regions of Knowledge" synergies will be sought with the EU's regional policy, in particular with regard to convergence and outermost regions.

The objective of the research potential component is stimulating the realisation of the full research potential of the enlarged Union by unlocking and developing the research potential in the EU's convergence regions and outermost regions,[49] and helping to strengthen the capacities of their researchers to successfully participate in research activities at EU level.

On science and society The objective of the science in society component is, with a view to building an effective and democratic European Knowledge society, to stimulate the harmonious integration of scientific and technological endeavour, and associated research policies in the European social web, by encouraging at European scale reflection and debate on science and technology, and their relation with society and culture. Contentious issues relating to emerging technologies should be addressed by society on the basis of well-informed debate leading to sound choices and decisions.

The actions proposed are varied and include, amongst others, reflection and debate on science and technology and their place in society; the creation of an environment which triggers curiosity for science in young people by reinforcing science education; helping scientists better communicate their work and supporting scientific information and media.

4. THE EXPECTED IMPACTS OF THE COMMISSION'S 7TH FRAMEWORK PROGRAMME PROPOSAL

Compared to its modest share of European public R&D funding, the Framework Proposal achieves significant impacts, especially in the long-term, mainly because of high crowding-in and economic multiplier effects. The doubling of funding foreseen in the European Commission's 7th Framework Programme proposal would boost Europe's economic growth. Depending on the rate of growth of the Framework Programme funding after the 7th Framework Programme, doubling the Framework Programme funding would generate at least 0.51 and up to 0.96 per cent of extra

[49] Convergence regions are those set out in Article 5 of the proposal for a Council Regulation laying down general provisions on the European Regional Development Fund, the European Social Fund, and the Cohesion Fund – COM(2004) 492. This includes "convergence" objective regions, regions eligible for funding from the Cohesion fund, and outermost regions.

GDP over and above the business-as-usual scenario of moderate growth in FP funding by the year 2030. In other words, assuming a GDP of 100 under the business-as-usual scenario for the year 2030, and given that the extra GDP generated by doubling the Framework Programme funding would amount to from 0.51 to 0.96 per cent of GDP by that same year, then total GDP would reach between 100.51 and 100.96 in the year 2030. Given the comparatively small size of the Framework Programme this constitutes a large impact.

Doubling Framework Programme funding would create extra jobs for European citizens (up to 900,000 extra jobs by the year 2030, of which up to 300,000 in research). It would also raise Europe's competitiveness (extra-European exports are increased by up to an extra 1.57 per cent by the year 2030, imports reduced by up to 0.9 per cent), and increase Europe's R&D intensity (the extra growth in Europe's R&D intensity could reach 0.2 per cent).

The proposed 7th Framework Programme would have large potential aggregate social impacts. It would contribute to the achievement of the Lisbon strategy and to addressing the main future social and political challenges of Europe. Through both thematic efforts in diverse areas as for example industrial technologies, energy, transport, ICT, food, agriculture, fisheries, maritime affairs, water management, life sciences, and so on, as well as through research that directly aims at the advancement of Social Sciences, the 7th Framework Programme would further enhance issues such as health and safety, social cohesion, human capital, well-being, governance, human rights and ethics, self-sufficiency, equity, and so on. Therefore, the new research effort in the 7th Framework Programme would enhance the impact of innovation and competitiveness, both on individual economic entities but ultimately also on the quality of life in the society as a whole. Research on ethics at European level is critical for arriving at a responsible approach towards S&T, which is consistent with the European Charter of Fundamental Rights and reflects public sentiment.

The Lisbon Agenda and the ERA clearly identify the need for innovative and competitive technological progress in line with environmental and socio-economic needs. Advances in knowledge and innovation further sharpen the competitive edge of societies which possess the know-how and capacities and have become key factors in decoupling economic development from adverse environmental impacts. To address the different challenges, research, and technological development affecting the environment in the 7th Framework Programme would aim to identify win-win technologies, improve natural resources management and services, and understand and predict the environment more precisely. Furthermore, the 7th Framework Programme should recognise the need for research activities for the analysis of sustainable development, scenario building, and impact assessment. Cross-cutting enabling technologies, such as nanotechnology, biotechnology, and industrial technologies could also have a positive environmental impact.

As far as the time dimension associated with these impacts is concerned, while showing significant results in the short term, investment in research shows its greatest impacts in the medium to long term as it takes time to transform research results into new products and processes.

As far as impacts on particular groups are concerned, care has been taken to make sure that all players in the European S&T system, including SMEs and the new Member States, would benefit to the maximum extent possible from the implementation of the Framework Programme. EU scientific achievements have a significant global impact – notably in developing countries, the Mediterranean, the Western Balkans, and newly independent countries of the former Soviet Union – and cover key areas such as agriculture, human health, food processing, post-harvest conservation, water treatment, erosion, and environmental protection.

CONCLUSION

As detailed in earlier chapters, a broad range of inputs helped shape the Commission's 7th Framework Programme proposal. Account was taken of the European societal challenges to be addressed, the role that S&T could play in addressing those challenges, the scope for public intervention, Europe's S&T weaknesses, the European (research) policy context since Lisbon, the lessons learned from the implementation of past Framework Programmes, and the views of stakeholders.

In addition, it was ensured that actions proposed for inclusion in the 7th Framework Programme pursued objectives such as (1) pooling and leveraging resources; (2) fostering human capacity and excellence in S&T; and (3) better integration of European R&D. In other words, it was ensured that these actions pursued objectives best pursued at European rather than at national/regional level, and were thus characterised by European added value.

A final input into the formulation of the Commission's 7th Framework Programme proposal consisted of assessing different policy options, in particular with regard to three issues: the maximisation of the Framework Programme's macro-economic impacts, overcoming the fragmentation of European basic research, and raising the competitiveness of European industry through research and innovation. The analysis pointed out that a much larger Framework Programme would have the largest macro-economic impacts, and that the establishment of a European Research Council and JTIs constituted the best means to overcome the fragmentation of European basic research and to raise the competitiveness of European industry through research and innovation.

On the basis of all of the aforementioned inputs, the Commission formulated its 7th Framework Programme proposal, centring on four Specific Programmes: cooperation, ideas, people, and capacities. The following chapter looks in more detail at the next stage in the legislative process for the 7th Framework Programme proposal: namely, what happened to the proposal in Council and Parliament after it was adopted by the Commission. In particular, it will describe how the European decision-making machinery responded to this proposal, and how the Financial Perspectives and other issues influenced the eventual outcome of this process.

CHAPTER 7

FROM PROPOSALS TO COMMITMENTS – THE 7TH FRAMEWORK PROGRAMME AND THE EUROPEAN DECISION-MAKING PROCESS

The previous chapters described how the European Commission's proposal for the 7th Framework Programme was prepared. But this is not the end of the story. The Commission does not have the power of decision; it simply drafts the proposal for new legislation, which it then presents to the European Parliament and the Council of Ministers. It is these institutions that will take the decision. They must examine the content of the proposal, propose their amendments, and arrive at an agreement on the final text – in other words, they transform the proposal into a legal decision. At the time of writing, the co-decision procedure is not yet completed and therefore the final shape of the 7th Framework Programme is not yet known.[1] Nevertheless, this chapter describes what has happened to date during this formal decision phase.

As will be seen, the Framework Programme is complex not only in terms of its preparation, but also in terms of institutional decision-making process through which it must pass. Greenhalgh and Russell describe decision-making as the "messy unfolding of collective action, achieved mostly through dialogue, argument, influence and conflict".[2] This chapter tries to do justice to the "messy unfolding" of the European decision-making process for the 7th Framework Programme, and focuses on two particular levels: the process itself and the content. It spotlights the hidden and the visible actors, explains how decisions are prepared and taken, and presents the main themes debated.

The analysis also attempts to link up with the issues set out in the previous chapters of this book, by exploring the extent to which decision-makers shared the Commission's analysis of the reasons for Europe's poor scientific and technological (S&T) performance, whether they agreed with the factors identified which prevent European S&T from realising its potential, and to what degree they supported the actions proposed by the Commission to overcome these weaknesses (Sections 2 and 3). Critically, the chapter also examines the tough compromises

[1] The manuscript of this book was submitted in August 2006, after the first reading of the Parliament and the Council's political agreement on the EC 7th Framework Programme. Any conclusions on the final decision are of a hypothetical nature. For the latest reference document, see: European Commission, Amended Proposal for a Decision of the European Parliament and the Council covering the 7th Framework Programme of the European Communities for Research, Technological Development and Demonstration Activities (2007–2013), COM (2006) 364, 28 June 2006.

[2] Trisha Greenhalgh and Jill Russell, Reframing Evidence Synthesis As Rhetorical Action in the Policy Making Drama, In: *Healthcare Policy* Vol. 1, No. 2, 2006, pp. 34–42.

that had to be made in order to reach a political consensus on the 7th Framework Programme.

But first it is important to understand how the EU decision-making machinery works in practice. This is the subject of the following section.

1. DECISION-MAKING IN THE FIELD OF RESEARCH POLICY

1.1. The legal basis: How is the Framework Programme decided?

The 1st EC Framework Programme (1984–1987) was adopted without an explicit legal basis.[3] Following the positive experiences with this pilot programme, a separate chapter on research and development (R&D) was included in the Single European Act in 1986.[4] Since then research is a Community responsibility with its own legal basis – and with the status of a fully fledged Community policy area.

The Maastricht Treaty further widened the Community role in R&D by highlighting its importance for Europe's industrial competitiveness and economic growth. In line with this valorisation, the Maastricht Treaty introduced the co-decision procedure for the adoption of the EC Framework Programme (art. 251 EC Treaty). Since then, the Parliament and the Council are placed on equal footing in this legislative process.[5] The 4th Framework Programme (1994–1998) was the first adopted under co-decision.

As the starting point of the co-decision procedure, the Commission proposal is sent to the Council and Parliament (on the organisation of the work in the institutions, see section 3).[6] The European Economic and Social Committee is also asked to set out its position. Parliament takes its *first reading* and sends its opinion to the Council.[7] On the basis of the Commission's proposal and Parliament's opinion and its own deliberations, the Council adopts a common position by qualified majority. The common position is then sent to Parliament for its *second reading*. The Parliament now has three months in which to do one of three things: Either, it accepts the Council's common position, then the proposal is deemed to have been adopted. Or it rejects the common position outright (by absolute majority), then the

---

[3] It was based on art. 235, now art. 308 of the Treaty Establishing the European Community (TEC), *Official Journal C325 of 24 December 2002*. The Council thus could initiate activities in the context of the internal market, on the basis of a Commission proposal after consultation of the European Parliament.
[4] Cfr. art. 163–173 TEC (OJ C 325 of 24.12.2002).
[5] It needs to be noted that the European Parliament is only equal with Council in the co-decision process for the *EC Framework Programme*. But the entire Framework Programme package is decided in three steps: The FP activities are elaborated by the *Specific Programmes* (art. 166 TEC). The Council by qualified majority approves the SPs after consulting the EP (advisory opinion). The SPs are implemented by detailed *Work Programmes*. These are adopted by Commission decision, after opinion from the management committee (the Parliament is not involved).
[6] Drawing on Klaus-Dieter Borchardt, *The ABC of Community Law,* Brussels, 2000, pp . 78ff.
[7] Theoretically, if Parliament does not amend the Commission's proposal, or the Commission accepts all amendments proposed by Parliament, the Framework Programme may be adopted at this stage of the procedure.

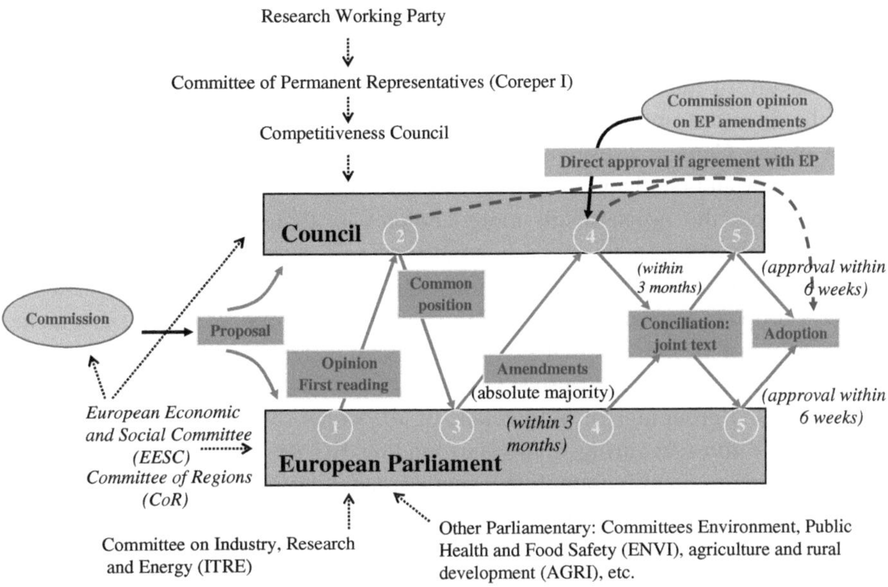

Fig. 7.1. A complex process – how the Framework Programme is decided
Source: DG Research

legislative process is at an end. Or it does not agree with the Council and amends the common position (again, by absolute majority), then the proposal is put before a conciliation committee, composed of equal numbers of the Council and Parliament representatives. Once this committee has reached an agreement, the text is sent again to the Parliament for the *third reading* and to the Council so that they can finally adopt it as law. If the conciliation procedure fails, the legislative process is then at an end (Fig. 7.1).

It will be apparent from this that the procedure is potentially very long, and its effectiveness depends entirely on the capacity of the three institutions to communicate continuously among themselves in order to reach a compromise.[8] Informal decision-making thus plays a great role in the process: much of the bargaining takes place in "informal trialogues" between representatives of the Council, the Commission, and the Parliament.

1.2. The Impact of co-decision: Consensus rules?

European decision-making on the Framework Programme involves many actors. It is not a simple matter of intergovernmental negotiation. The final shape of the Framework Programme arises from bargaining which is inter-institutional.[9] The

---

[8] Lucca Guzetti, *A Brief History of EU Research Policy*, Brussels, 1995, p. 154.
[9] John Peterson and Elizabeth Bomberg, *Decision-making in the European Union*, Basingstoke, 1999, p. 8.

process is complex and strongly influenced by external factors which constitute sources of unpredictability, such as politics and the political context, including the relative health of the European economy, the degree to which Member States are politically invested and committed to policy cooperation, the determination and finesse of the Council Presidency, and so on. Its outcome depends also on the relative power and charisma of individual personalities.

What makes the process still more complex is that, in reality, none of the institutional players is acting as a homogenous unit despite their Treaty status as single institutions. The European Commission, as well as the Council of Ministers and the European Parliament may seem to speak with one voice, but are in fact each made up of various internal assemblages with different functions and different policy concerns. This can be raised as striking or paradoxical, but is in fact a feature of all organised groupings of individuals, of all "organisations".

One of the most enduring myths surrounding EU decision-making is that the European Commission is a monolithic, single-minded institution. Seen from outside, it can indeed appear to be a homogenous body. However, as is the case for national administrations, the Commission is organised in sectoral departments (41 in total). Any political initiative brought forward needs to be based on an agreement of all services representing various groups of stakeholders with different and sometimes diverging interests. In relation to research policy, the Inter-Service Consultation on the 7th Framework Programme showed that – despite the overall agreement on the structure and priorities of the proposal – the services and Directorates-General involved sometimes, indeed, have different approaches to research policy, its design and focus.

The Member States are represented in the Council of Ministers. Their attitudes are shaped by national perceptions, the perceptions of 25 economically and culturally different countries. Nevertheless, they must also try to build a common view of what is needed for Europe as a whole. Member States themselves often suffer from internal divisions between ministries, not in the least between the Ministry in charge of Research and the Finance Minister, which can make it complicated for them to reach a single national position. The way in which Member States' positions are coordinated between domestic ministries before Council bargaining begins is another complicated variable in EU governance.

The European Parliament faces similar problems when it has to agree on a common position. Its 702 members stand for their different political groups. They also have to respect regional or constituency loyalties, while taking account of democratic lobbying. Acting individually but respecting these loyalties, they naturally wish to impose their particular stamp on the overall policy as they work towards the essential co-decision on the Framework Programme with the Council. Parliament is strongest in this process when its members exercise collective judgement about how far they can push the Council to accept its proposed amendments. The perceived need to stand up to the Council is a powerful incentive to strengthen cross-party cohesion in Parliament, and broad coalitions are found when it comes to defending and extending the Parliament's powers.[10] This is compounded by the legal need to achieve an absolute majority in second readings. Thus the fine print of co-decision has a significant influence on the degree of internal conflict within the European Parliament and its coherence against the Council.

If co-decision is to operate successfully, the Parliament and Council must demonstrate their capacity for compromise and direct their energies towards coming to an agreement. The process ensures on the one hand that decisions such as the Framework Programme are based on broad consensus. In a political system built on the assumption that European integration is, after all, a positive-sum process, it becomes imperative to ensure that outcomes produce as few visible losers as possible. The pressure for consensus makes it also important for the Commission to involve the relevant experts and stakeholders early in the policy design in order to build consent (see Chapter 5).[11] On the other hand, trying to incorporate all actors' concerns risks producing legislation of the "lowest common denominator" type. The resulting policy outputs may therefore reflect what is possible rather than the best regulatory option.

At the end of this complex and unpredictable process, how is it that co-decision regularly produces a strong and viable Framework Programme? This is probably because there is a large political agreement that research and innovation must be supported and that the Commission's proposals are generally pointing in the right direction. The following sections look in more detail at how this process unfolded in the case of the 7th Framework Programme.

2. RESEARCH POLICY AND THE LACK OF RESOURCES – THE NEGOTIATIONS ON THE FINANCIAL PERSPECTIVES AND THE BUDGET FOR EU RESEARCH

One of the main factors responsible for Europe's poor S&T performance is the lack of resources (Chapter 2). It is therefore not surprising that the initial negotiations on the 7th Framework Programme were dominated by a discussion of the size of the budget for EU research. As seen earlier, the Commission had proposed a doubling of Framework Programme funding. However, the money had to be found from the overall EU budget, which at the time of the Commission's proposal was still under heated discussion in the context of the so-called "Financial Perspectives".[12]

---

[10] See, for instance, Parliament's refusal of Council's budget deal of January 2006 which resulted not least in the fact that the Council had put forward its proposal without consulting the Parliament, and, MEPs recalled *that the conclusions of the European Council cannot overrule the powers of Parliament* (EP resolution ref.: B60049/2006 (see Section 2)).

[11] Taking account of such inputs and the needs of the research community not only ensures the technical quality of proposals but also guarantees a dialogue between science and civil society.

[12] The Financial Perspectives are the EU's long-term budgetary planning instruments which establish political priorities. They define the framework for all Community expenditure at 7 years intervals and set the ceilings and composition of planned expenditure. Thus, they largely determine the Community budget for R&D activities. This latest round of Financial Perspectives also had an impact on the duration

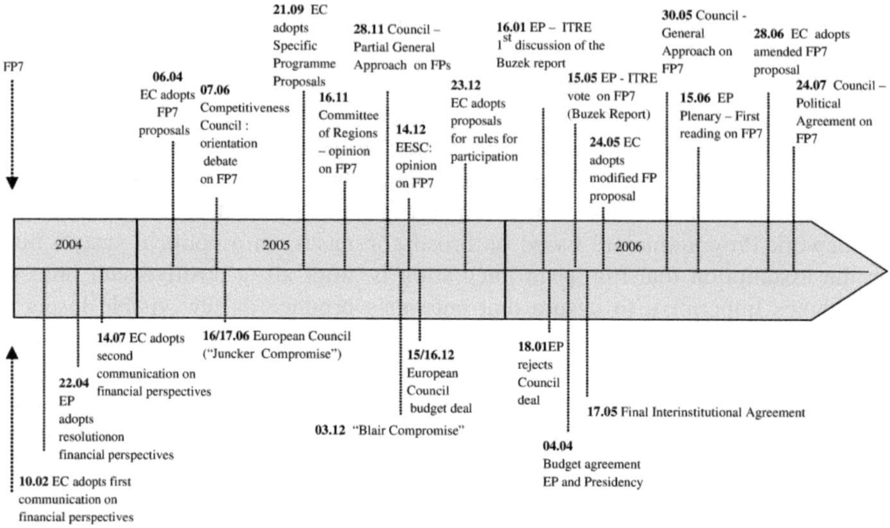

Fig. 7.2. Roadmap on the institutional decision-making process for the Financial Perspectives and FP7
Source: DG Research

The progress of the negotiations of both proposals in the institutions is illustrated in Fig. 7.2.[13] As will be seen, the budget question was eventually settled, but only after some hard negotiation and very tough decisions.

2.1. The Commission proposes a significant increase in the EU budget and a serious boost for research funding

In February 2004, the Commission presented a policy communication setting out the vision for the enlarged Union and its future budget, followed in July 2004 by a communication establishing a framework for the new Financial Perspectives for the period 2007–2013.[14]

The new budget proposed by the Commission reflected Europe's objectives and challenges in the twenty-first century. Above, it was shown that since the Lisbon Summit, and its strategy to bring the Union to the leading edge in the knowledge

---

of the 7th Framework Programme: It was for the first time extended to 7 years, so as to be aligned with that of the financial framework (from 2007 to 2013). Before, the duration of the FP was shorter than the Financial Perspectives: the Financial Perspectives 2000–2006 spun over FP5 (1998–2002) and FP6 (2002–2006).

[13] Figure 7.2 reflects the situation up to the time of writing (August 2006).

[14] European Commission, *Building Our Common Future: Policy Challenges and Budgetary Means of the Enlarged Union 2007–2013,* Communication from the Commission to the Council and the European Parliament, COM(2004) 101, 10 February 2004 and European Commission, *Financial Perspectives 2007–2013*, Communication from the Commission to the Council and the European Parliament, COM(2004) 487, 14 July 2004.

economy and society, Member States agreed on the need to boost competitiveness and growth.

The Commission argued that this required an average spending level of 1.21 per cent of EU-27 of Gross National Income over the 7-year period, compared with 1.12 per cent under the previous Financial Perspectives.[15] It proposed shifts among different spending priorities, and re-arranged the structure of the Financial Perspectives into five budgetary headings reflecting the Union's new priorities (Fig. 7.3). The position of the R&D post in the budget breakdown was a reflection of its political significance, with research and innovation, education and training featuring prominently under the first budgetary heading "Competitiveness for growth and employment" (sub-heading 1A). More than 12 per cent of the overall budget should go to this sub-heading (Fig. 7.3)[16] of which more than the half was foreseen for R&D.[17] This translated into a proposed doubling of the budget for EU research, or nearly € 73 billion spanning 7 years, a figure that was put forward in the 7th Framework Programme proposal in April 2005. Research and Innovation policy had never been higher on the European agenda.

These were the Commission's proposals. However, they still had to be agreed by the Council and the European Parliament.[18]

2.2. The hard reality of the budget negotiations

The Commission's intention to massively increase the EU funding for R&D received an enthusiastic welcome from the research community.

The European Parliament reacted positively too. In a resolution giving first orientations on the Commission's budget plans, it applauded the increase in research funding: "[Parliament] agrees... that the strengthening of the European effort in RTD constitutes a major objective [and] recalls... the importance of having appropriate funding."[19] In its final report on the Financial Perspectives, Parliament

---

[15] This increase may compare little since one would assume that the new priorities defined require a substantial raise of funding. However, there is the legal obligation that Member States' contributions must not exceed the ceiling of 1.24 per cent of GN (though in the last years, the EU annual expenditures only amounted to 1 per cent of GNI (0.98 per cent in 2004) and the ceiling of 1.24 per cent was never actually attained).

[16] € 121.687 billion over the period 2007–2013 (at 2004 prices, using a 2 per cent annual deflator).

[17] According to the Commission breakdown of expenditure, it was exactly 51.2 per cent.

[18] Decision-making on the financial framework follows a complex process owing to its soft law status: that is, it is not mentioned in the EC Treaty and is enacted through mutual agreement. The Inter-Institutional Agreement (IIA) establishes rules for the management of the financial framework over the seven-year period and for the setting up of annual budgets. It has to be agreed by the Council, Parliament, and the Commission. Without this agreement, financing is not available for the programmes proposed. After the new IIA was agreed on 17 May 2006, the legislative instruments relating to the various policy areas – such as FP7 – were revised to reflect the result of the negotiation. On 24 May 2006, the Commission adopted a revised package of EU programmes for the next financial framework, detailing funding for each programme and 31 proposals necessary for their implementation, among which were the modified proposals for FP7 (COM(2005) 119/3).

[19] Art. 32. Reference: A5-0268/2004.

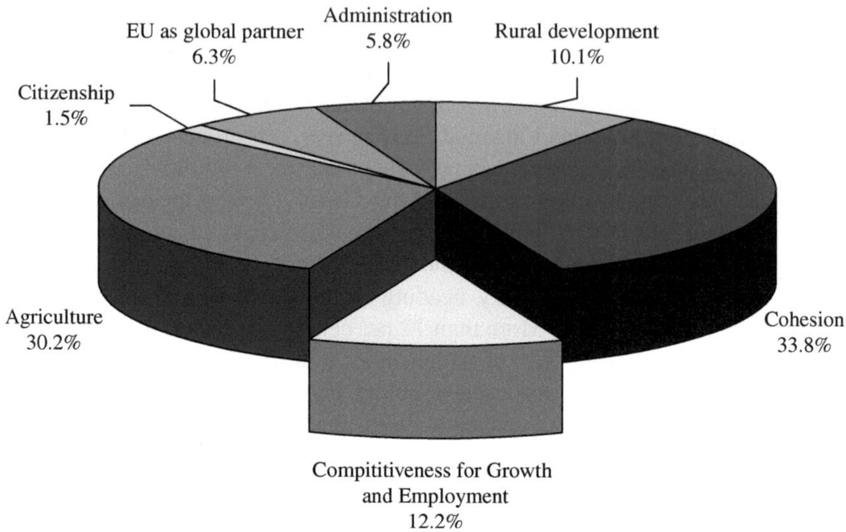

Fig. 7.3. A big slice of the cake for research – the proposal on the Financial Perspectives and the share of priority areas
Source: DG Research

stressed that "scientific research, technological development and innovation . . . are key factors for growth and sustainable development, the competitiveness of companies, employment, and to attain the objectives of the Lisbon Strategy; considers that research efforts should be boosted."[20] The Annex of this report goes on to give its full support to the proposed 7th Framework Programme budget.

However, the Commission's plans for the overall budget and for the R&D spending left the Council with mixed feelings. The debates on the Financial Perspectives turned out to be extremely difficult. These negotiations were carried out in a period of political tension: the French and Dutch had just rejected the Constitutional Treaty, Europe could not agree on the Union's future frontiers, and there was concern about the slow progress of the Lisbon strategy. Opinion polls reflected a sense of unease.[21] Furthermore, elections in several Member States were approaching, and national budgets were tight.

In this political context, the Brussels budget intentions were felt in some European capitals to be too ambitious. As early as December 2003, six of Europe's net payers to the budget addressed a joint letter to the then Commission President Prodi. They made it clear that they would not accept a massive increase of the

[20] Cfr. art. 16 of the so-called "Böge Report", adopted on 8 June 2005; Reference P6_TA_PROV(2005/0224).

[21] It is sobering that in April 2005 no fewer than 34 per cent of Germans and 41 per cent of French citizens believed that *the EU is responsible for me living less well.* Poll conducted of Financial Directors by the *Financial Times*, dated 11 April 2005.

EU budget: "In light of the painful cost-cutting efforts in the member states, our citizens would not understand why the EU budget should be exempt from this consolidation process", wrote the heads of state from Germany, France, the United Kingdom, the Netherlands, Austria, and Sweden.[22] This influential group agreed that the future EU expenditures should be capped at present level and should not go beyond 1 per cent of Gross National Income. For them, research funding could only be increased on condition that the 1 per cent ceiling was respected.[23] However, despite this early statement, the Commission proposed an average ceiling for the whole budget of 1.21 per cent of Gross National Income. The debate on the Member States' contributions to the budget promised thus to be difficult.

2.3. A new deal for the EU budget? Heading towards the knowledge economy or maintaining the emphasis on agriculture – 'Blair' versus 'Juncker'

In their coverage of the budget discussions in Council, the media identified two blocks of opposed Member States. On the one side, they saw the countries which supported an EU budget of a "traditional" shape with maintained emphasis on agriculture subsidies and regional aid.[24] On the other, they saw those Member States in favour of a modernised budget structure with more funding for the Lisbon-related issues such as research and less money spent on farms and regions.[25] This image of two opposed blocks is certainly simplistic – it overlooks the importance of the prior decision on agricultural funding up to 2013 and does no justice to the diversity in the enlarged Council, but it does help to sketch the principal lines of the debate.[26]

The two camps of Member States would find themselves represented by the succeeding Council Presidencies of the year 2005: the Luxembourg Presidency led by Prime Minister Juncker was followed in the second half of the year by the UK Presidency under Prime Minister Blair. The media communicated the image that Juncker was mainly supporting EU farmers and their communities, while Blair stood for a future, based on promoting research and innovation.

At the end of the Luxembourg Presidency, EU leaders were close to a budget deal. Prime Minister Juncker had presented a compromise proposal to the European Council in June 2005 which was for 1.06 per cent of Gross National Income (while the Commission had proposed 1.21 per cent). With regard to research, the compromise foresaw a cut of more than 40 per cent compared with the Commission's plans and a slight increase in agricultural subsidies.[27] Against

---

[22] http://www.bmdf.co.uk/blairlettereubudget.pdf. Last accessed on 28 August 2006.

[23] As seen in Chapter 5, these governments as well as Lithuania confirmed their adherence to the 1 per cent of GNI ceiling in their FP7 position papers.

[24] Among the prominent proponents of this approach were France, Poland, Spain, Germany, and Luxembourg.

[25] Represented by, among others, the UK and the Scandinavian countries.

[26] European Commission, *Financial Perspectives*, p. 4.

[27] Accordingly, a share of 8.2 per cent of the overall budget would be dedicated to research – decreasing from 12.2 per cent in the Commission proposal, while the agriculture share would increase from 30 to 33 per cent.

this background, Research Commissioner Potočnik addressed the Competitiveness Council in June 2005, urging Research Ministers to respect their commitment to the Lisbon and Barcelona goals by backing the original proposal. Eventually, the Juncker proposal failed to get backing at the summit owing to major differences of opinion between the EU Presidency and the net contributors to the budget (notably the Netherlands who had experienced a "no vote" in its referendum on the constitution a few weeks earlier, and Britain and France who disagreed over the continuation of the United Kingdom's yearly rebate).

The UK Council Presidency inherited the challenge of forging a deal, and this issue dominated the agenda of the Presidency. Under the UK's proposal for the Financial Perspectives the budget was further cut (to 1.03 per cent of Gross National Income). Prime Minister Blair insisted that his country would be unwilling to give up the rebate as long as the other countries opposed any further reform of the Common Agricultural Policy (CAP).[28] The United Kingdom argued that it is unjustified that CAP expenditures account for about 40 per cent of the EU budget whereas the farm sector only provides 5 per cent of jobs in the EU. They particularly underlined that they would like to see more money spent on issues related to Europe's competitiveness and growth to meet global challenges. A theoretical underpinning of this position can be found in the study by André Sapir of 2003 (see Chapter 3) who had qualified the budget structure as a "historical relic".[29] However, in his function as President of the European Council, Blair was also faced with the difficult task of tabling a compromise that all could agree to. It was therefore not surprising that in the end the Blair proposal followed broadly the same lines as Juncker's. Both compromises foresaw a considerable reduction of the overall budget – both at the expense of the "research" heading – while the share for agriculture was comparable. Table 7.1 provides detailed comparisons of the proposals.

2.4. EU summit budget deal: Rebuff for R&D

Eventually, at the European Council in December 2005, heads of State and Government found an agreement. They settled on an overall level of financial commitment over the 2007–2013 period, corresponding to 1.04 per cent of Gross National Income. This budget was higher than the totals which had been canvassed over months by the UK Presidency (1.03 per cent in the Blair proposal), but less than the June Juncker compromise of 1.06 per cent, and significantly below the Commission proposal (1.21 per cent). Since the Summit did not modernise the

---

[28] France had already persevered to agree a fixed arrangement for CAP spending that would not be changed until 2013. This agreement was reached as the result of an initiative by President Chirac and Chancellor Schröder at the European Council in October 2002. It involved only a small reduction in agricultural spending in real terms but it did make possible EU enlargement.

[29] *An Agenda for a Growing Europe. Making the EU Economic System Deliver* – Report of an Independent High-Level Study Group Established on the Initiative of the President of the European Commission, Brussels, 2003, p. 172.

budget structure – it could not reduce substantially agricultural expenses due to the prior agreement on agricultural funding until 2013 and it maintained the regional support post – items such as spending for research were severely amputated (heading 1A by more than 40 per cent, see Fig. 7.4).[30]

For the 7th Framework Programme, the agreement meant that its funding would be reduced by more than 30 per cent compared with the ambitious Commission proposal of €72 billion, bringing the budget down to €47 billion for the 7-year period.[31] Keeping a sense of perspective, it is important to note that despite this important cut the 7th Framework Programme's budget would still see a considerable increase (about 40 per cent on yearly average) with respect to the previous Framework Programme.[32]

Table 7.1. The Financial Perspectives (2007–2013): Comparison between Commission Proposal, Council's Compromise Suggestions, Council Agreement and Final Agreement

Table 7.1a. Comparison of expenditure levels 2007–2013

COMMITMENT APPROPRIATIONS 2007–2013 *(in million of €, at 2004 prices, incl. a 2% annual deflator)*	Commission proposal	Lux proposal	UK proposal	Council agreement	Final agreement
1a. Competitiveness for growth and employment	121,687	72,010	72,010	72,120	74,098
1b. Cohesion for growth and employment	336,308	309,594	296,900	307,619	308,041
2. Preservation and management of natural resources	400,294	377,800	367,464	371,244	371,344
3. Citizenship, freedom, security and justice	14,724	11,000	10,270	10,270	10,770
3a. Freedom, security and justice	9,210	6,630	6,630	6,630	6,630
3b. Citizenship	5,514	4,370	3,640	3,640	4,140
4. The EU as a global partner (excl. EDF)	62,770	50,010	50,010	50,010	49,463
5. Total administrative expenditure	57,670	50,300	49,300	50,300	49,800
6. Compensations BG/RO	800			800	800
Total commitments	994,253	871,514	846,754	862,363	864,316
in % of EU-27 GNI	1.21%	1.06%	1.03%	1.045%	1.05%

[30] Reduction to €72.12 billion (2004 prices), compared to the Commission proposal of €121.68 billion.

[31] Assuming a linear and proportional budget development between 2007 and 2013. Cfr. Internal Commission Calculations as of February 2006 (Fiche no. 94: Indicative breakdown of expenditure within individual headings and sub-headings, based on the European Council conclusions of December 2005/Fiche no.95: Impact of the financial framework foreseen by the European Council for 2007–2013).

[32] Exactly 44 per cent. Average annual funding of FP6 compared to the average annual funding of FP7. All calculations in real terms/constant prices.

Table 7.1b. Changes compared to Commission proposal

COMMITMENT APPROPRIATIONS 2007–2013 (*reduction of Commission proposal, in %*)	Lux proposal	UK proposal	Council agreement	Final agreement
1a. Competitiveness for growth and employment	-40.8	-40.8	-40.7	-39.1
1b. Cohesion for growth and employment	-7.9	-11.7	-8.5	-8.4
2. Preservation and management of natural resources	-5.6	-8.2	-7.3	-7.2
3. Citizenship, freedom, security and justice	-25.3	-30.2	-30.2	-26.9
3a. Freedom, security and justice	-28.0	-28.0	-28.0	-28.0
3b. Citizenship	-20.7	-34.0	-34.0	-24.9
4. The EU as a global partner (excl. EDF)	-20.3	-20.3	-20.3	-21.2
5. Total administrative expenditure	-12.8	-14.5	-12.8	-13.6
Total commitments	-12.3	-14.8	-13.3	-13.1

Source: European Commision

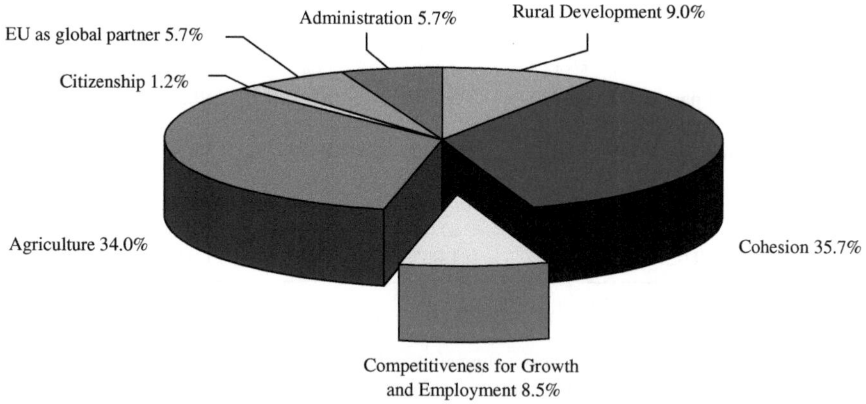

Fig. 7.4. Less cake for research – the decision on the Financial Perspectives and the share of priority areas
Source: DG Research

2.5. Disappointment on the side of the Parliament to the Council deal…

The Parliament rejected the budget deal as "unacceptably low" by an overwhelming majority.[33] With regard to R&D spending, it deplored the "unacceptable reduction in commitments for competitiveness, growth and employment, despite the emphasis given by all the EU institutions to the Lisbon strategy". In the wake of this resolution, Parliament threatened a return to an awkward system of setting annual spending rounds unless the Council agreed to spend more and reform the current budgetary procedure.

In its draft report on the 7th Framework Programme, Parliament openly opposed the budget deal.[34] The report was drawn up by the rapporteur Jerzy Buzek, member of the "Industry, Research and Energy" (ITRE) committee in Parliament which is responsible for research policy. Indeed Buzek had already made clear that there would be opposition to cuts in funding: "any reduction of the EU budget on research [is] unacceptable". Drastic cuts in the budget of the Framework Programme would constitute a "complete disaster".[35]

The European Economic and Social Committee called on the institutions "and, in particular, Member State heads of state or government to make the urgently needed R&D investments proposed by the Commission available in full and to ensure that these do not… fall victim to the negotiations on the EU's future overall budget".[36]

…and reaction from the research community Signatories of a "European petition for research and innovation" also asked the agreement to be revised: "Cutting the research budget… is damaging future growth, employment and development. [Member States should] align the Financial Perspectives with their political ambitions."[37]

Helga Nowotny, the Chair of the European Research Advisory Board (EURAB), a consultative committee to the European Commission on matters of research policy, indicated: "a drastic cut in the research budget… would be a severe blow to the goal of sustainable knowledge-based growth, hamper the desperately needed improvement in employment and engender dismal consequences for research and innovation in Europe".[38] The European University Association called

---

[33] EP resolution on the Council's position on the Financial Perspective and the renewal of the Interinstitutional Agreement 2007–2013 (12.01.2006); ref.: B6-0049/2006. In Parliament's view, the agreement would not guarantee a budget which could provide for the policies already decided by Council.

[34] Cfr. Draft Report on the FP7 proposal. Ref.: PE 360.033v01-00 of 15.12.2005 (ITRE), p. 64.

[35] Cfr. Second Working Document on the FP7 proposal. Ref.: PE 362.761v01-00 of 10.10.2005 (ITRE).

[36] Opinion of the EESC on the FP7 proposals. Ref.: EESC(2005)1484 of 14 December 2005.

[37] http://www.eurec.be/aboutEUREC/new.htm of 31.01.2006 (last accessed on 28 August 2006). The petition was initiated by EUREC, an agency of European renewable energy research centres. An earlier petition of July 2005 for increased Community research funding had been put forward by almost 20,000 researchers as a response to the Luxembourg proposal.

[38] http://europa.eu.int/comm/research/eurab/pdf/letter_1_hn_3_06_2005.pdf of 3 June 2005 (last accessed on 28 August 2006).

on the national rectors' conferences to put pressure on their national governments to maintain the initial Framework Programme budget and urged the national governments to demonstrate their commitment to their own Lisbon and Barcelona targets.[39]

2.6. Striking the deal: Parliament complies in the budget row

Nevertheless, a swift agreement on the overall EU budget was needed, and given the difficult political context of the negotiations, it can be assumed that Parliament wanted to avoid triggering a new crisis.[40] The importance of achieving an agreement can be measured in part by considering what would have been the consequences of failure in this period of political tension. In April 2006, the Parliament and the Austrian Presidency clinched a deal which was formalised by the renewed Interinstitutional Agreement in May 2006: the budget agreement was only slightly higher than what had been agreed at the European summit.[41] The result was disappointing when measured against the objective of Parliament – its members deemed that another €12 billion was necessary to finance policies with a potential high impact on the EU's jobs and growth strategy.[42] For the 7th Framework Programme, this extra money amounted to no more than €338 million.

2.7. Conclusion

The budget agreement was a vital achievement because it ensured that European programmes could be implemented as from January 2007, and put an end to any threat of EU political or financial paralysis. But the eventual compromise was not made in order to meet the Union's self-declared ambitions. The overall amount was far below the Commission's proposal.[43] In addition, the budget structure remained focussed on traditional priorities: about 80 per cent of the funding still went to farm subsidies and regional aid.[44] While the EU could possibly live with this budget, it

---

[39] Together with the European Federation of National Academies of Sciences and Humanities, European Academy of Sciences and Arts and Euroscience. Letter from the EUA President, Professor Georg Winckler of 8 June 2005. http://www.eua.be/eua/jsp/en/upload/FP7_budget_response_EUA.1118320894518.pdf (last accessed on 28 August 2006).

[40] There were early signs that Parliament was ready to approve the budget further to minor adjustments. In its January 2006 resolution, it had already declared its willingness to enter into constructive negotiations with the Council.

[41] Total amount of €864.4 billion. Member States gave €2 billion of 'new money', while another €2 billion was shuffled around within the accounting system (consisting €1.5 billion from moving emergency aid funding outside the formal budget structure and a €500 million reform of the EU institutions' pensions fund).

[42] Another controversial subject was the overall amount to be spent for the EU's CFSP and the 'flexibility instrument', to be put aside for reactions to unforeseen events such as natural disasters.

[43] More than €1.000 billion less.

[44] Out of the total of €864 billion, €371 billion was earmarked for rural subsidies, accounting for 43 per cent of all spending. Some €307 billion will be spent on regional support making up 35.5 per cent of the total.

could not be presented as a radical step to accelerate the Union's progress towards becoming the world's most competitive knowledge-based economy by 2010.[45]

As for the 7th Framework Programme, the final budget represented a reduction of approximately 30 per cent compared with the Commission's proposal. This remained well below the level of ambition originally proposed, although the spending for the new Framework Programme would be around 40 per cent higher on yearly average in real terms than for the previous one.[46] And compared to the past, each Specific Programme and Thematic Priority would benefit from substantially more financial resources (Figs. 7.5. and 7.6).

But as seen earlier, an increase in R&D spending as proposed by the Commission would have injected important new resources into the 7th Framework Programme and allowed for major new directions.[47] As Research Commissioner Potočnik commented: "Having an ambitious research policy with corresponding financial means is no longer a luxury. It is a necessity."[48]

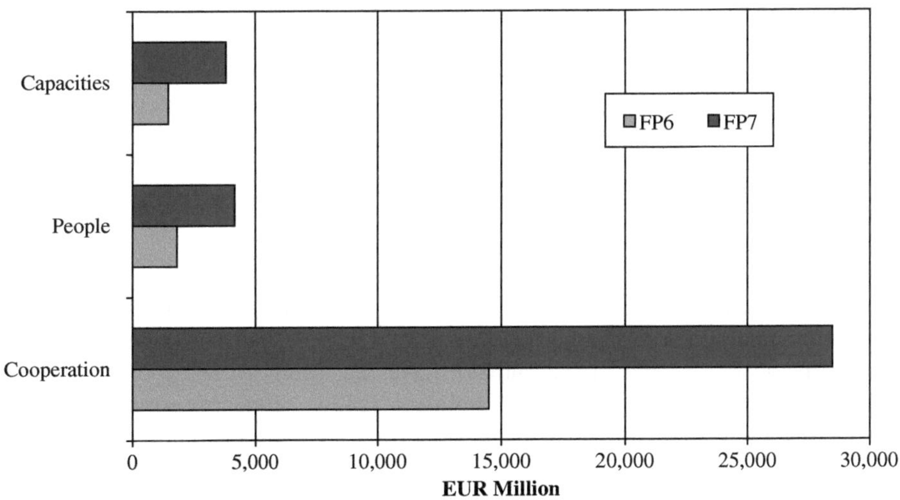

Fig. 7.5. Comparing the budget of some FP specific programmes/components between FP6 and FP7
Source: DG Research

[45] Not more than 5.5 per cent of its overall budget was dedicated to research funding.

[46] Exactly 44 per cent. Average annual funding of FP6 compared to the average annual funding of FP7. All calculations in real terms/constant prices.

[47] Achieving its potential economic, social, and environmental impacts would more than ever depend on each action reaching its minimum efficient budgetary scale, smooth implementation, careful monitoring and evaluation, and close coordination with other policies such as education and innovation. It would require stronger R&D efforts on the part of Member States, with more effective coordination between the EU, national, and regional levels.

[48] Speech given at the Toyota Motor Europe Technical Centre, 24.01.2006, Zaventem (Belgium).

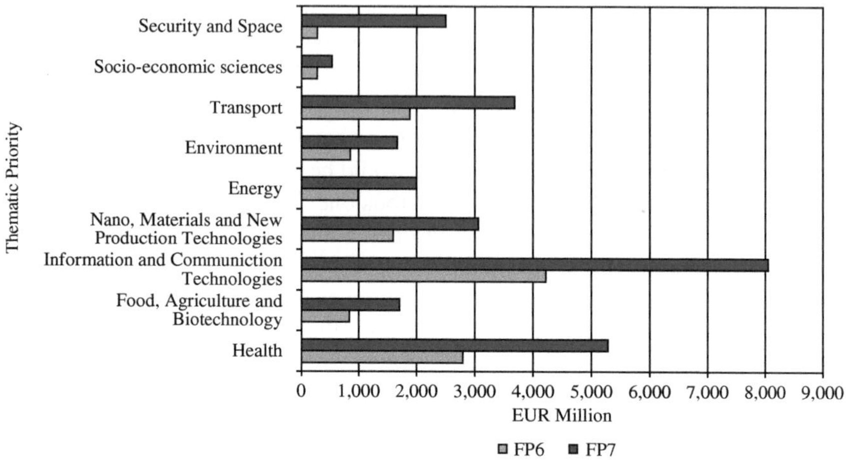

Fig. 7.6. Comparing the budget of the thematic priorities between FP6 and FP7 (2004 prices)
Source: DG Research
Note: FP6 and FP7 thematic priorities do not correspond exactly; thematic priority names in this graph are as under FP7; thematic priority names under FP6: Life sciences, genomics and biotechnology for health; Information society technologies; Nanotechnologies and -sciences, knowledge-based multifunctional materials and new production processes and devices; Aeronautics and space; Sustainable energy systems; Sustainable surface transport; Global change and ecosystems; Citizens and Governance in a knowledge-based society; Food quality and safety

At the end of this section, the time has come to ask why this gap between words and deeds continued to exist. Why was it that the decision-makers – who repeatedly underlined the importance of R&D and pleaded for strengthening of efforts in the field – did in the end not have the means to act upon their good intentions? The answer lies in the political reality of decision-making which leaves only little room for manoeuvre. At a time of low economic growth, tight budgets, and high unemployment in many parts of Europe, any kind of higher investment – as justified as it may be – is difficult to get accepted. On the one hand, the obligation to respect the Maastricht Treaty requirements places an important constraint on the Ministers' scope for action.[49] On the other hand, a large share of the budget was already determined due to the prior agreement reached on agricultural funding to 2013. Thus, while the need to boost R&D spending was widely agreed throughout the EU, political constraints still prevailed over the urge for reform. Indeed, the situation in the Union is similar to what happens at national level: earlier in this book we have seen that investment in research has barely increased in Member States despite the commitment taken in

[49] In this context, it is interesting to note that the idea to exempt R&D spending from the Maastricht public debt/deficit calculations was discussed during these debates.

Barcelona.[50] This situation had led Eska Aho, the chair of the "Aho-Group" and former Prime Minister of Finland, to describe the situation of research policy as "a gap of a political system that preaches the knowledge society and the reality of budgetary and other priorities that have shown little shift in preparing to engage with it".[51]

The debate on the Union's budget showed that the time may not yet be ripe to adapt the EU's funding priorities to the needs of the twenty-first century. Member States still use the EU budget primarily as a financial mechanism of redistribution. Under this angle of maximisation of net flows, it is only logical that the relatively more developed countries demand lower ceilings while the others would agree to higher amounts of total expenditure. It would seem that the real debate on the appropriate structure of the Union's budget has now been postponed until the revision of the Financial Perspectives in 2009. It is in this light that the European Council conclusion of December 2005 – indicating that "particular priority should be given to delivering a substantial and progressive enhancement of the EU's research effort" – should be interpreted.[52]

3. THE RESPONSE FROM DECISION-MAKERS TO THE NEW POLICY DESIGN – THE NEGOTIATIONS ON THE 7TH FRAMEWORK PROGRAMME

Money is important, but it is not the only issue. Simply boosting investment in S&T is not enough. Three other factors prevent European S&T from realising its potential: an inappropriate S&T governance model leading to fragmentation and dispersion, insufficient coordination with other policies, and a problematic relation of the public with S&T. This section will present the 7th Framework Programme discussions in Council and Parliament in relation to these three aspects.[53]

However, it is necessary first to say a few words about the way in which the work of Parliament and Council was organised.

3.1. Organisation of the work

Upon receiving the Commission's proposal in April 2005, the responsible forums of Parliament and Council started working on it (Fig. 7.2).

---

[50] Against this background it is ironic yet not surprising that, at the Informal Summit at London's Hampton Court in October 2005 – in the middle of the tough discussions on the budget – EU leaders underlined once again the overwhelming importance of R&D for the future of the EU growth and jobs. As a follow-up to the Council, the Aho Report was submitted in January 2006 on request of the Commission (see Chapters 2 and 3).

[51] *Creating an Innovative Europe*, Report of the Independent Expert Group on R&D and Innovation appointed following the Hampton Court Summit, Brussels, 2006.

[52] Cfr. European Council Conclusions of 15/16 December 2005, annex, §10. Ref.: Council doc. 15915/05 (Financial Perspectives agreement).

[53] At the time of writing (August 2006), decision-making is in full swing. All indications about the likely differences between the initial proposal and what will be the end-product are, therefore, provisional.

...in Parliament The co-decision procedure puts the ball first in the camp of Parliament which must prepare its amendments at first reading.[54] These were formulated in the "Industry, Research and Energy" committee (ITRE) before being debated and adopted in plenary session. This involved ITRE drafting a report (written by rapporteur Jerzy Buzek), giving its general appreciation of the 7th Framework Programme proposal.[55] In parallel, Members of Parliament (MEP) submitted specific amendments to the Commission proposal. In January 2006, Mr Buzek presented his draft report to ITRE for adoption in May 2006, and to plenary session for adoption in June 2006. Eight other committees provided further opinions.[56] The European Parliament also prepared more detailed reports on each of the Specific Programme proposals as part of the 7th Framework Programme package. These reports were expected to be voted on in late 2006.

...and Council The work in Council is highly segregated and fragmented. Decision-making is prepared by specialised working parties made up of experts from the Member States and chaired by the Member State holding the Council Presidency. In the *Research Working Party*, 25 national civil servants, normally the scientific counsellors, meet two to three times a week. They are usually based in Brussels in "permanent representations" headed by national ambassadors to the EU and act upon instructions from their capitals. By pulling in officials or just opinions from national capitals through the working groups, the number of actors involved in decision-making is expanded enormously. The scientific counsellors undertake the technical groundwork needed before the Council can decide, that is they discuss the details of the proposal. Their task is to deliver a technically uncontroversial text to the next level, the *(Deputy) Permanent Representatives Committee* (Coreper I for research policy).[57] Coreper I tackles the political issues which could not be solved by the Working Party. It is a powerful engine in the decision-making process: its members frequently must "construct" a national position in Council negotiations when instructions from national capitals are vague or reflect divisions between ministries. All remaining outstanding questions are taken to the third hierarchical level, to the Science and Research Ministers, meeting in the *Competitiveness Council* configuration (incorporating the former Research Council) two to three times per Presidency. This is the forum where the political decisions are taken (Fig. 7.7).

---

[54] As seen earlier, the introduction of co-decision significantly upgraded the EP's power in the legislative process. In particular, it enhanced the importance of the sectoral committees (such as the ITRE committee in charge of research policy) and particularly of the "rapporteurs" who conduct appraisals of specific proposals.

[55] Jerzy Buzek is the former Polish Prime Minister, member of the PPE-DE, and a professor of technical science. The "rapporteur" is chosen by a weighting system representative of the political groupings.

[56] The Committees for Budget, Environment, Public Health and Food Safety, Transport and Tourism, Regional Development, Agriculture, Fisheries, Legal Affairs, Women's Rights and Gender Equality.

[57] Romania and Bulgaria are present in Coreper as observers.

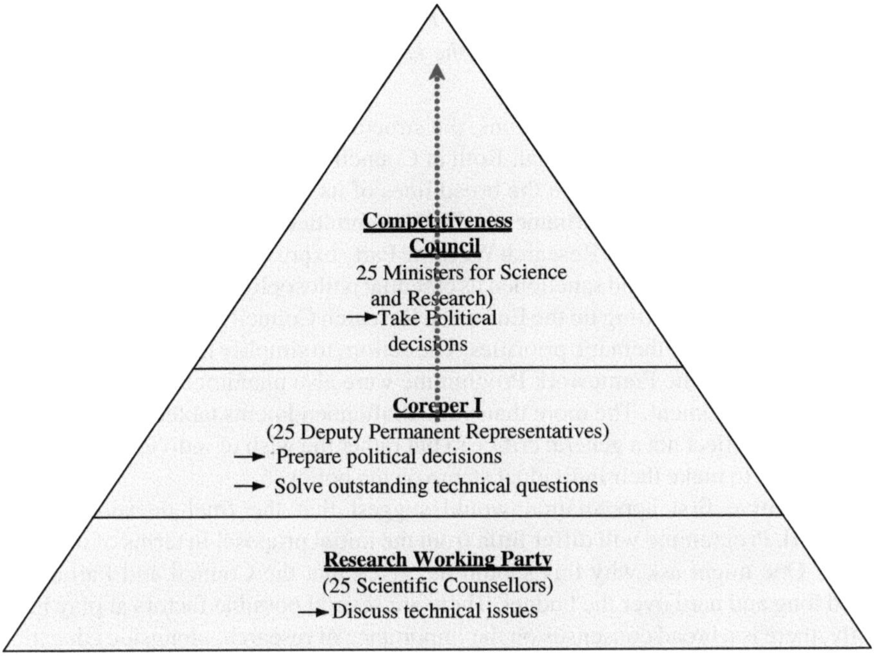

Fig. 7.7. Decision-making in Council at three levels
Source: DG Research

With regard to the Framework Programme, at the stage of the first reading, the work in Council ran concurrently with activity in the European Parliament. First position-taking in Council is always a cautious matter. Member States, with the Presidency in the lead, are anxious not to prejudge Parliament's opinion. However, on occasion, the Council may reach an agreement before Parliament ends its first reading, commonly termed a "general approach". This is what happened for the 7th Framework Programme:[58] Council agreed on a "general approach" in May 2006 which it formalised to a "political agreement" on all aspects of the Framework Programme in late July 2006, following Parliament's amendments in June 2006. At this stage in the process, one can expect a "second reading agreement" in autumn 2006 which would allow the 7th Framework Programme to be ready to be implemented in January 2007.

[58] Awaiting the EP opinion, the Council approved a *Partial General Approach* on the EC proposal in November 2005 (Council doc. 15602/05, 01.12.2005). This allowed for fixing the Council discussions on non-budgetary elements. It left open the possibility of adjusting agreed parts of the proposal following the opinion of Parliament in first reading and the outcome on the Financial Perspectives. The Council then moved to a *General Approach* in May 2006, after agreement was found on the overall budget. It went then from the general approach to a *Political Agreement* in July 2006 (Council doc. 11978/06 of 25.07.2006). In principle, it goes then to a common position after examining the EP's opinion. If the EP amendments coincide with the general approach, the act can be adopted at second reading.

3.2. *The proposal for the 7th Framework Programme and the decision-makers: How Council and Parliament rate the key factors accounting for Europe's S&T weaknesses*

From the beginning of the negotiations, the structure of the Framework Programme proposal has not been controversial. Both at Council and Parliament, there has been a large measure of agreement on the broad lines of its objectives and content.[59] The members of the European Parliament's ITRE committee[60] as well as the Scientific Counsellors in the Council's Research Working Party expressed overall support for the Commission's proposal and sanctioned its essential philosophy. The basic lines had the institutions' support: setting up the European Research Council, improving links with industry, the choice of thematic priorities. The actions to simplify the management and implementation of the Framework Programme were also unanimously welcomed by Council and Parliament. The more than 1200 draft amendments tabled by Parliament in April 2006 reflect not a general criticism but rather the wish of individual Members of Parliament to make their individual stamp on the policy.[61]

This positive first appreciation would suggest that the final decision on the Framework Programme will differ little from the initial proposal in terms of its broad content. One might ask why this should be, given that the Council and Parliament argued long and hard over the budget. There are several possible factors at play here. Firstly, there is a broad consensus on the importance of research, alongside education and innovation. Secondly, this may be evidence that the Commission proposal was well judged, and that the new approach to policy design described in this book has been successful. On the one hand the institutional actors perceive the 7th Framework Programme as a balanced draft which lives up to the expectations of the research community. On the other hand, it would appear that the stakeholders' consultation served its purpose: the absence of any general contestation confirming that the main inputs from the stakeholders were successfully taken on board. Thirdly, from an institutional perspective, the broad agreement on the content of the proposal may also affirm the importance decision-makers attach to the funding for the Framework

---

[59] The analysis draws on the following sources: For Parliament, the Buzek report is the first reference document. At the time of writing (August 2006), the reports on the Specific Programmes were not yet available. Positions within Council are mainly assessed on the basis of the debates at Ministers' level. This is done for two reasons: First, only the recordings of the Competitiveness Council are publicly accessible. Second, this is the forum where the most relevant political debates – beyond technical examination – take place.

[60] See the positive appreciation by the various versions of the draft Buzek report. In the version of 15 May 2006, the EP confirmed that "*the 7th Framework Programme is broadly accepted by the European Parliament*" (ref. 2005/0043(COD), p. 132). It also supported the selection of nine thematic areas for collaborative research, although ITRE – as well as did the Council – asked to split the area of "security and space" into two separate themes.

[61] At the ITRE vote on 15 May 2006, agreement between the political groups meant that their number was drastically reduced. Parliament eventually approved 257 amendments to the EC FP in the first reading, most of which did not change the Commission proposal. Many of the proposed changes were intended to increase the participation of SMEs, while others concerned the budget allocation, aimed to give the ERC more independence, specified particular scientific fields that should be included or the relationship between ETPs/ JTIs and FP7.

Programme. The research budget is perceived as the bottom line and, therefore, hard bargained over. It determines the content of the proposal, the technical details of which are often difficult to control by the institutions.[62] Table 7.2 visualises the key features of the decision-making process on the 7th Framework Programme.

Table 7.2. What goes in and what comes out – comparing Framework Programme features between proposal and decision

	Commission initial proposal[63]	Parliament first reading[64]	Council political agreement[65]	Likely decision[66]
Philosophy, objectives, structure		✓	✓	overall approach maintained
Budget: overall amount		decrease	decrease	decrease[67]
EC (2007–13)	€ 72,726 million	€ 50,521 million	€ 50,521 million	€ 50,521 million
EURATOM (2007–11)	€ 3,092 million	€ 2,751 million	€ 2,751 million	€ 2,751 million
Budget: allocation among the EC specific programmes	Cooperation 44,432 Ideas 11,862 People 7,129 Capacities 7,486	Cooperation[68] 32,492 Ideas 7,560 People 4,777 Capacities 3,944	Cooperation[69] 32,365 Ideas 7,460 People 4,728 Capacities 4,217	Cooperation 32,292 Ideas 7,460 People 4,727 Capacities 4,291

(Continued)

[62] The technical nature of the RTD sector is also confirmed by the fact that from the Council's point of view, research is the domain of Coreper I, not Coreper II, and thus is considered a technical, more than a political, arena for decision-making.
[63] European Commission, Proposal for a Decision of the European Parliament and of the Council concerning the Seventh Framework Programme of the European Community for Research, Technological Development and Demonstration Activities (2007–2013), COM(2005) 364, 6 April 2005.
[64] Ref.: Parliament doc. P6_TA-PROV(2006)0265 of 15.06.2006.
[65] Ref.: Council doc. 11978/06 of 25.07.2006.
[66] On 28 June 2006, the Commission presented its amended proposals for FP7, taking into account the discussions in Parliament and Council so far. Cfr. European Commission, Amended Proposal for a Decision of the European Parliament and of the Council concerning the Seventh Framework Programme of the European Community for Research, Technological Development and Demonstration Activities (2007 to 2013), COM(2006) 364, 28 June 2006.
[67] On 24 May 2006, the Commission presented its modified proposals for FP7, taking into account the agreement on the Financial Perspectives. Cfr. European Commission, Modified Proposal for a Decision of the European Parliament and of the Council concerning the Seventh Framework Programme of the European Community for Research, Technological Development and Demonstration Activities (2007–2013), (COM(2005) 119/3), 24 May 2005.
[68] The EP had wanted more for energy (renewable energies), socio-economic research and health. Less for ICT.
[69] The Council had wanted higher shares to all themes, except for food, socio-economic research and security. The areas which were most reduced in the Presidency proposal were security and infrastructures. Cfr. Council doc. 9187/06 of 18.05.2006.

Table 7.2. (Continued)

	Commission initial proposal[64]	Parliament first reading[65]	Council political agreement[66]	Likely decision[67]
	JRC (non-nuclear activities) 1,817	JRC (non-nuclear activities) 1,751	JRC (non-nuclear activities) 1,751	JRC (non-nuclear activities) 1,751
Thematic areas	9 areas proposed	✓ but: split security and space in 2 separate areas	✓ but: split security and space in 2 separate areas	✓ 10 areas (split security and space in 2 separate areas)
ERC (implementation modalities)		Start with executive agency; to be transformed in "independent structure" in co-decision after review	Start with executive agency; possibly transformation in "independent structure" in function of results of review	?
Ethics (funding stem cell research)		✓	✓	✓ In line with COM initial proposal

Source: DG Research

Notwithstanding this general agreement, the decision-makers did make a large number of important comments on the specific actions proposed in the new Framework Programme. The remainder of this section analyses their positions in relation to the ability of the Framework Programme to tackle the key weaknesses of European S&T.[70]

3.3. Research policy and the inappropriate S&T governance model

In Chapter 2, it was seen that Europe's research efforts are dispersed at regional, national, and European level, and characterised by overlap and fragmentation. As

[70] The choice of examples in the following section cannot claim to be exhaustive: The co-decision process is not yet completed at the date of writing (August 2006) and the institutions have not given their positions on all aspects of the proposal.

a result, efforts are not used efficiently. Often, different Member States pursue similar objectives in isolation, thus not reaching the necessary critical mass and partner mix.

Past Framework Programmes have tried to address these problems by promoting close S&T cooperation and coordination between Member States. Both the Council and the Parliament have always strongly supported such actions, for example the classical transnational collaborative research, mobility actions, or the ERA-NET scheme under the 6th Framework Programme. The new Framework Programme proposes stronger efforts in this direction, not only by increasing the funding for traditional actions, but also by introducing some new measures such as the Joint Technology Initiatives (JTIs) and the European Research Council (ERC). The ERC in particular represents a first step towards a completely new European S&T governance model based on supra-national measures.

"Traditional" cooperation and coordination actions: Full backing

The Council and the Parliament openly welcomed the 7th Framework Programme actions aimed at increasing cooperation. These include collaborative research projects and networks, the coordination of research programmes through a reinforced ERA-NET scheme,[71] as well as measures in the field of international cooperation between EU and third countries. The renewed emphasis given to the "people" programme – supporting training and career development of researchers through a further enhanced "Marie Curie" scheme – was particularly well received by the institutional players from the beginning of the debates.[72] Marie Curie was considered a highly successful initiative (as can be observed by the level of oversubscription in the previous Framework Programme) and the increased efforts proposed by the Commission received unanimous backing.[73]

"New" cooperation and coordination actions: More hesitations

European Technology Platforms and Joint Technology Initiatives. As new measures in the area of cooperation and coordination, the 7th Framework Programme foresees a set of actions, notably in its "ideas" programme: Besides the TPs in some selected areas, it proposes a more ambitious and concerted approach – the establishment of JTIs. These shall bring together European, national, and business resources into major research ventures corresponding to the technology-intensive markets of the future. Furthermore, the "capacities" programme delivers a strengthened infrastructures programme.

---

[71] In the "ERA-NET Plus module", the Commission provides an incentive to organising joint calls between national or regional programmes by "topping-up" joint transnational funding with Community funding.

[72] See draft Buzek report of 15 May 2006, ref. 2005/0043(COD), p. 132.

[73] Notwithstanding the ITRE amendments of May 2006 on the "people" part of FP7, asking for specific support for early stage researchers and for interregional mobility actions within a country.

From the preliminary discussions on the TPs and JTIs,[74] it was clear that there was a high level of interest and high expectations around these instruments. Parliament as well as Council underlined their potential. However, as was the case in the stakeholder consultation (Chapter 5), the institutional actors expressed concern about the implementation modalities. While TPs enjoyed support in principle, main concerns related to JTIs.[75] Parliament felt that it was not obvious from the proposal how the different new initiatives (i.e. TP/JTIs, ERC, and large infrastructures) would interact with each other and pointed to the risk of poor integration with other efforts at European, national, and regional levels. It suggested that special mechanisms should enhance their coordination, such as expanding the role of Networks of Excellence as a tool to foster networking and integration with national programmes and corresponding ERA-NETs. Further particular issues of concern expressed by Members of European Parliament and Scientific Counsellors regarded the selection criteria of JTIs, their effectiveness, and the involvement of SMEs.[76] As for the Council, pending agreement on the Financial Perspectives and awaiting concrete Commission proposals on specific Technology Initiatives, discussion was postponed to a later stage.

The European Research Council – Strong support from the very first day As seen earlier, the creation of an autonomous body to boost European frontier research across all scientific and technological fields is a genuine innovation in EU research policy. With this idea, the Community moves from its traditional funding rationale – transnationality – to a new concept: promoting excellence by fostering Europe-wide competition between individual researchers and teams. The implied change in the definition of the European added value and in the European governance model is important.

This proposal has been extremely well received within the scientific community, by Member States' governments, and by the Parliament members. The idea of a European Research Council has never been contested in itself. On the contrary, as signalled during the stakeholders' consultation, there was broad agreement that the EU should fund the brightest and best of European research. In addition, there was consensus from the very beginning on the overarching principles of this organisation such as autonomy, accountability, transparency, and excellence.

However, the implementation of the European Research Council was exposed to several contestations. Questions raised included how autonomy and accountability would work in practice, what criteria would be used to select projects, and who would do the selecting.

---

[74] First debates took place under Luxemburg Council Presidency (Spring 2005) and at the moment of the drafting the report on the "capacities" Specific Programme in Parliament (Spring 2006).

[75] For example, they would jeopardise competition, lead to stakeholders grouping together and shutting themselves off from others in the research community or elsewhere. As said Jerzy Buzek at CORA on 10 May 2006: "They will be big enough not to have contact with anybody. We must avoid the fragmentation of our research."

[76] Amendment 87 of the Buzek report (15 June 2006) added criteria (e.g. genuine societal need and commitment from industry; added value in terms of excellence and synergies obtained through cross-border cooperation; the capacity to encourage entrepreneurship; agenda for the training of researchers involved).

The European Research Council's selection criteria: Excellence or transnationality? The beginning of the Council debate was marked by one of the recurrent controversies in research policy. Culminating in the Competitiveness Council in March 2004 – the last Council before enlargement – Member States found themselves caught in a debate about whether the principle of excellence should prevail over that of transnationality. One group considered that basic research, undertaken with Community funding, should be highly competitive.[77] Projects should be selected based on the single criterion of excellence. Some, mainly southern Member States,[78] while supporting the principle of excellence, thought that it could not be the sole criterion, and that account should be taken of the cooperative and transnational context of European research. This would imply the collaborative participation of teams from different Member States. Indeed, Italy argued that the European added value of EU research was its very transnationality. This debate was settled quickly afterwards, with the opponent Member States accepting the prevalence of the excellence principle in the light of an overwhelming majority. The discussion then turned to the governance structure of the European Research Council. Should it be an executive agency or a body following article 171 EC Treaty?[79]

The overall architecture of the European Research Council: How independent should it be? The Member States, as well as Parliament, remained divided for a long time over the status of the ERC. The Commission had proposed setting up the Research Council as an executive agency – a legal body ultimately answerable to the Commission. European scientists, however, feared that this could lead to political criteria determining scientific priorities; they voiced their particular concern that the Commission would be able to ignore advice from the ERC's Scientific Council.[80] Therefore, some EU research ministers[81] favoured establishing the ERC as a legal entity entirely separate from the Commission according to article 171 EC Treaty, as did the Parliament in the Buzek report. This idea was supported by several scientific organisations from France, Germany, Poland, Spain, and the United Kingdom. However, according to the Commission, an autonomous, self-regulating ERC would be vulnerable to the principle of 'juste retour', meaning a structure ignoring scientific merit and distributing grants according to national quotas or contributions to the budget. The essence of the Research Council which is a pan-European competition on the basis of scientific excellence might thus be undermined.

---

[77] Denmark, Germany, France, Austria, Netherlands, Sweden, UK, and the Commission.

[78] Italy, Spain, Greece, and Portugal (Slovakia joined this group later).

[79] Art. 171 TEC states: *The Community may set up joint undertakings or any other structure necessary for the efficient execution of Community research, technological development and demonstration programmes.*

[80] As an independent body, the Scientific Council (made up of 22 scientists) should have the role to determine the ERC's scientific strategy and ensure that its operations are conducted according to principles of excellence. At the March 2006 Competitiveness Council, Ministers agreed on the autonomy of the ERC, with authority of the Scientific Council over scientific strategy and management, and proper accountability via the Commission.

[81] Those from Hungary, Poland, Sweden, Portugal, and Denmark. Situation as reflected at the Competitiveness Council of 30 May 2006.

The Council's Partial General Approach and the Parliament's Buzek report helped to clarify the ERC's overall structure: Agreement was found that the European Research Council should, in a first time, be based on an Executive Agency. It was consented that, after an independent review in the context of the interim evaluation of the Framework Programme, there would be the possibility to create a structure based on the article 171 EC Treaty.[82]

3.4. Research Policy and the insufficient coordination with other policies

Earlier in this book, it was demonstrated that until 2000 and as inscribed in the Treaty, research was characterised as a policy area complementary to national competencies and at the service of other Community objectives (e.g. industrial competitiveness). With Lisbon, this model was no longer valid. The Lisbon agenda placed research at the heart of Community policies which needed to be reassessed so as to maximise their contribution to the knowledge-based economy. Against this background, efficient horizontal coordination of research policy with the other components of the knowledge triangle (namely education and innovation policies) and between the knowledge triangle policies and other Community policies became increasingly important. However, as seen in Chapters 2 and 3, the insufficient coordination of European S&T with other policies has been one of the factors contributing to the slow progress in achieving the Lisbon goals.

Looking into the institutional debate on the Framework Programme proposal, we notice that the main focus has been on its size and content, but little has been said on the subject of policy coordination. The institutional actors may have expressed concerns about the coordination gap and its negative impacts on the achievements of the Lisbon objectives, but they have not analysed possible solutions, and their interventions remained rather general. For instance, Parliament's Buzek report acknowledged the need for closer horizontal coordination between Community policies in general terms. It recommended paying attention to the relation of the Framework Programme notably with the "Competitiveness and Innovation Programme",[83] but

[82] However, amendments 239 and 249 of the Buzek report of 15 June 2006 suggested that the ERC would in the future make the transition to an independent structure established under co-decision procedure.

[83] Not under the FP7 umbrella, the CIP is an initiative regarding coordination of Community policies: It brings together programmes relevant to innovation and competitiveness under one roof (such as the entrepreneurship and innovation programme, the information and communications technology support programme, and the intelligent energy-Europe programme). FP7 and CIP are complementary (the CIP focuses on downstream activities such as market take-up of technologies and supports the development of technology transfer networks. FP7 focuses on the technological needs of industry and supports innovation-related activities within research projects). At the March 2006 Competitiveness Council, a partial general approach was reached on the CIP, the main features of which are: clear articulation of the complementarities between CIP, FP7, and the Structural Funds; clarification and higher visibility in respect to eco-innovation; improvement of the monitoring and evaluation aspects of implementation of the CIP.

also the "Education and Training Programme" and the "Structural Funds".[84] It also stressed the importance of vertical coordination between Framework Programme actions and activities in Member States. For its part, the Council acknowledged on several occasions the need for coordination between EU research policy and other policies, and for complementarity between regional funds and other Community measures. In its "Partial General Approach", agreed in November 2005, the Council highlighted the importance of such integrated approach to advancing European research (notably in the vertical sense but also touching upon horizontal coordination). It added a separate title in the Capacities Programme: "Support to the coherent development of research policies", stating as its objectives "*enhancing the effectiveness and coherence of national and Community research policies and their articulation with other policies, improving the impact of public research and its links with industry*". The Commission, in its amended proposal of June 2006,[85] accepted this insertion. However, even if the issue of horizontal coordination was brought up in the debates, the statements of all institutional actors tended to be of general principle rather than detailed debates with concrete follow-up actions. Why?

This is not entirely surprising. Coordination across policies is difficult, and research decision-makers have plenty to discuss in order simply to arrive at an EU-wide agreement on the Framework Programme. Such meta-debates might be more productively handled at a level that transcends the EU's policy process, for example the European Council. This is the forum where global decisions such as on the budget are taken and where coordination issues are best discussed. The debate on the Union's Financial Perspectives, described earlier in this chapter, serves as an illustration of such overarching decisions on EU policy balance. The Commission had proposed to reshuffle policy priorities and to put research at the centre of European policies. This would have given a new importance to coordination of research policy with other policies. But we have seen that in the end this proposal was not adopted. Political constraints, conflicts between competing policy areas, and prior commitments impeded a re-balancing. It would appear that the real debate on the Union's political and financial priorities and thus horizontal policy coordination was postponed to a later stage, possibly to the agreed revision of the Financial

---

[84] FP7 includes two specifically targeted schemes to support regions as important players in the transition to the knowledge-based economy: the "Regions of Knowledge" scheme, to support existing regional research-driven clusters and "Unlocking the research potential" to develop the research potential in the EU's less-favoured regions. These schemes not only aim to promote research in the broad sense. They also aim to improve synergies of the FP with European Regional Policy in the sense that more use is made of the Structural Funds to support R&D activities. At the same time, it needs to be acknowledged that while being in favour of the reinforcement of complementarities and synergies, the Commission cannot interfere in the decisions of the national/regional authorities concerning the use of the Structural Funds.

[85] Without dedicating a budget line to this heading. European Commission, Amended Proposal for a Decision of the European Parliament and of the Council concerning the Seventh Framework Programme of the European Community for Research, Technological Development and Demonstration Activities (2007 to 2013), COM(2006) 364, 28 June 2006.

Perspectives in 2009. In other words, the time has not yet seemed to be ripe. This may also explain why the conclusions of the European Council, when they touch upon the subject of coordination of research policy with other policies tend to be rather broad brush and do not set clear and binding targets.[86]

In addition to the 'timing problem', the absence of a real examination of the coordination trap points at a gap in the decision-making process itself: The institutional translation of Lisbon and of this shift of priorities is lagging behind. The system has not yet adapted to the new European objectives set; the Council is still no exchange forum for best practices at national level; the Framework Programme is still and almost exclusively perceived as an instrument to implement research projects. It appears as if policy coordination is a less immediate – even if not necessarily less important – priority. This leads back to Chapters 2 and 3, which touched upon the difficulty of efficiently coordinating policies, especially research and innovation policies which are transversal and systemic in nature. Indeed, as is the case at national level, policy coordination is a complicated matter and even more difficult at EU level. Here, another layer of complexity is added since its ambitious targets require coordination not only across different policy areas, but also, as seen earlier, between different levels of governance. Therefore, it has proved difficult to come up with concrete measures (some of these would be much longer term, relating to institutions and governance generally).

3.5. Research policy and the difficult relationship with the public

While knowledge frontiers have proven to be endless, the same cannot be said for public support for research. Scientists are constantly discovering new frontiers, most importantly ethical ones, which force the research community and the policy-making community to take a position. In some cases, these ethical issues are complex and public opinion may be cautious, sceptical, or even fearful. These facets have been discussed in Chapter 1.

The 7th Framework Programme starts from an awareness on the side of the policy-shapers and decision-makers of the ethical dimensions of research. It makes clear that all research supported must be carried out in compliance with ethical principles, and that any activity forbidden in a Member State will not be funded there. The fact that the media has brought the activities in which researchers engage in to the attention of the "lay" people has enabled civil society to express its opinion. The theme "Science and Society" proposes a variety of concrete measures to bridge the gap between S&T and society at large, including improved communication between the scientific world and the public. In their discussions on the 7th Framework

---

[86] Cfr. for example the Conclusions of the European Spring Council of March 2006 which endorsed the relaunched Lisbon Strategy: *Action undertaken under the Research Framework Programme should be better coordinated with other European and national actions* (Art. 19); *Links should be strengthened between R&D, innovation systems and business environment* (Art. 22) and *The search for excellence and innovation at all levels of education and training, in particular through better links between higher education, research and enterprises is crucial* (Art. 23). Reference: 7775/1/06 REV 1.

Programme, the decision-makers welcomed this approach.[87] But the debate on ethical principles has not been without some controversy, most notably in relation to the funding of research involving the use of embryonic stem cells.

Research and embryonic stem cells The issue of stem cell research and embryonic stem cell research in particular is one of the most controversial ethical themes dominating the debate in Europe, in fact worldwide.[88] Stem cell research offers the prospect of developing therapies for serious diseases such as Alzheimer. At the same time, it raises difficult and controversial questions which are reflected in a variety of moral perspectives and national regulatory regimes. These vary from total prohibition to qualified authorisation. In the context of the EU decision-making, the issue fervently discussed concerns the conditions under which Community funding is granted to projects involving the use of embryonic stem cells. This subject which had dominated the negotiations on the 6th Framework Programme, re-emerged under the 7th Framework Programme.

In its new proposal, the Commission set out the same pragmatic approach as taken in the previous Framework Programme: restrictions on the research should be rare, well-justified, and flexible enough to accommodate changes in the quickly evolving field; scientists should be free to receive EU funds to implement this kind of research in other EU countries if it is banned in their own country. Article 6 of the proposal provides that "*All the research activities carried out under 7th Framework Programme shall be carried out in compliance with fundamental ethical principles.*" Applying this provision for embryonic stem cell research, the Commission proposes the following: Certain "no go areas" not be eligible for Community funding would be specified; there would be an ethical evaluation for each project concerned; and any funding for such projects would be decided through the regulatory procedure involving the programme committee.

Within the EU, national legislation and practices differ. Consequently, two main groups of Member States emerged during the discussions on the 7th Framework Programme in Council: the "liberal" and the "restrictive" Member States. Interestingly, both camps referred to the subsidiarity principle to back their respective positions.

The more restrictive Member States[89] opposed Framework Programme funding for research activities which include the consumption of embryonic stem cells. They requested that further "no go areas" should be added to those already excluded

[87] In the plenary debate in June 2006, several MEPs underlined the necessity to have an open and well-informed debate on ethical aspects inherent in research policy. ITRE had also asked, for instance, that the Commission, in case of ethically sensitive research projects, systematically examines the projects and report back to Council and Parliament on a yearly basis (cfr. draft Buzek report of 30 May 2006).

[88] See the recent Eurobarometer: George Gaskell *et al.*, *Europeans and Biotechnology in 2005: Patterns and Trends*, Eurobarometer 64.3. A Report to the European Commission's Directorate-General for Research, Brussels, 2006, p. 5 C13.

[89] Notably Austria, Germany, Lithuania, Malta, Poland and Slovakia.

from Community support. Following their interpretation of subsidiarity, funding should be based on the ethical standards dictated by the most restrictive country. This would imply that only research that is accepted in all Member States should be eligible for Community funding.[90] Countries, in which embryonic stem cell research is legal, should fund it from their national budgets.[91] The more liberal Member States[92] accepted the Commission proposal as a minimum compromise solution vis-à-vis the national approaches, and asked for no further restrictions to be imposed. Within this group, the least restrictive countries preferred not to list "no go areas" explicitly, considering the provisions proposed by the Commission to be sufficient. For the liberal countries the exclusion of research activities forbidden in one Member State would contradict the subsidiarity principle. In their view, countries should retain responsibility for deciding whether human embryonic stem cells may be used in their respective territories. In their opinion, this was in full respect of subsidiarity and of the fundamental ethical principles.[93]

Eventually, Ministers reached an agreement in late July 2006 on the modalities of Community-funded stem cell research by a small majority,[94] broadly accepting the Commission proposal.[95] Some of the restrictive countries, the critics of such research,[96] agreed once there was an explicit pledge that European funding would not be used for the phase of work that involves the destruction of human embryos. This solution made possible the Council's political agreement on the 7th Framework Programme proposal. Encouragingly, the international press took great interest in this agreement, commenting that the United States was risking a brain drain of disillusioned scientists in this field of research to Europe wanting to make the most of Europe's more liberal rules.[97]

---

[90] Such as somatic and umbilical cord stem cell research.

[91] See also the European Parliament resolution on egg cell trade of 10.03.2005 (p6_TA_PROV(2005)0074).

[92] Notably the Czech Republic, Cyprus, Denmark, Hungary, France, Slovenia, Sweden, Belgium, Netherlands, Portugal, UK, Finland, Spain, and Estonia.

[93] Arguing that the 7th Framework Programme proposal guarantees anyhow that no research activity will be funded in a country where that research is forbidden. To this end, a group of Member States submitted at the Council meeting in March 2006 a declaration to the Council minutes, calling for decisions on "fundamental ethical principles" to be "left to national legislators in full accordance with the principle of subsidiarity".

[94] The Austrian, Lithuanian, Maltese, Polish, and Slovakian delegations voted against.

[95] That is, stem cell research may continue to be financed under the case-by-case approach used in the 6th Framework Programme; research projects will only be considered for funding from Member States where the research is legal; no activity will be funded that is forbidden in all Member States.

[96] It was Germany and Italy. As wrote the German Minister Schavan to the Presidency on 20 July 2006 according to Reuters: *The EU research programme should not be used to give financial incentives to kill embryos.*

[97] The Guardian, 25.07.2006 (http://www.guardian.co.uk/genes/article/0,,1828069,00.html). In July 2006, US President Bush had limited federal funds for this area, using a presidential veto against a bill that would have expanded such work in the United States. Shortly afterwards, however, a scientific discovery appeared to get round this basic ethical objection to stem cell research: Scientists in the United States were reported to have created human embryonic stem cells without destroying embryos. Cfr. *Financial Times*, Europe Edition, 24 August 2006.

The clash between Member States echoed the disruptions in the Parliament: the debate on stem cell research and Community funding divided the house into supporters and opponents across party group lines. Eventually, as was the case in Council, at the plenary vote on the Buzek report in June 2006, Parliament declared itself in line with the Commission and its members passed the provision on stem cell research with a majority of 284 votes.[98]

CONCLUSION

The decision-making process for the Framework Programme is complex, and to the outsider it appears to be something of a black box. It is a multi-layer, multi-player, multi-interest system in which all actors strive for compromise. What is perhaps most striking is its disaggregated nature. Given the multitude of institutional actors and sectoral interests, there is no homogeneous opinion building and decision-making. Despite their treaty status as a single institution, the Commission, Council, and Parliament are far from homogeneous entities with one clear opinion. It is not simply "the Council" confronting "the Parliament". Nor does the Commission act throughout as a monolithic body. In addition to this internal fragmentation, there is the added complexity of the different levels and layers in which the work of the institutions is organised. By pulling in officials, experts, or just opinions from outside Brussels, the number of actors involved in decision-making is expanded enormously. These factors make the process more difficult to predict – and to recount. At the same time, the legal requirement – the Framework Programme falls under co-decision – obliges these heterogeneous entities to find a compromise.

Turning to the content: How did the decision-makers react to the new policy design as presented in the 7th Framework Programme proposal? At this stage in the process, after the first reading, the institutional players appear to share in broad terms the Commission's analysis of the weaknesses of European S&T. The main features of the proposal will survive. Both in Council and Parliament, there has been a large measure of agreement on the broad lines of the structure of the 7th Framework Programme, on its S&T objectives and content. With regard to particular problems, the agreement is less unanimous. The negotiations on the Union's overall budget confirm the ambivalence of current European R&D policy: the funding for the 7th Framework Programme will be 30 per cent less that initially proposed by the Commission, although representing an increase in average terms vis-à-vis the previous Framework Programme of approximately 40 per cent. This increase in funding is a success of sorts, admittedly, but one which could have

[98] In form of the amendment 66 of the Buzek report of 15 June 2006. This amendment suggested to clarify which areas should not receive EU funding which was in line with the practice under FP6 (i.e. human cloning; heritable modifications of the human genome; and production of human embryos solely for stem cell procurement), added that such research may be financed in countries where the legal system permitted it, and in the framework of strict licensing conditions. It was adopted with 284 votes in favour, 249 against, and 32 abstentions.

been more resounding. In spite of the shared agreement on the role of R&D for the Union's future, there are numerous political and financial constraints which limit the room for manoeuvre of decision-makers. Regarding the fragmentation and dispersion of research efforts in Europe and the proposed solutions, the institutional actors are positive, at least as far as the classical activities proposed are concerned: there is strong support for continuing and reinforcing traditional cooperation and coordination activities.[99] However, when it comes to new activities to enhance collaboration and to move beyond mere cooperation towards a true supranationalism, the problems are obvious. While there was generally strong support for the creation of an ERC, it is no surprise that it was exactly in the discussions on its implementation that the tensions about the organisation of future European S&T became clearest.[100] But with the acceptance of the general idea to establish an ERC an important first step has been taken towards a new model of European S&T governance. Finally, the debate on the issue of stem cells provides yet another time a striking example of the difficulties of research facing new frontiers. Neither policy nor decision-makers can solve this problem – the only way out seems to be a policy design in full awareness of the ethical dimensions of research activities and an informed and continuous dialogue between the scientific world and the public.

---

[99] Such as foreseen in the collaborative research projects, the Marie Curie scheme, in ERA-NET, and so on.

[100] As seen, the supporters of the "new" governance model proposed awarding ERC funding on the basis of excellence only, and making the ERC scientifically independent but accountable to the Commission for management issues. The other group proposed tempering the excellence criterion with the requirement for transnational collaboration. This way, the ERC would be independent from the Commission but also vulnerable to the promotion of national interests.

CONCLUSION AND PERSPECTIVES

THE 7TH FRAMEWORK PROGRAMME AND A NEW DEAL FOR EUROPEAN S&T GOVERNANCE

The final chapter of this book starts out by summarising in Section 1 the analysis of the design of the 7th Framework Programme. In accordance with its impact assessment guidelines, the Commission developed its proposal for the new Framework Programme on the basis of a wide range of inputs.

The focus of Section 2 of this chapter is less on the design of the 7th Framework Programme than on its content, examining its expected impacts and the conditions for its success.

Section 3 offers some reflections on what is needed beyond a successful 7th Framework Programme to arrive at a more effective European research policy. The discussion centres on the concept of the European Research Area. In particular, ways in which the European Research Area can be achieved as originally conceived and proposed will be discussed. At the same time, it seems necessary that the concept is redefined and revitalised into a "New Deal" for an effective European research policy.

1. SUMMARISING THE DESIGN OF THE 7TH FRAMEWORK PROGRAMME

Chapter 1 explored some of the important challenges Europe is facing, the expectations held by the public of S&T in addressing these challenges, and the role that S&T could potentially play. The 7th Framework Programme was designed against the background of Europeans feeling anxious because the continent is experiencing a number of major economic, social, and environmental challenges – or indeed against the background of a Europe turning from a positive into a negative "exception" at global level. Economic growth is slow. Europe's competitive position is feeble. There are not enough jobs, and not enough of them are in high-skilled and knowledge-intensive growth industries. Europe is still characterised by significant poverty and regional inequality. An important demographic challenge is emerging. Europeans' health is affected by serious lifestyle and contagious diseases. And the environment is being degraded. Expectations of S&T have never been higher than they are now. Such expectations held of S&T are partially justified. S&T can indeed play an important role in addressing societal economic, social, and environmental challenges. They are the engine of economic growth and competitiveness. The employment effects of S&T are positive. They can play a major role in addressing the consequences of ageing, and the cohesion, public health, and environmental challenges. S&T are part and parcel of our lives, be they framed in a Knowledge Society or otherwise, and they are the linchpin of a Knowledge Society's emergence.

Chapter 2 showed that if Europe wants to be able to meet the expectations of S&T held by citizens and governments, it needs to tackle a number of key S&T weaknesses currently preventing Europe from achieving its full S&T potential. It not only needs to spend more on research, but also needs to deploy these scarce resources more efficiently and to greater effect through more cooperation and coordination of S&T policies in the EU. This will require further improvement in the EU system of governance for research policy, as well as closer coordination between research policy and other related policies. Furthermore, the diverse relations between citizens and S&T must be recognized and taken into account if Europe is to advance in certain key fields.

Chapter 3 traced how, while the aforementioned S&T weaknesses are by no means new or sudden, the growing perception that something needed to be done emerged from a new policy context, which put the spotlight on knowledge, research, education, and innovation. The Lisbon European Council of March 2000 and its aftermath created a new European research policy context, within which it was possible to pursue a Framework Programme much more ambitious in scale and scope than any past Framework Programme. At Lisbon, European policy-makers recognised unambiguously that to overcome its societal challenges, Europe would have to turn itself into a knowledge-based society. At the same time, it was acknowledged that putting knowledge at the centre would require more and better investment in the knowledge triangle of research, education, and innovation. The objective to invest more in research centred on the 3 per cent objective, while the objective to invest better in research centred on the concept of the European Research Area. After 2000, however, little progress was made towards the achievement of these objectives. This was mainly due to weaknesses in the governance model initially used to implement the Lisbon strategy. This resulted in 2005 in a reform of the open method of coordination, and in a general revaluation of the role of the Union in the achievement of the Lisbon objectives. This recognition of the relative effectiveness of action through the Union spilled over into the debate on Community funding programmes up for renewal, and opened the way for an ambitious 7th Framework Programme.

Having taken stock of these "exogenous" factors – Europe's challenges and the role S&T can play in addressing them, the weaknesses of Europe's S&T system, and, finally, the new policy context – the next three chapters of the book looked at how the Framework Programme was actually conceived so as to respond to these issues. Chapter 4 took stock of the experience gained through the implementation of past Framework Programmes. The accumulated evidence convincingly shows that the Framework Programme has had a clear positive impact on participants' immediate competitive position, but also on their capabilities, behaviour, and ability to compete. It has improved Europe's scientific, technological, and innovative performance. And it has strengthened Europe's human capital, integrated European research infrastructures, and enhanced the coordination of Member State research policies and actions. It has also generated wider economic, social, and environmental benefits.

Chapter 5 explained how in the development of the Commission's 7th Framework Programme proposal, account was taken of the views of outside stakeholders. The extensive stakeholder consultations launched by the Commission were a valuable tool for capturing opinions from a wide range of actors. The resulting stock of information helped to design the proposal around key areas of consensus and to make it responsive to the practical concerns of actors on the ground. For the Commission, the consultations proved how challenging it is to balance the different, often opposing views and to formulate a proposal acceptable to the vast majority of stakeholders.

The focus of Chapter 6 was on European added value, by which all actions included in the 7th Framework Programme have to be characterised, and on the analysis of policy options. It was ensured that all actions proposed for inclusion in the 7th Framework Programme pursued objectives best pursued at European rather than at national/regional level such as (1) pooling and leveraging resources; (2) fostering human capacity and excellence in S&T; and (3) better integration of European R&D. Different policy options were analysed with regard to three issues: the maximisation of the Framework Programme's macro-economic impacts, overcoming the fragmentation of European basic research, and raising the competitiveness of European industry through research and innovation. The analysis pointed out that a much larger Framework Programme would have the largest macro-economic impacts, and that the establishment of a European Research Council and Joint Technology Initiatives constituted the best means to overcome the fragmentation of European basic research and to raise the competitiveness of European industry through research and innovation.

On the basis of all of these inputs, the Commission formulated its 7th Framework Programme proposal, centring on four Specific Programmes. Chapter 7 looked in detail at the inter-institutional legislative process: namely, what happened to the proposal in Council and Parliament after it was adopted by the Commission. In particular, it described how the European decision-making machinery responded to this proposal, and how the Financial Perspectives and other issues influenced the eventual outcome of this process. At this stage, after the first reading, the institutional players appear to share in broad terms the Commission's analysis of the weaknesses of European S&T. In both Council and Parliament, there has been a large measure of agreement on the broad lines of the structure of the 7th Framework Programme, and on its S&T objectives and content. With regard to particular problems, the agreement is less unanimous. The negotiations on the Union's overall budget confirm the ambivalence of current European R&D policy: the funding for the 7th Framework Programme will be 30 per cent less than that initially proposed by the Commission, although representing an increase in average terms vis-à-vis that of the previous Framework Programme of approximately 40 per cent. Regarding the fragmentation and dispersion of research efforts in Europe and the proposed solutions, the institutional actors are positive, at least as far as the classical activities proposed are concerned. However, when it comes to new activities to move beyond mere cooperation towards a true supra-nationalism, the problems are obvious. While there was strong support for the creation of an ERC, it is no surprise that it was in the discussions on its implementation that the tensions about the organisation of future European S&T became clearest. But with the acceptance of the idea to establish an ERC an important first step has been taken towards a new model of European S&T governance.

2. PERSPECTIVES ON THE 7TH FRAMEWORK PROGRAMME – ITS IMPACTS AND THE CONDITIONS FOR ITS SUCCESS

2.1. The expected impacts of the 7th Framework Programme

The expected impacts of the 7th Framework Programme, though smaller than those of the Commission proposal, are substantially larger than those of a business-as-usual Framework Programme.[1] A good insight into those expected impacts can be obtained by extrapolating on the basis of the actual impacts of past Framework Programmes, and through econometric modelling.

The substantially larger budget for collaborative research will accommodate many more participations by firms, universities, and public-sector research institutes from a wide variety of regions and Member States in a significantly larger number of tightly integrated collaborative research networks and projects of "critical mass". High quality research that otherwise would not be carried out will be supported under a number of predefined thematic priorities, while sufficient flexibility is retained to respond to emerging scientific trends and societal emergencies. This will strengthen participants' research capabilities by enhancing their knowledge base, and improving the skills of their staff, while also permanently changing their behaviour (e.g., networking and collaboration) resulting in a better ability to compete. The 7th Framework Programme will also have a large impact on Europe's scientific, technological, and innovative performance. The extension of the knowledge base will be codified and made explicit through large numbers of high-quality, and mainly intra-European, collaborative scientific publications. The programme will enhance the development and use of new tools and techniques; the design and testing of models and simulations; the production of prototypes, demonstrators, and pilots; and other forms of technological development. The collaborative research projects funded by the 7th Framework Programme will also generate a large number of patents. As a result of their participation in such research projects, firms will be able to realise increased turnover and profitability, enhanced productivity, improved market shares, access to new markets, reorientation of a company's

---

[1] Although the co-decision process on the Framework Programme is not yet completed at the time of writing, there was already agreement on its budget following the deal on the Financial Perspectives of April 2006. Cfr. European Commission, *Modified Proposal for a Decision of the European Parliament and of the Council Concerning the Seventh Framework Programme of the European Community for Research, Technological Development and Demonstration Activities (2007 to 2013)*, COM(2005) 119/3) 24 May 2005.

commercial strategy, enhanced competitiveness, enhanced reputation and image, and reduced commercial risks.

Similar to the case of collaborative research, the budget will also be boosted to an important extent for human resources actions. Thus many more researchers will be able to participate in top trans-national teams and high-level projects, and benefit from training and knowledge sharing. This will ultimately lead to better R&D, and have a positive impact on the attractiveness of the EU as a place to pursue a scientific career for European and third-country researchers. The actions will provide more researchers than before with an (first) international academic or industrial experience they would not get otherwise. They will gain international research experience, have dedicated time to carry out research, develop research skills, produce scientific publications and patents, and contribute to the development of new technologies. They will generally return home to more responsible positions, taking with them the networks established and reinforced through their fellowship.

Research infrastructure actions will benefit from more resources as well under the 7th Framework Programme. This will provide more researchers with access to unique or outstanding infrastructures in a broad range of disciplines – access they would not have otherwise – and thus contribute to the production of high-quality research. It will also help develop interdisciplinary approaches between several areas, and cooperation between European researchers, while giving a European dimension to several national facilities and, in several cases, organising more efficiently their networking. The programme will also contribute to improving the technical performance of several categories of infrastructures, and thus to enhancing European scientific competitiveness in several fields.

The 7th Framework Programme will also generate wider economic, social, and environmental impacts. Limiting the discussion here to macro-economic impacts, for instance, the 7th Framework Programme can be expected to increase GDP growth, create research and other jobs, raise exports and reduce imports, and increase Europe's R&D intensity (Table C.1).

Beyond the 7th Framework Programme's substantial direct and sometimes quantifiable impacts discussed in the previous paragraphs, it is important to point

Table C.1. 7th Framework Programme macro-economic impacts (deviation from business-as-usual scenario)

	2010	2015	2020	2025	2030
GDP	0.01%	0.05%	0.12%	0.23%	0.37%
Total employment	21.7	57.5	90.7	159.8	285.4
Research employment	14.9	35.8	55.1	80.0	111.6
Extra-EU exports	0.00%	0.04%	0.18%	0.42%	0.72%
Extra-EU imports	0.01%	0.00%	−0.09%	−0.25%	−0.42%
R&D intensity	2.06	2.15	2.24	2.35	2.46

Source: DG Research
Note: For instance, in the year 2030, total GDP will be 0.37 per cent larger when the 7th Framework Programme is implemented than when a business-as-usual Framework Programme is implemented.

also to its considerable indirect and hardly if ever quantifiable effects. In spite of its being a relatively small player in the European research system in monetary terms, the Framework Programme has a large restructuring effect on the European research system, for instance, via its acting as a point of reference for and thus guiding the reorientation of public and private research agendas. A similar restructuring effect can be expected from new Framework Programme initiatives such as the European Research Council and the Joint Technology Initiatives.

2.2. The conditions for the success of the 7th Framework Programme

The various impacts estimated are the impacts that can be achieved if the new Framework Programme is implemented successfully. Yet, its full success is not guaranteed. In order for the 7th Framework Programme to tap its full potential, a number of conditions have to be met.

First and foremost, the partnership and commitment of all actors involved is needed. Of course, the Commission plays an important role when it comes to managing the Framework Programme and implementing the ongoing simplification efforts. But it is not only the Commission which will determine whether the new Framework Programme will achieve the maximum impacts. Its success will also depend on the research community itself – on its readiness to master the fundamental legal requirements for Framework Programme participation; on the industry – on its awareness of the opportunities offered by the Framework Programme; and finally, on the national and regional authorities which collaborate with the Commission to construct conducive framework conditions.

Second, the ongoing initiatives for simplification of Framework Programme application and participation administrative requirements have to be carried through. This will reduce Framework Programme barriers-to-entry, especially for small and medium enterprises and for participants from the new Member States. It will bring into the Framework Programme new capabilities and ideas, and reduce the concentration of Framework Programme participation and the stability across calls and Framework Programmes of networks. It will have a positive impact on dissemination. And it will reverse the alienation of a sizeable share of the scientific community who participated in past Framework Programmes. A first 7th Framework Programme mid-term evaluation is planned for the year 2009. A key issue that will be looked into at that time is programme implementation. Simplification will have to show results by that time.

A third condition concerns programme management. The various management arrangements proposed for the 7th Framework Programme have to deliver. The Commission has successfully managed past Framework Programmes, but it has never had to manage a Framework Programme of such a scale and such a scope. The management of the European Research Council will be externalised. The challenge here is to build up a new executing agency from scratch – with all that it entails in terms of locating premises, hiring staff, establishing procedures, and so on – and at the same time meet the high expectations of the European scientific community by delivering professional results quickly. The scientific management of other Framework Programme components will be kept in-house. The challenge here is how to safeguard the widely appreciated high quality of scientific management given a much larger budget and a less than proportionate increase in the number of project officers. The partial or complete externalisation of back-office functions under review may help meet this challenge.

A fourth condition concerns the programme's output and impact orientation. The efforts being made to further strengthen this programme dimension have to be carried through. The existence of research and technological uncertainty means that the outcome of research projects cannot be predicted, and that some research projects will not succeed in generating new knowledge. But given the right incentives, the search for new knowledge will be perceived as more urgent and productive by research project participants. Increased efforts will be made to disseminate new knowledge and valorise it by converting it into and capitalising upon scientific and technological outputs. At the same time, actions have to have critical mass, and feedback effects between projects have to be supported. In the end, the concern is with the programme's impact, and time and experience will tell what the most appropriate degree of externalisation is to achieve maximum programme impact.

A fifth condition concerns the programme's responsiveness and the adaptability of the programme's scientific content. The 7th Framework Programme will run over seven years, a very long period of time in the world of S&T. New societal challenges may emerge, and so may new scientific disciplines, thematic priorities, and topics. Content-related flexibility is built into the Framework Programme. But being able to make the correct choices at the most appropriate moments will depend on having at one's fingertips the required strategic intelligence. That means investing in the establishment and maintenance of linkages with the scientific community and society at large, but that also means developing a strong internal monitoring and analytical capability.

A sixth condition concerns horizontal policy coordination. As discussed in Chapter 3, direct support programmes in the field of research should be coordinated with policy initiatives concerning research framework conditions. Research policy should be coordinated with education and innovation policies, that is, with the other "knowledge triangle" policies. And knowledge triangle policies should be coordinated with other, for example structural, policies. The main challenge in the case of the 7th Framework Programme concerns the coordination with innovation policy and that with non-knowledge triangle policies. The Framework Programme used to contain its own innovation activities. That will no longer be the case under the 7th Framework Programme. Innovation activities are now part of a newly developed Competitiveness and Innovation Programme to be managed by the Commission's Enterprise Directorate-General. The Research Directorate-General was involved in the conceptualisation of the new programme. But effective coordination should be ensured during the policy implementation stage too. This is not self-evident, and will have to be ensured in a proactive manner. Even more of a challenge will be the coordination between knowledge and other policies. As we have discussed in the third chapter, the horizontal policy coordination arrangements originally put in place by Lisbon were quite weak. A serious effort is now being made to strengthen these arrangements. Their effectiveness still remains to be seen, however.

A key role in the mitigation of implementation risk can be played by the Framework Programme monitoring system. In view of the implementation of the 7th Framework Programme, this is being revised. First, a renewed effort will be made towards the better collection of data on Framework Programme applicants and participants, and on Framework Programme proposals and projects. Second, the annual monitoring exercise, which used to be carried out by external experts often far removed from the reality of the Framework Programme will be internalised and carried out by high-level Commission officials.

The success of the 7th Framework Programme will have to be judged on the basis of a thorough evaluation. This requires an ambitious and strong Framework Programme evaluation system matching the ambition of the Framework Programme itself. Initiatives being taken in this regard have to be achieved. At the same time a methodological reflection is required to avoid the risk of undervaluation.[2]

3. PERSPECTIVES BEYOND THE 7TH FRAMEWORK PROGRAMME – TOWARDS A NEW DEAL FOR AN EFFECTIVE EUROPEAN RESEARCH POLICY

If the 7th Framework Programme is a success, a significant contribution will have been made towards remedying the European scientific and technological weaknesses discussed in Chapter 2, that is, the lack of R&D investment, the S&T fragmentation and dispersion, and the lack of horizontal policy coordination. This will put European S&T in a substantially better position to meet the economic, social, and environmental challenges analysed in Chapter 1.

However, even a successful 7th Framework Programme will still only represent a small fraction of Europe's overall public research efforts, and will not by itself eliminate all Europe's S&T weaknesses. To achieve this, a more coherent and effective European research policy is required, encompassing all government efforts across the EU, be they at national, regional, or EU level.

But what is the best way to arrive at a more effective European research policy? The Commission gave an answer to this question in 2000 with the launching of a European Research Area. This visionary concept had two key dimensions: first, its focus on integration, and, next, its emphasis on excellence. Greater coordination

---

[2] At an April 2006 conference on the evaluation of research, Luke Georghiou referred to "the risk of undervaluation". He argued that present research evaluation techniques have a tendency to undervalue the contribution of R&D, and that the failure of research ex-post evaluators to appreciate the full extent of both the private and the social returns to R&D is a key reason for the international decline in government investment in research. He even raised the possibility that industry itself may underestimate its own benefits from R&D, particularly under shareholder pressure to show short-term profits. Georghiou's concern was mainly with research evaluation techniques. He argued that present techniques, and particularly surveys (often used in the case of ex-post evaluations of the Framework Programme too), focus on predictable and linear effects while missing or recording only perfunctorily a series of other effects which may be of equal or greater value. He also argued that even where effects are properly identified and/or calculated there is a tendency to underestimate the contribution of public intervention.

and cooperation had to be achieved throughout Europe. More links had to be established between the different players (public authorities, firms, universities, research institutes) at all policy levels (regional, national, Community, inter-governmental) in the European research system. In addition, excellence was supposed to become the core principle on the basis of which all these players would operate and research funds would be allocated.[3]

Furthermore, the European Research Area initiative identified a list of specific actions required to promote the integration of European S&T and improve its performance. Private research investment had to become more dynamic. The environment for private research investment, R&D partnerships, and high technology start-ups had to be improved by using tax policies, venture capital, and EIB support. Effective tools had to be developed to protect intellectual property, so that innovation and ideas would be adequately rewarded within the new knowledge-based economy. Greater mobility of researchers had to be achieved in Europe. Obstacles to the mobility of researchers had to be removed, and a European dimension introduced into scientific careers. A greater place and role for women in research had to be achieved, and the young had to be given a taste for research and careers in science. Europe had to offer attractive prospects to the best brains, so that high-quality research talent would be retained and attracted. European S&T centres of excellence had to be networked, and virtual ones created. A European approach had to be defined to research facilities. Regions had to play a more important role in the European research effort. And the scientific communities of Western and Eastern Europe had to be better integrated.

Seven years after the launch of the European Research Area, we recognise that progress has been made in terms of the two key dimensions (coordination and excellence). But much remains to be done. Some steps forward were indeed taken towards the integration of European S&T via the Open Method of Coordination, the mutual opening up of national research programmes, and Framework Programme instruments such Networks of Excellence and ERA-Nets. Excellence was indeed accepted as the core principle on the basis of which to work. But the efforts made so far to promote greater coordination and cooperation between S&T players are just a start, and the excellence principle still has to be applied more widely and more rigorously.

While the original European Research Area objectives remain valid today, taking stock of the current situation, a more fundamental observation can be made. As originally conceived, the European Research Area was a rather static concept concerning mainly the S&T input side. It essentially accepted the existing architecture of the European S&T system – the established public authorities and institutions, each with their own roles and responsibilities – and simply focussed on building links between them and funding excellent S&T players, under the assumption that this

[3] For instance, in the evaluation of FP proposals, a larger weight was assigned to scientific excellence as an evaluation criterion.

would lead to a higher societal impact of research than under a situation of complete S&T fragmentation and dispersion.

Yet the urgency of realising the Lisbon Agenda is greater than ever. This calls not for a piecemeal raising of effectiveness and impact, but for making effectiveness and impact the key priorities. As a key precondition this means that action should be taken where most effective. This means a reallocation of responsibilities and assumes going beyond the existing structure of the European research system and redesigning, even reinventing, it. A more dynamic approach is needed towards the realisation of the European Research Area.

This calls for a frank and thorough exploration of subsidiarity issues, and of the notion of "added value". It is important to determine clearly what kind of action is best taken at Community level, and what kind of action is best at regional, national, or intergovernmental level. The debate so far has been confined to European added value. In other words, so far it has been the Union which – confronted with Member States striving to retain their national powers unless given a compelling reason to do otherwise – has had to demonstrate that it is better to implement some actions at Community level than at other policy levels. In order to spend our scarce R&D resources more effectively, there is an urgent need to broaden this debate. The burden of proof needs to be more evenly divided. In addition to European added value, an honest discussion is needed on regional added value, national added value, and intergovernmental added value. In fact, the discussion of different kinds of added value is systemically linked. One cannot hope to really understand European added value if one does not have a clear insight into the added value of other policy levels, the one being the flipside of the other.

This debate on which kind of action to implement at which level, that is, the systematic exploration of the added value, must be evidence-based and must be based on solid data. More generally, the evidence-based approach to policy-making, which has already been embraced by the Union through the implementation of its Better Regulation Action Plan and the application of ex ante impact assessment to major legislative initiatives, could be usefully extended to other policy levels.

In order to explore added value and achieve a clear division of labour and true complementarity between different policy levels, it is important to have a clear insight into what is presently being done at each level. At present, however, there is no complete overview of research measures in Europe. While Community research policies are well known in the Member States, and often analysed and assessed there, more complete and consolidated information is needed on the S&T policies implemented by regional and national authorities, which still account for over 90 per cent of Europe's public R&D expenditure. Some useful initiatives have been taken in this regard (e.g., TrendChart),[4] but this still gives only a partial picture, and a more exhaustive inventory is needed.

---

[4] The European TrendChart on Innovation, a product of the European Commission, measures innovation performances across the European Union. It includes, for example, annual country reports, information on national policy measures and government decisions.

Next, comparative information is needed on the effectiveness of policies at different policy levels. But how to measure this? Existing S&T indicators and datasets cannot provide this kind of information. Within the framework of the S&T benchmarking initiatives launched within the context of the Open Method of Coordination, a wealth of S&T information has been collected over the past few years. New (e.g., composite) S&T indicators have been developed. A serious effort has been made to collect comparable data at the regional, national, and aggregate European levels for both existing and newly developed indicators. And a great multitude of S&T indicator reports have been published at regional, national, and Community levels. However, the purpose of these indicators has been mainly to track progress in terms of S&T inputs or performance. They do not provide us with information on the relative efficiency and effectiveness of implementing an action at one policy level versus another.

That means that a completely new data effort is needed. The first step in this new-generation data effort should consist of the development of new and improved ex-post evaluation methodologies. These should seek to link a particular public research support input with all possible S&T and societal outputs and impacts.[5] This can only be done if rapid progress is made towards addressing long-standing methodological challenges in the field of ex-post evaluation concerning, for instance, additionality and crowding-in/crowding-out, and attribution. These new evaluation methodologies have to be action-specific. For example, ex-post evaluation methodologies for research infrastructure actions would differ from those for human resources actions. Efforts towards the development of new evaluation methodologies should also be broad-based, involving the Member States as well as the Community so that the results obtained will be credible, legitimate, and broadly supported.

The next step has to consist of both the Member States and the Community systematically applying the newly developed ex-post evaluation methodologies to their complete project portfolios, and then capturing, via a common set of indicators, the relative effectiveness and efficiency of project implementation. By facilitating for the first time a comparison of information across policy levels, this evidence-base would provide a solid foundation for an improved distribution of effort between the different levels.

If successful, this approach would turn research evaluation from an all-too-often bureaucratic exercise of ex-post justification into a real strategic planning tool. Two things are needed to make this a reality. The first is the integration of the European research ex-post evaluation community. A "European evaluation area" has to be achieved, in which both Member States and Community try to make real methodological progress in the field of research ex-post evaluation, and to arrive at broadly supported evaluation approaches. In addition, a further revision is needed of the policy cycle. The role of ex-post evaluation has to be strengthened and its link with policy development tightened.

[5] For instance, tracing to a particular research project all its scientific and technological outputs and wider impacts.

What is needed is a pan-European knowledge community, not only in the area of ex-post evaluation, but also in socio-economic and scientific and technical foresight. One cannot hope to restructure the European research system and achieve a true European Research Area if there is no common vision of how major societal and S&T challenges will evolve, how scientific disciplines will develop, and how the research system as such will progress. Europe has a long history of activity in this field and a large community of excellent players has built up a wealth of experience and knowledge on these issues. But European foresight suffers from some of the same problems as the European research system. Foresight efforts are fragmented and dispersed, and there is substantial overlap among national initiatives and between national and Community projects. Tackling this requires more effectively linking the different players in this field, and perhaps creating a common institution.

Once pan-European evaluation and foresight communities have been established, and a credible evidence base has been built up, it will be possible to draw solid conclusions, and take bold and courageous political decisions – as indeed Europe has always done at key junctures over the last fifty years.

These are the foundations for a "New Deal" between the Member States and the Community. Its aim is to develop a more coherent and effective European research policy which can have a greater societal impact, and help Europe to tackle the many existing and emerging challenges it is facing. This "New Deal" would involve a more ambitious approach towards the realisation of the European Research Area. It builds upon the solid foundation of the European Research Area concept launched in 2000, and shares its concern about the fragmentation and dispersion of European S&T. But it seeks to go much further by re-designing the architecture of the European research system. Unlike the original ERA initiative, it does not simply promote the establishment of links between existing European S&T players, each with their own existing roles and responsibilities. Instead it proposes a more dynamic and impact-oriented approach towards the completion of the European Research Area. It puts on the agenda the redesign, even the reinvention, of the European research system. It opens the possibility of a significant reallocation of research policy responsibilities between policy levels should available evidence point into that direction. Within this context, it calls for an in-depth debate by the Community and the Member States on the issues of subsidiarity and added value based on solid evidence of relative policy effectiveness and a common vision for the future. It is, above all, important to keep an open mind about the outcome of this debate and inquiry. It could mean an expansion of national or regional activities in some areas. It could lead to an increase in EU-level actions in others. It may even result in a need to build new, common European S&T institutions, whether supranational (in the vein of the ERC) or intergovernmental (in the vein of CERN). The New Deal would mean preparing these decisions together based on solid, shared evidence, and bravely facing the facts about what may need to change. Europe has risen to such challenges in the past. Its ability to do so again could herald a new phase in European S&T.

BIBLIOGRAPHY

Aghion, Philippe and Peter Howitt, *Appropriate Growth Policy: A Unifying Framework*, 9 August 2005.

Alesina, Alberto and Enrico Spolaore, *The Size of Nations*, Cambridge, MA, 2003.

Allum, Nick *et al.*, *A Meta-Analysis of Science Knowledge and Attitudes Across Cultures*, Presentation at the Annual Meeting of the American Association for the Advancement of Science, Washington, DC, February 2005.

American Electronics Association, *Losing the Competitive Advantage? The Challenge for Science and Technology in the United States*, Washington, DC, 2005.

An Agenda for a Growing Europe. Making the EU Economic System Deliver, Report of an Independent High-Level Study Group Established on the Initiative of the President of the European Commission, Brussels, 2003.

Applica *et al.*, *Network on Social Inclusion and Income Distribution – Final Report*, European Observatory on the Social Situation (SSO) – Contract No. VC/2004/0462, December 2005.

Arnold, Erik, *What the Evaluation Record Tells Us about Framework Programme Performance*, May 2005.

Banthien, Henning *et al.*, *Governance of the European Research Area: The Role of Civil Society*, Conference Proceedings, Brussels, 20 October 2003.

Basu, Aparna, Using ISI's "Highly Cited Researchers" to Obtain a Country Level Indicator of Citation Excellence, In: *Scientometrics*, Vol. 68, No. 3, 2006, pp. 361–375.

Baumol, William J., *The Free-Market Innovation Machine. Analyzing the Growth Miracle of Capitalism*, Princeton, NJ and Oxford, 2002.

Beck, Ulrich, *Risk Society: Towards a New Modernity*, London, 1992.

Beck, Ulrich, *World Risk Society*, Cambridge, 1999.

Beffa, Jean-Louis, *Pour une Nouvelle Politique Industrielle*, Paris, 2005.

Bibliometric Analysis for Participants under the 5th European Framework Programme for Research and Technical Development (Contract No. EVA6-CT-2005–00001) – Final Report.

Biotechnology: Europeans Start to Believe the Health Benefits, In: *Europe Information Service*, European Report No. 3108, 20 June 2006.

Blind, Knut *et al.*, *New Products and Services: Analysis of Regulations Shaping New Markets*, Fraunhofer Institute Systems and Innovation Research, Luxembourg, 2004.

Borchardt, Klaus-Dieter, *The ABC of Community Law*, Brussels, 2000.

Branscomb, Allan, Scenario Development, In: Hulse, David *et al.* (Eds.), *Willamette River Basin Planning Atlas – Trajectories of environmental and Ecological Change*, Corvallis, 2002.

Breschi, Stefano and Lucia Cusmano, *Unveiling the Texture of a European Research Area: Emergence of Oligarchic Networks under EU Framework Programmes*, CESPRI Working Paper No. 130, July 2002.

von Bubnoff, Andreas, Asia Squeezes Europe's Lead in Science, In: *Nature*, Vol. 436, 21 July 2005, p. 314.

Building Ireland's Knowledge Economy. The Irish Action Plan for Promoting Investment in R&D to 2010, Report to the Inter Departmental Committee on Science, Technology and Innovation, July 2004.

Bush, Vannevar, *Science – The Endless Frontier*, A Report to the President, Washington, DC, 1945.

Callon, Michel and Bruno Latour, Unscrewing the Big Leviathan: How Actors Macro-Structure Reality and how Sociologists Help Them to Do So, In: Knorr-Cetina, Karin and Cicourel, Aaron V. (Eds.), *Advances in Social Theory and Methodology: Towards an Integration of Micro and Macro Sociologies*, London, 1981.

Caracostas, Paraskevas and Ugur Muldur, *Society – The Endless Frontier*, Luxembourg, 1998.

CEPII-CIREM, *European Industry's Place in the International Division of Labour*, Report Prepared for DG Trade of the European Commission, July 2004.

Centre for Urban and Regional Development Studies, University of Newcastle Upon Tyne, *Regional Science Policy and Instruments: An Overview of British Developments, Background Paper Prepared for the "Strategy for Success" Scrutiny and Policy Development Panel*, North East Assembly, 2004.

Chellaraj, Rajkumar *et al.*, *The Contribution of Skilled Immigration and International Graduate Students to US Innovation.* World Bank Policy Research Working Paper 3588, May, 2005.

Cohen, Daniel and Michèle Debonneuil, *Nouvelle Économie*, Paris, 2000.

Committee on Ensuring the Best Presidential and Federal Advisory Committee Science and Technology Appointments – Committee on Science, Engineering, and Public Policy – National Academy of Sciences, National Academy of Engineering and Institute of Medicine of the National Academies, *Science and Technology in the National Interest, Ensuring the Best Presidential and Federal Advisory Committee Science and Technology Appointments*, Washington, DC, 2005.

Conseil d'Analyse Économique, *Politiques Environnementales et Compétitivité*, Paris, 2004.

Construire L'Europe Politique. 50 Propositions pour l'Europe de Demain, Dominique Strauss-Kahn, Président de la Table Ronde "Un Projet Durable pour l'Europe de Demain" Établie à l'Initiative du Président de la Commission Européenne, Rapporteur Olivier Ferrand, April 2004.

Cooke, Philip, *Towards Regional Science Policy? The Rationale from Biosciences*, Presented to Conference on Rethinking Science Policy: Analytical Frameworks for Evidence-Based Policy, Science Policy Research Unit, University of Sussex, 21–23 March 2002.

Council for Science and Technology Policy Japan, *Japan's Science and Technology Basic Policy Report*, 27 December 2005.

Court of Auditors, *Special Report No. 1/2004 on the Management of Indirect RTD Actions under the Fifth Framework Programme (FP5) for Research and Technological Development (1998 to 2002), Together with the Commission's Replies (Pursuant to Article 248(4) Second Subparagraph EC) (2004/C 99/01)*, 23 April 2004.

Creating an Innovative Europe, Report of the Independent Expert Group on R&D and Innovation Appointed following the Hampton Court Summit, Brussels, 2006.

Databank Consulting *et al.*, *IST Impact Study – Microelectronics & Microsystems, Health, Mobile Communications – Draft Final Report D.6*, 4 August 2004.

De Mooij, Ruud and Paul Tang, *Four Futures of Europe*, Den Haag, 2006.

Decisia *et al.*, *Final Report – Assessment of the Impact of the Actions Completed under the 3rd and 4th Community Framework Programmes for Research; Survey for the Five-Year Assessment of Community Research Activities (1999–2003)*, 2 March 2004.

Denis, Cécile *et al.*, *The Lisbon Strategy and the EU Structural Productivity Problem*, European Economy, European Commission, Directorate-General for Economic and Financial Affairs, Economic Papers, No. 221, February 2005.

Dierkes, Meinolf *et al.* (Eds.), *Between Understanding and Trust: The Public, Science and Technology*, Amsterdam, 2000.

Discours du Premier Ministre Guy Verhofstadt prononcé devant le Parlement européen. Bruxelles, le 31 mai 2006.

Dosi, Giovanni *et al.*, *Evaluating and Comparing the Innovation Performance of the United States and the European Union*, Expert Report Prepared for the TrendChart Policy Workshop 2005.

Dratwa, Jim, Social Learning at the European Commission and the Codex Alimentarius, In: Reinalda, Bob and Verbeek, Bertjan (Eds.), D*ecision Making within International Organizations*, London and New York, 2004, pp. 215–229.

Dratwa, Jim, Taking Risks with the Precautionary Principle: Food (and the Environment) for Thought at the European Commission, In: *Journal of environmental Policy and Planning*, Special Issue on Risk and Governance, Vol. 4, No. 3, 2002, pp. 197–213.

Drucker, Peter F., *The Age of Discontinuity. Guidelines for a Changing Society*, London, 1969.

DTI, *Competing in the Global Economy: The Innovation Challenge*, Innovation Report, London, December 2003.

DTI, *Targeted Review of Added Value Provided by International R&D Programmes*, London, May 2004.

DTI, *The Impact of the EU Framework Programmes in the UK*, July 2004.

DTI, *2003 R&D Scoreboard,* London, 2004.

Durieux, Luc and Gilbert Fayl, The Scheme Used for Evaluating the European Research and Technological Development Programmes, In: OECD (Eds.), *Policy Evaluation in Innovation and Technology – Towards Best Practices*, Paris, 1997.

Ede, Andrew and Lesley B. Cormack, *A History of Science in Society – From Philosophy to Utility*, Toronto, 2004.

E-Government Strategic Support Unit (Office of the Deputy Prime Minister, UK), *Engaging the Community in E-government*, 2004.

Elzinga, Aant, The Growth of Science: Romantic and Technocratic Images [Nobel symposium "Progress in Science and Its Social Conditions", 15–19 August 1983, Lidingö, Sweden], In: *Essays on Scientism, Romanticism and Social Realist Images of Science*, Göteborg, 1984, pp. 51–90.

Enlarging the European Research Area: Identifying Priorities for Regional Policy Focusing on Research and Technological Development in the New Member States and Candidate Countries (FP6 contract No. COP6-CT-2004–00001).

Esterle, Laurence and Ghislaine Filliatreau (Eds.), *Indicateurs de Sciences et de Technologies Édition 2004*, Rapport de l'Observatoire des Sciences et des Techniques, Paris, 2004.

EU-Chinese Co-operation Pinpoints Potential Treatments for SARS, In: *European Research Headlines*, 29 July 2005.

European Commission, DG INFSO, *Impact of ICT on Sustainable Development*, 2004.

European Commission, DG JRC – Institute for Prospective Technological Studies, *E-Government in the EU in the Next Decade: Vision and Key Challenges*, 2004.

European Commission, *Action Plan "Simplifying and Improving the Regulatory Environment"*, COM(2002) 287, 10 June 2002.

European Commission, *Action Plan to Improve Communicating Europe by the Commission*, SEC(2005) 985 final, 20 July 2005.

European Commission, *Amended Proposal for a Decision of the European Parliament and of the Council Concerning the Seventh Framework Programme of the European Community for Research, Technological Development and Demonstration Activities (2007 to 2013),* COM(2006) 364, 28 June 2006.

European Commission, *BRITE-EURAM – Making a Lasting Impression on Europe*, Luxembourg, 2002.

European Commission, *Building Our Common Future: Policy Challenges and Budgetary Means of the Enlarged Union 2007–2013*, Communication from the Commission to the Council and the European Parliament, COM(2004) 101, 10 February 2004.

European Commission, *Building the Europe of Knowledge*, Proposal for a Decision of the European Parliament and of the Council Concerning the Seventh Framework Programme of the European Community for Research, Technological Development and Demonstration Activities (2007 to 2013) – Proposal for a Council Decision Concerning the Seventh Framework Programme of the European Atomic Energy Community (Euratom) for Nuclear Research and Training Activities (2007 to 2011), COM(2005) 119 final, 6 April 2005.

European Commission, *Building the Knowledge Society: Social and Human Capital Interactions*, Commission Staff Working Paper, SEC(2003) 652, 28 May 2003.

European Commission, *Clean, Safe and Efficient Energy for Europe – Impact Assessment of Non-Nuclear Energy Projects Implemented Under the Fourth Framework Programme – Synthesis Report*, Luxembourg, 2003.

European Commission, *Commission Interim Working Document on the Implementation of a Funding Mechanism for Basic Research*, 29 September 2004.

European Commission, *Commission Staff Working Document in Support of the Report from the Commission to the Spring European Council, 22–23 March 2005, on the Lisbon Strategy of Economic, Social and environmental Renewal*, SEC(2005) 160, 28 January 2005.

European Commission, *Communication from the Commission to the Council and to the European Parliament on the Presentation of a Proposal for a Directive and Two Proposals for Recommendations on the Admission of Third-Country Nationals to Carry Out Scientific Research in the European Community*, COM(2004) 178, 16 March 2004.

European Commission, *Delivering on Growth and Jobs: A New and Integrated Economic and Employment Co-ordination Cycle in the EU*, Companion Document to the Communication to the Spring European Council 2005, COM(2005) 24, Working Together for Growth and Jobs, A New Start for the Lisbon Strategy, SEC(2005) 193, 3 February 2005.

European Commission, DG Health and Consumer Protection, *Public Health, Disease and Conditions Information Sheets: Cardiovascular Diseases.*

European Commission, DG Health and Consumer Protection, *Public Health, Disease and Conditions Information Sheets: Diabetes.*

European Commission, *Employment in Europe 2001 – Recent Trends and Prospects*, Brussels, 2001.

European Commission, *Employment in Europe 2003*, Luxembourg, 2003.

European Commission, *Employment in Europe 2004*, Luxembourg, 2004.

European Commission, *Employment in Europe 2005*, Luxembourg, 2005.

European Commission, *Enhancing Access to Research Infrastructures – Improving the Human Research Potential and the Socio-Economic Knowledge Base (1998–2002)*, Technical Review of IHP-ARI Contracts – Panel's Technical Review Report, January 2003.

European Commission, *Enlargement, Two Years After – An Economic Success*, Communication from the Commission to the Council and the European Parliament, COM(2006) yyy final, 3 May 2006.

European Commission, *Enlargement, Two Years After: An Economic Evaluation*, European Economy, Occasional Papers, No. 24, May 2006.

European Commission, *ERA-NET – Networking of National Research Programmes in the European Research Area – Project Synopses*, Luxembourg, 2004.

European Commission, *Eurobarometer 2005, Social Values, Science and Technology*, Luxembourg, 2005.

European Commission, *Europe and Basic Research*, COM(2004) 9, 14 January 2004.

European Commission, *European Competitiveness Report 2004*, Luxembourg, 2004.

European Commission, *Europeans, Science and Technology*, Special Eurobarometer, June 2005.

European Commission, *Financial Perspectives 2007–2013*, Communication from the Commission to the Council and the European Parliament, COM(2004) 487, 14 July 2004.

European Commission, *Five Year Assessment Report Related to the Specific Programme Competitive and Sustainable Growth Covering the Period 1995–1999*, June 2000.

European Commission, *Five-Year Assessment of the European Community RTD Framework Programmes – Report of the Independent Expert Panel Chaired by Viscount E. Davignon and the Commission's Comments on the Panel's Recommendations*, Luxembourg, 1997.

European Commission, *Five-Year Assessment of the European Union Research and Technological Development Programmes, 1995–99*, Luxembourg, 2001.

European Commission, *Five-Year Assessment of the European Union Research Framework Programmes 1999–2003*, 15 December 2004.

European Commission, *Five-Year Assessment of the European Union Research Framework Programmes 1999–2003. Executive Summary.*

European Commission, *Five-Year Assessment Report Related to the Specific Programme: Quality of Life and Management of Living Resources Covering the Period 1995–1999.*

European Commission, *Five-Year Assessment: 1999–2003 – Research and Technology Development in Information Society Technologies – Interim Panel Report*, June 2004.

European Commission, *FP7 Simplification – Feedback from Actors and Stakeholders on Commission Staff Working Document*, 2005 (Internal Commission document – DG Research (A3)).

European Commission, *Frontier Research: The European Challenge – High-Level Expert Group Report*, Luxembourg, February 2005.

European Commission, *General Budget of the European Union for the Financial Year 2006 – The Figures*, Luxembourg, 2006.

European Commission, *General Principles and Minimum Standards for Consultation of Interested Parties by the Commission*, COM(2002) 704, 11 December 2002.

European Commission, *Green Paper "Confronting demographic Change: A New Solidarity between the Generations"*, Communication from the Commission, COM(2005) 94 final, 16 March 2005.

European Commission, *Green Paper on Greenhouse Gas Emissions Trading within the European Union*, Presented by the Commission, COM(2000) 87 final, 8 March 2000.

European Commission, *Growth and Jobs – A New Start for the Lisbon Strategy*, Luxembourg, 2005.

European Commission, *Health in Europe – Data 1998–2003*, Luxembourg, 2005.

European Commission, *Human Capital in a Global and Knowledge-based Economy – Final Report*, Luxembourg, 2003.

European Commission, *Impact Assessment Guidelines*, SEC(2005) 791, 15 June 2005.

European Commission, *Implementation Report 2004 on "A Mobility Strategy for the European Research Area" and "Researchers in the ERA: One Profession, Multiple Careers"*, Commission Staff Working Document, SEC(2005) 474, 6 April 2005.

European Commission, *Independent External Monitoring and Evaluation of Community Activities in the Area of Research and Technological Development*, Communication from the Commission to the Council and the European Parliament, COM(96) 220 final, 22 May 1996.

European Commission, *Innovation Policy: Updating the Union's Approach in the Context of the Lisbon Strategy*, Communication to the Council, the European Parliament, the European Economic and Social and the Committee of the Regions, COM(2003) 112 final, 11 March 2003.

European Commission, *Investing in Research: An Action Plan for Europe*, Communication from the Commission, COM(2003) 226 final/2, 4 June 2003.

European Commission, *Key Figures 2005: Towards a European Research Area: Sciene, Technology and Innovation*, Luxembourg, 2005.

European Commission, *Main Report: Overall Summary – Impact Assessment and Ex Ante Evaluation*, Commission Staff Working Paper, Annex to the Proposal for the Council and European Parliament Decisions on the 7th Framework Programme (EC and Euratom), SEC(2005) 430, 6 April 2005.

European Commission, *More Research for Europe – Towards 3 Percent of GDP*, Communication from the Commission, COM(2002) 499 final, 11 September 2002.

European Commission, *Progress Towards the 3 Percent Target on the Basis of the Information in the NRPs*, December 2005 (Internal Commission Document, DG Research/Directorate M).

European Commission, *Proposal for a Directive of the European Parliament and of the Council Establishing a Scheme for Greenhouse Gas Emission Allowance Trading within the Community and Amending Council Directive 96/61/EC*, COM(2001) 581 final, 23 October 2001.

European Commission, *R&D Investment Snapshot*, Brussels, September 2004.

European Commission, *Raising EU R&D Intensity – Improving the Effectiveness of Public Support Mechanisms for Private Sector Research and Development – Direct Measures*, Report to the European Commission by an Independent Expert Group, 2003.

European Commission, *Report – Evaluation of the Effectiveness of the New Instruments of Framework Programme VI*, Report of a High-level Expert Panel Chaired by Professor Ramon Marimon, 21 June 2004.

European Commission, *Report on European Technology Platforms and Joint Technology Initiatives, Fostering Public-Private R&D Partnerships to Boost Europe's Industrial Competitiveness*, Commission Staff Working Document, SEC(2005) 800, 10 June 2005.

European Commission, *Science and Society Action Plan*, COM(2001) 714, 4 December 2001.

European Commission, *Science and Technology, the Key to Europe's Future – Guidelines for Future European Union Policy to Support Research*, COM(2004) 353 final, 16 June 2004.

European Commission, *Science, Society and the Citizen in Europe*, Commission Working Document, SEC(2000) 1973, 14 November 2000.

European Commission, *Second European Report on S&T Indicators, 1997 – Report*, Luxembourg, 1997.

European Commission, *Simplification in the 7th Framework Programme*, SEC(2005) 431, 6 April 2005.

European Commission, *Special Eurobarometer – The Attitudes of European Citizens Towards Environment*, April 2005.

European Commission, *Stimulating Technologies for Sustainable Development: An environmental Technologies Action Plan for the European Union*, Communication from the Commission to the Council and the European Parliament, COM(2004) 38 final, 28 January 2004.

European Commission, *Towards a European Strategy for Nanotechnology*, Communication from the Commission, COM(2004) 338 final, 12 May 2004.

European Commission, *Technology Platforms – From Definition to Implementation of a Common Research Agenda*, Luxembourg, September 2004.

European Commission, *The Commission in Close-up – How Decisions Are Made*, Brussels, 2005.

European Commission, *The Economic Costs of Non-Lisbon – A Survey of the Literature on the Economic Impact of Lisbon-Type Reforms*, Commission Staff Working Document, SEC(2005) 385, 15 March 2005.

European Commission, *The Overall Socio-Economic Dimension of Community Research in the European Framework Programme*, Luxembourg, 2003.

European Commission, *Third European Report on Science and Technology Indicators*, Luxembourg, 2003.

European Commission, *Towards a European Research Area – Science, Technology and Innovation – Key Figures 2003–2004*, Luxembourg, 2003.

European Commission, *Towards a European Research Area – Science, Technology and Innovation – Key Figures 2002*, Luxembourg, 2002.

European Commission, *Towards a European Research Area*, Communication from the Commission to the Council, the European Parliament, the Economic and Social Committee and the Committee of the Regions, COM(2000) 6 final, 18 January 2000.

European Commission, *Unity, Solidarity, Diversity for Europe, Its People and Its Territory. Second Report on Economic and Social Cohesion*, Luxembourg, 2001.

European Commission, *White Paper on a European Communication Policy*, COM(2006) 35 final, 1 February 2006.

European Commission, *Working Together for Growth and Jobs, A New Start for the Lisbon Strategy*, Communication to the Spring European Council 2005, COM(2005) 24, 3 February 2005.

European Environment Agency, *Greenhouse Gas Emission Trends and Projections in Europe 2004*, Luxembourg, 2004.

European Foundation for the Improvement of Living and Working Conditions & PREST (Manchester Business School), *The Knowledge Society Delphi: EUFORIA Project Report*, 2004.

European Research Advisory Board (EURAB), *Report on European Technology Platforms*, January 2004.

European Research Advisory Board (EURAB), *Report On: "SMEs and ERA" – Report and Recommendations*, 2006.

Evidence Ltd *et al.*, *The Role of Selectivity and the Characteristics of Excellence*, Report to the Higher Education Funding Council for England: A Consultancy Study within the Fundamental Review of Research Policy and Funding, October 2000.

Ezrahi, Yaron, *The Descent of Icarus: Science and the Transformation of Contemporary Democracy*, Cambridge, MA, 1990.

Facing the Challenge. The Lisbon Strategy for Growth and Development, Report from the High Level Group Chaired by Wim Kok, Brussels, 2004.

Fahrenkrog, Gustavo *et al.* (Eds.), *RTD-Evaluation Toolbox Assessing the Socio-Economic Impact of RTD-Policies*, JRC-IPTS and Joanneum Research, Strata Project HPV 1 CT 1999–00005, August 2002.

Federal Ministry of Education, Science and Culture, together with Federal Ministry of Transport, Innovation and Technology, and Federal Ministry of Economics and Labour, *Austrian Research and Technology Report 2004*, Vienna, 2004.

Forfás, *The 4th Framework Programme in Ireland – An Evaluation of the Operation and Impacts in Ireland of the EU's Fourth Framework Programme for Research and Development*, 2001.

Fougeyrollas, Arnaud *et al.*, *The Némésis Model: New Econometric Model for Environment and Sustainable Development Implementation Strategies*, Brussels, 2002.

French, Howard W., As China Ages, a Shortage of Cheap Labor Looms, In: *The New York Times*, 30 June 2006.

García-Quevedo, José, Do Public Subsidies Complement Business R&D? A Meta-Analysis of the Econometric Evidence, In: *Kyklos*, Vol. 57, No. 1, pp. 87–102.

Gaskell, George *et al.*, *Europeans and Biotechnology in 2005: Patterns and Trends*, Eurobarometer 64.3. A Report to the European Commission's Directorate-General for Research, Brussels, 2006.

Georghiou, Luke *et al.*, *"Making the Difference" – The Evaluation of "Behavioural Additionality" of R&D Subsidies*, Institute for the Promotion of Innovation by Science and Technology in Flanders (IWT), IWT Studies, 2004.

Georghiou, Luke, Assessing The Framework Programmes – A Meta-Evaluation, In: *Evaluation*, Vol. 1, No. 2, 1995.

Geuna, Aldo, *University Participation in Community Programmes: How Does the Selection Process Work?* (Prepared for the Conference "New Research Findings: The Economics of Scientific and Technological Research in Europe", University of Urbino, 24–25 February, 1995), February 1995.

Gibbons, Michael *et al.*, *The New Production of Knowledge: The Dynamics of Science and Research in Contemporary Societies*, London, 1994.

Glänzel, Wolfgang *et al.*, A Bibliometric Analysis of International Scientific Cooperation of the European Union (1985–1995), In: *Scientometrics*, Vol. 45, No. 2, 1999, pp. 185–202.

Glänzel, Wolfgang National Characteristics in International Scientific Co-Authorship Relations, In: *Scientometrics*, Vol. 51, No. 1, 2001, pp. 69–115.

Gómez, Isabell *et al.*, Collaboration Patterns of Spanish Scientific Publications in Different Research Areas and Disciplines, In: M.E.D. Koenig and A. Bookstein (Eds.), *Proceedings of the Biennial Conference of the International Society for Scientometrics and Informetrics,* Medford, NJ, 1995, pp. 187–196.

Greenhalgh, Trisha and Jill Russell, Reframing Evidence Synthesis As Rhetorical Action in the Policy Making Drama, In: *Healthcare Policy*, Vol. 1, No. 2, 2006, pp. 34–42.

Guellec, Dominique and Bruno van Pottelsberghe de la Potterie, *R&D and Productivity Growth: Panel Data Analysis of 16 OECD Countries*, STI Working Papers 2001/3, 14 June 2001.

Guy, Ken and Wolfgang Polt, *Strategic Options for the Evaluation of the R&D Programmes of the European Union*, Plattform – Technologie – Evaluierung, No. 8, April 1999.

Guy, Ken *et al.*, *Strategic Options for the Evaluation of the R&D Programmes of the European Union*, Final Report Prepared for STOA, Brighton, 1999.

Guzetti, Lucca, *A Brief History of EU Research Policy*, Brussels, 1995.

Hajer, Maarten A., Ecological Modernisation as Cultural Politics, In: Lash, Scott *et al.* (Eds.), *Risk, Environment and Modernity: Towards a New Ecology*, London, 1996.

Hajer, Maarten A., *The Politics of Environmental Discourse: Ecological Modernization and the Policy Process*, Oxford, 1995.

Hilliard, Rachel and Roy Green, Governance and Institutional Change in Ireland, In: *Governance of Innovation Systems: Vol. 2. Case Studies in Innovation Policy*, Paris, 2005.

International Development Research Centre, *Information and Communication Technologies for Development in Africa: Trends and Overview*, 2001.

Institute for Fiscal Studies, *Issues in the Design and Implementation of an R&D Tax Credit for UK Firms,* Briefing Note No. 15, London, 2001.

International Monetary Fund, *Euro Area Policies*, Washington, DC, 2 July 2004.

International Monetary Fund, *World Economic Outlook April 2006*, Washington, DC, 2006.

Joanneum Research Forschungsges.m.b.H *et al.*, *Evaluation of Austrian Participation in the 4th EU Framework Programme for Research, Technological Development and Demonstration*, 2001.

Kaiser, Robert and Heiko Prange, A New Concept of Deepening European Integration? – The European Research Area and the Emerging Role of Policy Coordination in a Multi-Level Governance System, In: *European Integration Online Papers*, Vol. 6, No. 18, 2002.

King, David A., The Scientific Impact of Nations. What Different Countries Get for Their Research Spending, In: *Nature*, Vol. 430, 15 July 2004.

Kremer, Michael, *Creating Markets for New Vaccines Part I: Rationale, Part II Design Issues*, NBER Working Papers, 2000.
Kuhlmann, Stefan and Jakob Edler, *Governance of Technology and Innovation Policies in Europe: Investigating Future Scenarios.*
Kuhlmann, Stefan and Jakob Edler, Scenarios of Technology and Innovation Policies in Europe: Investigating Future Governance, In: *Technological Forecasting & Social Change*, Vol. 70, 2000, pp. 619–637.
Landler, Mark, The Energy Challenge – Europe's Green Image Clashes with Reliance on Coal, In: *The New York Times*, 20 June 2006.
Latour, Bruno, *Science In Action: How to Follow Scientists and Engineers Through Society*, Cambridge, MA, 1987.
Latour, Bruno, *We Have Never Been Modern*, Cambridge, MA, 1993.
Levidow, Les and Susan Carr (Eds.), Precautionary expertise for EU Agbiotech Regulation, In: *Science & Public Policy* (Special Issue), Vol. 32, No. 4, 2005, pp. 257–332.
Lewenstein, Bruce V., Science and the Media, In: Jasanoff, Sheila *et al.* (Eds.), *Handbook of Science and Technology Studies*, London, 1995/2002, pp. 343–388.
Lundvall, Bengt-Åke (Ed.), *National Systems of Innovation: Towards a Theory of Innovation and Interactive Learning*, London, 1992.
Maddison, Angus, *The World Economy: A Millennial Perspective*, Paris, 2001.
Martin, Ben *et al.*, *The Relationship between Publicly Funded Basic Research and Economic Performance*, A SPRU Review, Report Prepared for HM Treasury, July 1996.
Merton, Robert K., The Matthew Effect in Science, II – Cumulative Advantage and the Symbolism of Intellectual Property, In: *ISIS*, Vol. 79, 1988, pp. 606–623.
Merton, Robert K., The Matthew Effect in Science, In: *Science*, Vol. 159, No. 3810, 1968, pp. 56–63.
Ministry of Education, Culture, Sports, Science and Technology, *Science and Technology and Society in the Future*, White Paper on Science and Technology 2004, Tokyo, 2004.
National Institute of Science and Technology Policy, Ministry of Education, Culture, Sports, Science and Technology, *The 2001 Survey of Public Attitudes Toward and Understanding of Science & Technology in Japan*, Tokyo, 2002.
National Science Board, National Science Foundation, Division of Science Resources Statistics, *Science and Engineering Indicators 2004*, Arlington, VA, 2004.
National Science Board, National Science Foundation, Division of Science Resources Statistics, *Science and Engineering Indicators 2006*, Arlington, VA, 2006.
Nelson, Richard R. (Ed.), *National Innovation Systems: A Comparative Analysis*, New York and London, 1993.
News Alert - Commission Releases New Funds for Avian Flu Research, 17 January 2006.
NIFU *et al.*, *Evaluation of Norway's Participation in the EU's 5th Framework Programme*, 2004.
Nobel Excellence in EU Research Projects, In: *European Research Headlines*, 9 December 2005
Observatoire des Sciences et des Techniques, *Indicateurs de sciences et de technologies, Edition 2004 – Rapport de l'OST*, Paris, 2004.
OECD and IEA, *World Energy Outlook 2004*, Paris, 2004.
OECD, *Benchmarking Industry-Science Relationships*, Paris, 2002.
OECD, *Citizens as Partners*, OECD Handbook on Information, Consultation and Public Participation in Policy-Making, Paris, 2001.
OECD, *Economic Survey of the Euro Area 2005: Outlook and Challenges*, Paris, 12 July 2005.
OECD, *Environment and Employment: An Assessment*, Paris, 2004.
OECD, *Governance of Innovation Systems – Volume 1: Synthesis Report*, Paris, 2005.
OECD, *Government R&D Funding and Company Behaviour – Measuring Behavioural Additionality*, Pre-publication Copy, Paris, 2006.
OECD, *Science, Technology and Industry Outlook*, Paris, 2005.
OECD, *Science, Technology and Industry Scoreboard 2005 – Towards a Knowledge Based Economy*, Paris, 2005.
OECD, *Governance of Public Research*, Paris, 2003.

OECD, *Integrating Multiple Interests into Policy, Meeting of Senior Officials from Centres of Government on Integrating Multiple Interests into Policy* (30 September–1 October 2006), PUMA/MPM(96) 2, 2006.

OECD, *Promise and Problems of E-Democracy – Challenges of Online Citizen Engagement*, Paris, 2003.

OECD, *Turning Science into Business – Patenting and Licensing at Public Research Organisations*, Paris, 2003.

Office of Science and Technology Policy – Executive Office Of The President, *Promoting Innovation*, Final 1.31.03.

Olson, Robert L. and Jessica Biamonte, *New Global Agenda: An Analysis of Major Overview Studies of the Global Environment*, A White Paper Prepared For The Foresight And Governance Project, Woodrow Wilson International Center For Scholars, Washington, DC, August 2004.

O'Mahony, Mary and Bart van Ark (Eds.), *EU Productivity and Competitiveness: An Industry Perspective – Can Europe Resume the Catching-up Process*?, Luxembourg, 2003.

Oughton, Christine *et al.*, The Regional Innovation Paradox: Innovation Policy and Industrial Policy, In: *Journal of Technology Transfer*, Vol. 27, No. 1, 2002, pp. 97–110.

Parliamentary Office of Science and Technology, *Debating Science*, Postnote No. 260, March 2006.

Peterson, John and Elizabeth Bomberg, *Decision-making in the European Union*, Basingstoke, 1999.

Pluem, Gesche, *Analysis of Impact Variables Resulting from Research Projects in the Fisheries and Aquaculture Domain of the European Commission's Fair Programme (1994–1998)*, 2003.

Porter, Michael E. and Christian H.M. Ketels, *UK Competitiveness: Moving to the Next Stage*, DTI Economics Paper No. 3, London, May 2003.

Potočnik, Janez, *Improving Research Policies through the Open Method of Coordination*, 18 May 2006.

Pouillot, Didier and Alain Puissochet, R&D Spending on ICT, In: *Communications & Strategies*, No. 48, 4th quarter 2002.

Presidency Conclusions Barcelona European Council 15 and 16 March 2002.

Presidency Conclusions Lisbon European Council 23 and 24 March 2000.

Prest *et al.*, *Assessing the Socio-Economic Impacts of the Framework Programme*, June 2002.

Pronéos in Collaboration with Corvinus University of Budapest, Comenius University and Observatoire des Sciences et des Techniques, *Private Sector Interaction in the Decision Making Processes of Public Research (Draft)*, Study for the European Commission – DG Research, 14 March 2006.

Ravetz, Jerome R. and Silvio O. Funtowicz, A New Scientific Methodology for Global environmental Issues, In: Costanza, Robert (Ed.), *Ecological Economics: The Science and Management of Sustainability*, New York, 1991.

Regional Analysis of Framework Programme Participants from 1987 (FP2) until 2002 (FP5) (FP6 contract No. EVA6-CT-2005–000002).

Remøe, Svend Otto, Governing Fragmentation: The Case of Norway, In: *Governance of Innovation Systems: Vol. 2. Case Studies in Innovation Policy*, Paris, 2005.

Report for the Dutch Presidency and Environment Ministry, *Towards a More Sustainable EU: The Need for Investments that Benefit Economy and Environment Alike*, September 2004.

Rising Above The Gathering Storm: Energizing and Employing America for a Brighter Economic Future, Statement of Norman R. Augustine, Retired Chairman and Chief Executive Officer Lockheed Martin Corporation and Chair, Committee on Prospering in the Global Economy of the 21st Century – Committee on Science, Engineering, and Public Policy – Division on Policy and Global Affairs, The National Academies, before the Committee on Science, US House of Representatives, October 20, 2005.

Rodríguez-Pose, Andrés, *The Dynamics of Regional Growth in Europe: Social and Political Factors*, Oxford, 1998.

Rye, Mette, Evaluating the Impact of Public Support on Commercial Research and Development Projects, In: *Evaluation*, Vol. 8, No. 2, 2002, pp. 227–248.

Science and Technology Policy Council, *Finland: A Knowledge-based Society*, Helsinki, 1996.

Science and Technology Policy Council, *Towards an Innovative Society: A Development Strategy for Finland*, Helsinki, 1993.

Scott, James C., *Seeing Like a State: How Certain Schemes to Improve the Human Condition Have Failed*, New Haven, 1998.

Siune, Karen, *Convergence in European Research Policy: Strength or Weakness?*, Note from The Danish Centre for Studies in Research and Research Policy 2005/2, Aarhus, 2005.

Soete, Luc, The Challenges and the Potential of the Knowledge-based Economy in a Globalised World, In: Rodrigues, Maria João (Ed.), *The New Knowledge Economy in Europe. A Strategy for International Competitiveness and Social Cohesion*, Cheltenham, 2002, pp. 28–53.

Tassey, Gregory, Policy Issues For R&D Investment In A Knowledge-Based Economy, In: *Journal of Technology Transfer*, Vol. 29, 2004, pp. 153–185.

Tiits, Marek *et al.*, *Competitiveness and Future Outlooks of the Estonian Economy – R&D and Innovation Policy Review*, Tallinn, 2003.

Technopolis, *An International Review of Methods to Measure Relative Effectiveness of Technology Policy*, Amsterdam, 2001.

Technopolis, *Policy Instruments for Sustainable Innovation*, Amsterdam, 2004.

UK Treasury, Dept. of Trade and Industry and Dept. of Education, *Science & Innovation Investment Framework 2004–2014*, London, 2004.

UNAIDS, *AIDS Epidemic Update: December 2005 – Global Summary of the AIDS Epidemic*, Geneva, December 2005.

UNICEF, *The State of the World's Children 2005*, New York, 2005.

United Nations, *The World Investment Report 2003 – FDI Policies for Development: National and International Perspectives*, New York and Geneva, 2003.

United Nations Research Institute for Social Development, *Information and Communication Technologies and Social Development in Senegal: An Overview/Les Technologies de l'Information et de la Communication et le Développement Social au Sénégal: un Etat des Lieux*, 2000.

Uotila, Marjo *et al.*, *Finnish Participation in the EU Fifth Framework Programme and Beyond*, 2004.

US Slide in World Share Continues as European Union, Asia Pacific Advance, In: *Science Watch*, Vol. 16, No. 4, July/August 2005.

van de Sande, Daphne *et al.*, *Impact Assessment of the Marie Curie Fellowships Under the 4th and 5th Framework Programmes of Research and Technological Development of the EU (1994–2002)*, IMPAFEL 2 – Contract nr. IHP-D2–2003–01, June 2005.

Wagner, Caroline S. *et al.*, *ERAnets – Evaluation of NETworks of Collaboration Among Participants in IST Research and Their Evolution to Collaborations in the European Research Area (ERA) – Final Report*, March 2005.

WHO, *The World Health Report 2004*, Geneva, 2004.

World Bank, *World Development Report 2000/2001*, Washington, DC, 2000.

Wynne, Brian, May the Sheep Safely Graze? A Reflexive View of the Expert-Lay Knowledge Divide, In: Lash, Scott *et al.* (Eds.), *Risk, Environment and Modernity: Towards a New Ecology*, London 1996, pp. 44–83.

GLOSSARY

Absorptive (absorption) capacity: Absorptive capacity is the ability of a firm to recognize the value of new, external information, assimilate it, and apply it to commercial ends. Absorptive capacity is a function of the firm's prior knowledge. The accumulation of prior knowledge enhances the ability to acquire new knowledge; the diversity of prior knowledge facilitates novel associations and linkages and helps deal with uncertainty. The effect of absorptive capacity is cumulative since it enhances the firm's ability to assimilate and exploit new knowledge and to evaluate the importance of new technological advances. The term was first introduced by Cohen and Levinthal in their 1990 article entitled "Absorptive capacity: A new perspective on learning and innovation" (In: Administrative Science Quarterly, Vol. 35, Issue 1, pp. 128–152).

Applied research: Original investigation undertaken in order to acquire new knowledge. Contrary to *basic research,* it is directed primarily towards a specific practical aim. The results of applied research are intended to be valid for a single or limited number of products. The knowledge or information derived from it is often patented but may also be kept secret. (Definition based on Frascati manual, Source: Eurostat)

Article 169 (TEC): It enables the Community to participate in research programmes undertaken jointly by several Member States. The "European and Developing Countries Clinical Trials Partnership" is the first case of application of article 169.

Article 171 (TEC): It allows the Community to set up joint undertakings or any other structure necessary for the efficient execution of Community RTD programmes. Article 171 was applied for the first and only time for the GALILEO Joint Undertaking.

Associated country: A country which has an international agreement with the Community on the basis of which it makes a financial contribution to all or a part of the Framework Programme. In exchange, its institutions participate and get funding on similar conditions as organizations from Member States. Currently the following countries are associated: five Candidate Countries (Bulgaria, Romania, Turkey, Croatia, and Former Yugoslav Republic of Macedonia), as well as Iceland, Israel, Liechtenstein, Norway, and Switzerland. (http://europa.eu/abc/eurojargon/index_en.htm)

Basic research: Experimental or theoretical work undertaken primarily to acquire new knowledge of the underlying foundations of phenomena and observable facts, without any particular application or use in view (contrary to *applied research*). The results of basic research are not generally sold but are usually published in scientific journals. Basic research can be split into two categories: 1) Pure basic research, which is carried out for the advancement of knowledge, with no positive efforts being made to apply the results to practical problems. 2) Oriented basic research, which is carried out with the expectation that it will produce a broad base of knowledge likely to form the background to the solution of recognised or expected current or future problems or possibilities. (Definition based on Frascati manual, Source: Eurostat)

Business enterprise sector: It includes all firms, organisations, and institutions the primary activity of which is the market production of goods or services (other than higher education) for sale, and the private non-profit institutes mainly serving them. The core of the sector is made up of private enterprises. Among these may be found some firms for which R&D is the main activity (commercial R&D institutes and laboratories). It also includes non-profit institutions which are market producers of goods and services other than higher education. (Definition based on Frascati manual, Source: Eurostat)

BRIC-countries: Brazil, Russia, India, and China.

Call for proposals: A legal text calling interested parties to submit proposals for Community-funded projects. The text defines the necessary specifications to prepare and submit a proposal, that is, thematic priorities, *instruments* used, address and other technical means for submission, deadlines, and so on. Calls are published in the *Official Journal* of the EU in all Community languages.

Candidate countries: Romania, Croatia, Bulgaria, Turkey, and Former Yugoslav Republic of Macedonia.

"Capacities" Specific Programme: It elaborates in detail the part of the 7th Research Framework Programme which will enhance research and innovation capacities throughout Europe and ensure their optimal use. This aim will be achieved through optimising the use and development of research infrastructures, supporting the development of regional research-driven clusters, horizontal actions and measures in support of international co-operation, and so on.

Co-decision procedure (article 251 TEC): Introduced by the Treaty of Maastricht for the adoption of the EC Framework Programme, it gives the European Parliament equal power with the Council of Ministers in the legislative process. The procedure comprises one, two, or three readings. (http://europa.eu/scadplus/glossary/index_en.htm)

Collaborative Projects: Support to Framework Programme–funded research projects carried out by consortia with *participants* from different countries. The size, scope, and internal organisation of projects can vary from field to field and from topic to topic. Projects can range from small- or medium-scale focused research actions to larger integrating projects which mobilise a significant volume or resources for achieving a defined objective. (http://cordis.europa.eu/fetch?ACTION=D&SESSION=&DOC=1&TBL=EN_DOCS&RCN=6797&CALLER=FP7_LIB)

Consortium: All the *participants* in the same project. The majority of FP6 projects are multi-partner projects requiring a minimum number of participants from different *Member States* or *Associated Countries.*

Consortium agreement: An agreement that *participants* conclude amongst themselves for the implementation of a research activity. It allows the participants to determine the detailed administrative and management provisions necessary to carry out their research project but it cannot negate the provisions established by the EC contract or the *rules for participation.*

Constant prices: Values that are expressed in such a way as to enable comparisons to be made across a period of years. To measure real national income, economists value total production in each year at a constant set of prices that applied in a chosen base year. (Source: http://www.bized.ac.uk/cgi-bin/glossarydb/search.pl)

"Cooperation" Specific Programme: It elaborates in detail the part of the 7th Research Framework Programme which will support transnational cooperation on policy-defined themes. It will provide support to cooperation at every scale across the European Union and beyond, in a number of thematic areas corresponding to major fields of the progress of knowledge and technology.

Co-operative research projects (CRAFT): Framework Programme–funded projects whereby a number of Small or Medium-Sized Enterprises (SMEs) assign scientific and technological research activities to RTD performers (e.g. to research centres or universities). These activities may also be carried out by innovative and high-tech SMEs in cooperation with research centres and universities. The SMEs retain ownership of the results.

Coordination Actions: One of the *instruments* to implement Framework Programme–funded projects. They are intended to promote and support the networking and coordination of research and innovation activities. They will cover the definition, organisation, and management of joint or common initiatives as well as activities such as the organisation of conferences, meetings, the performance of studies, exchange of personnel, the exchange and dissemination of good practices, setting up common information systems and expert groups.

Coordinator: One *participant* appointed by all participants of a *consortium* and accepted by the Commission, having specific additional obligations arising out of the contract, such as distributing Community funds to other participants, reporting to the Commission, and so on.

CORDIS: The Community Research and Development Information System (CORDIS) (http://www.cordis.lu) is a huge Internet information system comprising information on past and ongoing projects, current calls for proposals, partner search facilities, an electronic proposal submission system (EPSS), and many more features.

COST: An intergovernmental framework for European cooperation in the field of S&T (http://cost.cordis.lu/src/home.cfm), allowing the coordination of nationally funded research on a European level. COST actions cover basic and pre-competitive research as well as activities of public utility.

CREST: The Scientific and Technical Research Committee (CREST), composed of representatives of Member States, is a high-level advisory board to the Commission and the Council in the field of RTD.

Coreper: The French acronym for the Permanent Representatives Committee. Within the Council of Ministers, it is the forum of the Member States' Ambassadors to the European Union ("Permanent Representatives"). It is responsible for preparing the decision-making of Ministers themselves. Research items are dealt with by Coreper I, which consists of the Deputy Permanent Representatives. Coreper II is composed by the Permanent Representatives themselves. (http://europa.eu/scadplus/glossary/index_en.htm)

Current prices: Values that are expressed in the prices that exist at present. (Source: http://www.bized.ac.uk/cgi-bin/glossarydb/search.pl)

Directorate General (DG): An administrative unit of the Commission, comparable with national ministries. Currently the Commission is divided into about 40 DGs (and comparable services). Five of them are involved in the management of FP6: DG Research (RTD), DG Information Society (INFSO), DG Transport and Energy (TREN), DG Enterprise (ENTR),and DG Fisheries (FISH).

Direct action: A research activity carried out by the *JRC* directly on behalf of and in the interest of the institutions of the EU. Results are owned by the European Communities.

EFTA-countries: Iceland, Norway, Switzerland, and Liechtenstein.

ERA-NET: The principal means for the FP6 to support the coordination of national and regional research programmes. The scheme is financed as a part of the *specific programme* "Integrating and strengthening the European Research Area".

EU-15 Member States: Before 1 May 2004, the European Union consisted of 15 Member States (Austria, Belgium, Denmark, Finland, France, Germany, Greece, Ireland, Italy, Luxembourg, the Netherlands, Portugal, Spain, Sweden, and the United Kingdom).

EU-25 Member States: The current 25 Member States of the European Union are Austria, Belgium, Cyprus, Czech Republic, Denmark, Estonia, Finland, France, Germany, Greece, Hungary, Ireland, Italy, Latvia, Lithuania, Luxembourg, Malta, the Netherlands, Poland, Portugal, Slovak Republic, Slovenia, Spain, Sweden, and the United Kingdom.

Eurobarometer surveys: Since 1973, the European Commission has been monitoring the evolution of public opinion in the Member States, thus helping the preparation of its legislation and the evaluation of its work. The surveys and studies address major topics concerning European citizenship such as enlargement, health, culture, information technology, environment, the Euro, and so on. A Standard Eurobarometer survey consists of approximately 1000 face-to-face interviews per Member State. Reports are published twice yearly. (http://ec.europa.eu/public_opinion/index_en.htm)

EURATOM: The European Atomic Energy Community (EURATOM) is one of the building blocks of the EU. In relation to Community research policy, the EC Framework Programme is complemented by an EURATOM Framework Programme under the Euratom Treaty which covers training and research activities in the nuclear sector.

EUREKA: A pan-European network for market-oriented, industrial R&D. EUREKA supports the competitiveness of European companies through international collaboration, in creating links and networks of innovation. The objective is to bring high quality research and development efforts to the market and to use the multiplying effects of cooperation. (http://www.eureka.be/)

European Research Advisory Board (EURAB): An independent advisory committee created by the Commission to provide advice on the design and implementation of EU research policy. EURAB is made up of 45 top experts from EU countries and beyond.

European Research Area (ERA): A general concept proposed by the Commission and endorsed by the European Parliament and Council in 2001 to overcome the fragmentation of European research and innovation efforts. The concept comprises organising cooperation at different levels, coordinating national or European policies, networking teams and increasing the mobility of individuals and ideas. FP6 is an important tool in support of the ERA, alongside national efforts and other European cooperative research activities.

European Research Council (ERC): Introduced in FP7, it will be the first pan-European funding agency for *frontier research*. Early stage as well as fully established investigators from across Europe will be able to compete for grants with scientific excellence as the sole criterion for funding. The independent Scientific Council will direct the ERC's scientific operations and ensure that its support is in accordance with the highest standards of science and scholarship. (http://ec.europa.eu/erc/index_en.cfm)

European Space Agency (ESA): Established in 1975, it is an intergovernmental organisation dedicated to the exploration of space, currently with 17 Member States. Its mission is to shape the development of Europe's space capability. By coordinating the financial and intellectual resources of its members, it can undertake programmes and activities far beyond the scope of any single European country. (For more information see: http://www.esa.int/export/esaCP/index.html)

Financial perspective: It forms the framework for Community expenditure over a period of several years. It is the product of an inter-institutional agreement between the European Parliament, the Council, and the Commission, and indicates the maximum volume and the composition of the foreseeable Community expenditure. It is adjusted annually by the Commission to take account of prices and the development of Community GNP. To date, four inter-institutional agreements of this type have been concluded, the first in 1988, the second in 1992, the third in 1999, and the fourth in 2006: the 1988–92 financial perspective (Delors I package), the 1993–99 financial perspective (Delors II package), the 2000–06 (Agenda 2000), and the 2007–13 financial perspective. (For more information see http://ec.europa.eu/budget/documents/multiannual_framework_en.htm) (http://europa.eu/scadplus/glossary/index_en.htm)

Framework Programme (FP): Since 1984, research and innovation activities of the EU are grouped in one big multi-annual programme, the Framework Programme for Research and Technical Development. While FP1–FP6 were conceived for a period of four years, FP7 is synchronised with the duration of the EU's *financial perspective* and covers the period 2007–2013. The FPs are elaborated and proposed by the Commission and have to be adopted by the European Parliament and the Council in co-decision. They are further elaborated by Specific Programmes which are implemented in detail by *Work Programmes*. (For further information see http://europa.eu.int/comm/research/why.htm)

Framework Programme ex-post evaluation system: Introduced in 1994, its two main components are yearly monitoring exercises and five-yearly in-depth assessments (the so-called "Five-Year Assessments") carried out overall and usually also at specific programme level. The annual monitoring exercise is intended to enable a quick response to issues arising from ongoing programme implementation.

The objective of the Five-Year Assessments is to provide input for policy formulation on the basis of feedback obtained from programme implementation. Both annual monitoring and Five-year Assessment rely to a great extent on panels of independent experts making use of studies carried out by professional evaluators.

Frontier research: The term reflects a new understanding of *basic research*, introduced in FP7. On the one hand it denotes that basic research in science and technology is of critical importance to economic and social welfare, and on the other that research at and beyond the frontiers of understanding is an intrinsically risky venture, progressing in new and most exiting research areas, and is characterised by an absence of disciplinary boundaries. (http://ec.europa.eu/erc/index_en.cfm?p=3_glossary)

Government budget appropriations or outlays on R&D (GBAORD): All appropriations allocated to R&D in central government budgets. Data on government R&D appropriations therefore refer to budget provisions, not to actual expenditure, that is, GBAORD measures government support for R&D using data collected from budgets. (Source: Eurostat)

Gross domestic expenditure on R&D (GERD): Total intramural expenditure on R&D performed on the national territory during a given period. GERD includes R&D performed within a country and funded from abroad but excludes payments made abroad for R&D. (Source: Eurostat)

Gross Domestic Product (GDP) at market prices: This aggregate represents the result of the production activity of resident producer units. It corresponds to the economy's output of goods and services, less intermediate consumption, plus taxes linked to imports. The sum of the regional values of the GDP at market prices might differ from the national values for some countries. (Source: Eurostat)

Gross National Income (GNI): GDP less net taxes on production and imports, less compensation of employees and property income payable to the rest of the world plus the corresponding items receivable from the rest of the world. In other words, GDP less primary incomes payable to non-resident units plus primary incomes receivable from non-resident units. (Source: Eurostat)

Higher education sector: It is composed of all universities, colleges of technology, research institutes, and other institutes of post-secondary education. While this definition describes the general coverage of the sector, it is difficult to provide clear guidelines which ensure internationally comparable reporting of data because it is not backed by the System of National Accounts (SNA). As its definition is based on mixed criteria, it is particularly susceptible to varying interpretation resulting from national policy preoccupations and definitions of the sector. (Source: Eurostat)

"Ideas" Specific Programme: It elaborates in detail the part of the 7th Research Framework Programme which deals with investigator-driven research based on initiatives of the research community. The "Ideas" Specific Programme will create the *European Research Council*, and provide its source of funding.

Impact Assessment: Since 2003, all draft EU legislation has to be accompanied by a report which assesses the expected impacts of the proposal in question on the economy, society, and the environment. Impact Assessment (IA) is part of the so-called "Better Regulation Package" which is designed to cut red-tape, to tackle excessive regulation, and to help strike the right balance between the costs and the benefits of legislation. As a method, it identifies the main options for achieving the objective and analyses their likely impacts in the economic, environmental and social fields. It outlines advantages and disadvantages of each option and examines possible synergies and trade-offs. The 7th Framework Programme is the first FP to be accompanied by an in-depth IA report. (http://ec.europa.eu/governance/impact/index_en.htm).

Indirect action: Research and other activities carried out by legal entities (e.g. enterprises, research institutions, universities, etc.) in their own interest, with funding by the EU *Framework Programme*. Most of the indirect actions are based on shared costs, that is, the Community contribution covers a percentage of the overall costs. Results are, in general, owned by the institutions carrying out the actions. The bulk of FP6 is implemented by indirect actions (for comparison see *direct action*).

Instrument: The different mechanisms for indirect Community intervention to implement FP-funded activities (in other words, the different types of projects). The range of instruments in FP6 covers research and development, demonstration and innovation activities, integrating activities, special actions for SMEs, fellowships for individuals, support for access to large research infrastructures, and so on.

Integrated Project (IP): One of the new *instruments* of FP6. An Integrated Project is a transnational multi-partner project with the main aim to generate knowledge in the priority thematic areas of FP6. Integrated Projects have to include R&D activities and can in addition have demonstration, training, and innovation-related activities.

Intellectual Property Rights (IPR): They cover all aspects of owning, protecting, and giving access to knowledge and pre-existing know-how.

Interactive Policy Making initiative (IPM): An initiative by the European Commission to collect feedback from citizens, consumers, and businesses in order to better understand how they perceive and what they expect from Community policies. It consists of two Internet-based instruments (Feedback mechanism and Online Consultation Mechanism) which are both available via the "Your Voice in Europe" webportal, developed in the context of IPM and is the Commission's "single access point" for consultations. (http://ec.europa.eu/yourvoice/ipm/index_en.htm)

Joint Research Centre (JRC): As a service of the European Commission, the mission of the JRC is to provide customer-driven scientific and technical support for the conception, development, implementation, and monitoring of EU policies. It functions as a reference centre of science and technology for the Union. The JRC has a network of research institutes in different Member countries (Belgium, Germany, Italy, the Netherlands, and Spain). Its activities are financed by the *Framework Programme* via the *direct actions*.

Labour productivity: Labour productivity is defined as GDP (in PPS) per hour worked. According to the growth accounting methodology, labour productivity can be decomposed into capital deepening and multi-factor productivity. (Source: Eurostat)

Lisbon strategy: At the European Council Meeting in Lisbon in March 2000, the EU's political leaders set a new goal for the EU in order to be able to compete with other major world players: to become, within a decade, "the most competitive and dynamic knowledge-based economy in the world, capable of sustainable growth with more and better jobs and greater social cohesion". The "Lisbon strategy" covers such matters as research, education, training, Internet access, and online business. It also covers reform of Europe's social protection systems, which must be made sustainable so that their benefits can be enjoyed by future generations. Every spring the European Council meets to review progress in implementing the Lisbon strategy. (http://europa.eu/abc/eurojargon/index_en.htm)

Marie Curie Actions: The main objective of the FP's Marie Curie Actions is to strengthen training, the career prospects, and mobility of European researchers in order to provide support for the development of world-class human resources. Under FP7, special focus will be given to skills and career development, increasing mobility between university and industry, and strengthening links with national systems.

NEMESIS: The NEMESIS model is a large-scale econometric model at the macro-level and sectoral level, which has been built by a Community-funded *consortium* of European research institutes. It comprises roughly 70,000 equations. The model can be used for several purposes, which include the assessment of structural (mainly R&D and environmental) policies, the study of the short- and medium-term consequences of a wide range of economic policies, short- and medium-term forecasting (up to 8 years) at the macro-level and sectoral level, and building long-term baseline scenarios (up to 30 years). More information on NEMESIS can be found on http://www.nemesis-model.net.

Network of Excellence (NoE): A new *instrument* introduced in FP6 to overcome the fragmentation of the European research landscape with the objective to strengthen European excellence in a given area. Its purpose is to reach a durable restructuring/shaping and integration of efforts and institutions (laboratories, departments, units, teams, etc.) in areas where this is necessary.

New instruments: The specific aim of FP6 is not just to fund good research but also to have a structuring and coordinating effect on the European Research Area (see *ERA)*, and requires the application of new types of projects (new mechanisms for indirect Community intervention) bringing together a critical mass of resources and leading to lasting integration of research capacities. The three new instruments are *Integrated Projects*, *Networks of Excellence* and the application of *Article 169*.

One-stage procedure: Within this procedure of proposal submission and evaluation in FP6, a full proposal has to be submitted immediately and will be the basis for evaluation and selection of projects to be funded (see also *two-stage procedure*).

Open method of coordination (OMC): A relatively new and intergovernmental means of governance in the EU, based on the voluntary cooperation of Member States. It rests on soft law mechanisms such as guidelines and indicators, benchmarking and sharing of best practice, not on official sanctions for laggards. Rather, the method's effectiveness relies on a form of peer pressure and naming and shaming, as no Member State wants to be seen as the worst in a given policy area. The OMC was first applied in EU employment policy, as defined in the Amsterdam Treaty of 1997. It was officially defined and endorsed at the *Lisbon* Council for the realm of a broader fields of policies, among which is research. Since then it has been applied in the European employment strategy, pensions, immigration, education and culture, and asylum, and its use has also been suggested for health as well as environmental and research affairs (such as the 3% objective).

Participant: Legal entities in FP-funded projects contributing to an *indirect action* and having rights and obligations with regard to the Community and to one another under the terms of the *Rules for Participation*. Under the contract with the Community, participants are referred to as contractors.

Peer review: The *evaluation* of proposals with the help of independent external experts (peers). For FP6, the procedures for the evaluation of proposals are described in detail in a Commission decision on "Guidelines on proposal evaluation and selection procedures".

"People" Specific Programme: It elaborates in detail the part of the 7th Research Framework Programme which deals with support of individual researchers. The "People" Programme will strengthen the human potential in R&D in Europe by putting into place a set of *Marie Curie* actions, addressing researchers at all stages of their career.

Proportionality principle: Alongside *subsidiarity* a major principle governing the exercise of powers within the Community. By virtue of this principle, action taken by the Union, in terms of its form and content, does not exceed what is required to achieve the objectives set out in the Treaties. (http://europa.eu/scadplus/constitution/subsidiarity_en.htm)

Project evaluation: The process of assessing the quality of proposals and arranging them in a ranked list to prepare the decision on the selection of projects to be funded. For FP6, evaluation procedures are described in the "Guidelines on Proposal Evaluation and Selection Procedures".

Purchasing Power Standards (PPS): Financial aggregates are sometimes expressed in Purchasing Power Standards, rather than ecu/euro based on exchange rates. PPS are based on comparisons of the prices of representative and comparable goods or services in different countries in different currencies on a specific date. (Source: Eurostat)

Research and experimental development (R&D): R&D comprises creative work undertaken on a systematic basis in order to increase the stock of knowledge, including knowledge of man, culture, and society and the use of this stock of knowledge to devise new applications. This term covers three activities: *basic research, applied research* and experimental development (Definition based on Frascati manual, Source: Eurostat)

Researchers: Professionals engaged in the conception or creation of new knowledge, products, processes, methods, and systems, and in managing the projects concerned. Included in this are managers and administrators engaged in the planning and management of the scientific and technical aspects of a researcher's work, as well as postgraduate students engaged in R&D. (Source: Eurostat)

R&D intensity: Gross Domestic Expenditure on R&D (GERD) expressed as a percentage of Gross Domestic Product (GDP). (Source: Key Figures, 2005)

Rules of Participation for the Framework Programme: They set out the framework that governs the relationship between the Commission and the institutions that *participate* in the programme, covering aspects such as procedures for *calls for proposals*, types of grants, levels of financing, *consortia* composition, the *evaluation* process, financial management of projects, and dissemination of project results. The Rules of Participation are adopted by the European Parliament and the Council in co-decision upon a proposal from the Commission (art. 167 TEC).

Small and medium-sized enterprises (SMEs): Enterprises having fewer than 250 employees and with either an annual turnover of no more than ECU 40 million or a balance sheet total of no more than ECU 27 million. (Source: Eurostat)

Specific Support Actions (SSA): As an *instrument* of FP6, they are intended to support its implementation and may also be used to help prepare for future EU research policy activities. Within the priority thematic areas, they involve, for example, conferences, studies, scientific awards, working groups, dissemination activities, or a combination of these. They are also intended to stimulate the participation of small research teams, SMEs as well as organisations from the candidate countries – in the activities of the priority thematic areas, especially through *networks of excellence* and *Integrated Projects.*

Specific Targeted Innovation Projects (STIP): As multi-partner innovation projects, their purpose is to support activities exploring, validating, and disseminating new innovation concepts and methods at European level. The Community contribution is paid as a *grant to the budget* (percentage of total costs of the project).

Specific Targeted Research Projects (STREP): As multi-partner research or demonstration projects, their purpose is to support RTD activities of a more limited scope than *Integrated Projects.* The Community contribution is paid as a *grant to the budget* (percentage of total costs of the project). STREP are used in implementing the priority thematic areas, in areas anticipating S&T needs, in specific international cooperation research activities, and in research activities developing harmonious relations between science and society.

Stability and Growth Pact: To be seen against the background of the third stage of economic and monetary union (EMU), which began on 1 January 1999. Its aim is to ensure that the Member States continue their budgetary discipline efforts once the single currency has been introduced. In the medium term the Member States have undertaken to pursue the objective of a balanced or nearly balanced budget. Along the same lines, States not taking part in the third stage of EMU are required to submit a convergence programme. The Stability and Growth Pact opens the way for the Council to penalise any participating Member State which fails to take appropriate measures to end an excessive deficit. (http://europa.eu/scadplus/glossary/index_en.htm)

Stakeholder: Any person or organisation with an interest in or affected by EU legislation and policy-making is a "stakeholder" in that process. The European Commission makes a point of consulting as wide a range of stakeholders as possible before proposing new legislation or new policy initiatives. (http://europa.eu/abc/eurojargon/index_en.htm)

Subsidiarity principle: Alongside *proportionality,* a major principle governing the exercise of powers within the Community. It means that EU decisions must be taken as closely as possible to the citizen. In other words, the Union does not take action (except on matters for which it alone is responsible) unless EU action is more effective than action taken at national, regional, or local level. (http://europa.eu/abc/eurojargon/index_en.htm)

Third country: A country which is neither a Member State nor an Associated country and which falls in any of the following categories: Candidate countries; countries neighbouring the EU; Mediterranean partner countries; Western Balkans and the Newly Independent States; developing countries; emerging economies.

Technology Platforms: Introduced in FP7, they will bring together companies, research institutions, the financial world, and regulatory authorities at European level to define a common research agenda which should mobilise a critical mass of – national and European – public and private resources. This approach has been, or will be, adopted in areas such as energy (hydrogen technology, photovoltaic solar energy), transport (aeronautics), mobile communications, embedded systems, and nanoelectronics. Often, it will be possible to implement the agenda by means of *Integrated Projects*. A possible framework for their implementation is a structure based on *article 171 TEC* (*Joint Undertaking*).

Thematic priorities/areas: Every Framework Programme provides support for transnational cooperation at every scale across the EU and beyond, in a number of thematic areas corresponding to major fields of the progress of knowledge and society, where research must be strengthened to address European social, economic, environmental, and industrial challenges. The nine themes determined for EU action in the FP7 proposal are (a) Health; (b) Food, Agriculture, and Biotechnology; (c) Information and Communication Technologies; (d) Nanosciences, Nanotechnologies, Materials and new Production Technologies; (e) Energy; (f) Environment (including Climate Change); (g) Transport (including Aeronautics); (h) Socio-economic Sciences and Humanities; and (i) Security and Space.

Triadic Patents: The set of patented inventions for which protection has been sought at all three major patent offices (the European Patent Office – EPO, the US Patent and Trademark Office – USPTO, and the Japanese Patent Office – JPO). The country of origin is defined as the country of the inventor. The advantage of triadic patents is that they can eliminate the "home advantage effect". They may also be associated with patents of a higher expected commercial value, since it is costly to file through three patent systems. However, it is also likely that they tend to reflect the patenting activity of larger companies who seek, and can afford, broader international protection (Source: Key Figures, 2005)

Two-stage procedure: This submission and evaluation procedure of FP6-funded projects includes a first step where a short outline proposal is to be submitted and evaluated, followed by a second step of submission and evaluation of a full proposal only for the outline proposals evaluated positively. The application of this procedure will be announced in the *work programmes* and in the *calls for proposals* (see also *one-stage procedure*).

Work Programme: Plans drawn up by the Commission for the implementation of the *specific programmes* of the Framework Programme. They comprise detailed descriptions of the activities (thematic priorities, *instruments* used, evaluation procedures and criteria, deadlines, roadmaps) and are revised at least annually. They provide all information necessary to launch calls for proposals.

INDEX

MIX
Papier aus verantwortungsvollen Quellen
Paper from responsible sources
FSC® C105338

If you have any concerns about our products,
you can contact us on
ProductSafety@springernature.com

In case Publisher is established outside the EU,
the EU authorized representative is:
Springer Nature Customer Service Center GmbH
Europaplatz 3, 69115 Heidelberg, Germany

Printed by Libri Plureos GmbH
in Hamburg, Germany